KU-335-721

International Turf Management

HANDBOOK

International Turf Management

HANDBOOK

EDITED BY

D.E.Aldous

Department of Environmental Horticulture and Resource Management,
University of Melbourne, Burnley College,
Richmond, Victoria

INKATA PRESS
An imprint of Butterworth Heinemann

INKATA PRESS
A DIVISION OF BUTTERWORTH-HEINEMANN

AUSTRALIA	BUTTERWORTH-HEINEMANN	22 Salmon Street, Port Melbourne, Victoria 3207
SINGAPORE	REED ACADEMIC ASIA	
UNITED KINGDOM	BUTTERWORTH-HEINEMANN Ltd	Oxford
USA	BUTTERWORTH-HEINEMANN	Woburn, Massachusetts

National Library of Australia Cataloguing-in-Publication entry

International turf management handbook.

Includes index.
ISBN 0 7506 8954 4.

1. Turf management—Handbooks, manuals, etc. I. Aldous, David E., 1946– .

635.9642

Edited by Brenda Hamilton
Designed by Pauline McClenahan, Captured Concepts
Typeset in 10.5/13.5 Meridian by Post Pre-press Group
Production by Prose Editorial Services
Printed by The Bath Press

CONTENTS

FOREWORD

Sports turf, and thus turf culture, is a comparatively modern idea. The roar of suburban turf was heard for the first time at the turn of the present century. It only reached a crescendo in the last fifty years or so, as motorised mowers and edging machines, aerators and scarifiers made their appearance even in small gardens and at every park, golf course and playing field. When I started work, not that long ago — not that long anyway — advertisements could still be found seeking scythe hands, even though the mowing machine was invented in 1830. Now forty years later, the skill which allowed the scythe to leave a surface as fine and even as a mowing machine, is almost unknown. The pace of change is still accelerating as new grass varieties are bred, new equipment is evolved, and new techniques, fertilisers, weed killers and pesticides are developed.

Turf culture only started to aspire to its present level of sophistication when people had the security and leisure to start gardens and to take part in sports. It is true that Pliny the Younger who lived from 61 AD until 113 AD mentioned a lawn at his villa in Tuscany, but it seems to have been a small affair. There was little or nothing written about lawns before that and not much to indicate that they existed on any extensive scale between then and mediaeval times. Boccacio has some of his characters, in the *Decameron*, putting grassy meades to uses which the modern Park Superintendent would easily recognise, but he was thinking of them as meadows filled with wild flowers and not as well kept, neatly mown, turf. If you look closely at paintings of the period you may see the corroborating evidence, grass, yes, but probably only scythed a couple of times a year and thus with a rich thatch of herbs as a major component of the sward. It would have lacked the durability given by modern methods of cultivation and new varieties of grasses, but it would in its season have looked pretty in a way that we have tended to forget.

As it happens, in central London we have now re-established three hundred acres of meadowland. Once familiar herbs like the meadow geranium, the harebell and scores of other wildflowers are shyly starting to appear in it, after decades of absence. It is within living memory that Hyde Park was kept trim by sheep which grazed it, and Richmond and Bushy Parks still have deer as the principle groundkeepers, just as they have been for half a millennium. Both permit a wide flora to flourish along with the grass. But tastes change. Neatly mown grass was until recently *de rigeur* in urban parks and gardens, at least in Britain, and a shift to a more relaxed style of upkeep for extensive areas of informal grass is a newish trend. Dealing with the problems that arise will be the next set of conundrums for management. There are plenty of questions to ask.

How do you keep an urban meadow clean? What do you do about the problems caused, and left, by dogs? How do you stop one or two species from dominating the sward at the expense of others? Should you deliberately introduce esteemed herbs or ought you to be patient enough to allow them to reappear naturally even though it may take many years? What kind of herb should be introduced, and ought its provenance to be considered important? There are plenty of problems to resolve. Nor are meadows a cheap option. They will not look after themselves. They must be kept clear of litter and debris just as a lawn has to be, and rubbish is more trouble to find and pick out of long grass than short. In the British climate meadows must be mown in the autumn after flowering, and again in the spring. Without this, they would turn into spinneys and after a titanic struggle between warring species into woodland, which is the destiny of all plantations in our climate, just as the bush would no doubt encroach in Australia.

But conservationists are wrong when they refer to a mown lawn as a green desert, though these days they are often bold enough to do so. It is in fact a green carpet on which people, sometimes in large numbers, can play, sit, walk, sunbathe and disport themselves. It can be highly decorative in its own right and it is the perfect foil for flowers. It is the most durable of all living surfaces.

Sports turf shares the need to be tough. But where the skill of the player can be subverted by the nature of the playing surface, the care and precision with which it is made, and then kept, is a matter of key importance. It deserves the greatest care and study. The greenkeeper who has faced a cricketer bowled out because of a crumbling wicket, or a golfer who has missed a putt because of the alleged quality of the green, or a bowler whose perfect shot has just finished in the ditch, will not wish to repeat the experience frequently. It makes no difference whether the complaint was justified or not, in an extreme case the sportsman's irritation may not be far removed from road rage and is to be avoided in so far as skill can do it. If it can't then the biblical observation that 'The soft answer turneth away wrath' has much truth in it, and the Koran has an even blunter view of the matter 'When fools speak, say peace'.

This book has an international band of contributors. They come from Australia, Canada, China, New Zealand and the United States of America as well as from the United Kingdom. Techniques in all of these countries vary, different grass species may be used, and methods of cultivation must address the particular circumstances of climate and soil. But the basic principles of managing turf are abiding ones and apply everywhere. The purpose of the turf has to be known and understood. The type and frequency of its use must be assessed. The way the lawn or grassland is constructed has to be worked out so that it will provide an enduring surface, which is fitted for its chosen role and use. And then the great toil commences. The area has to be kept, perhaps for generations or even centuries, in ways that satisfy its purposes in life and give pleasure to those who use it or simply admire its appearance. Throw in the old, but newly relabelled, concepts of sustainability and biodiversity, and the frugal use of the earth's resources, and this subject can be seen for what it is; a highly complex, ever developing technology. I hope that this book, in drawing on the experience and wisdom of many contributors, will assist in the understanding and practice of turf management. It is fundamental to enjoying most sports, and essential to the beauty and usefulness of parks and gardens everywhere in the world.

David Welch
The Royal Parks
London

LIST OF CONTRIBUTORS

Aldous, D.E.
Institute of Land and Food Resources, University of Melbourne, Burnley College, Swan Street, Richmond, Victoria 3121, Australia

Baker, Sz.W.
The Sports Turf Research Institute, St. Ives Estate, Bingley, West Yorkshire BD16 1AU, England

Beehag, G.W.
Globe Australia Pty Ltd, Miranda, NSW 2229, Australia

Brereton, J.S.
Institute of Land and Food Resources, University of Melbourne, Burnley College, Swan Street, Richmond, Victoria 3121, Australia

Cockerham, S.T.
Agricultural Operations Department, University of California, Riverside, California 92521, United States of America

Connellan, G.J.
Institute of Land and Food Resources, University of Melbourne, Burnley College, Swan Street, Richmond, Victoria 3121, Australia

Chivers, I.H.
Racing Solutions Pty. Ltd., P.O. Box 133, Sandringham, Victoria 3191, Australia

Haining, V.A.
Parks and Recreation, The City of Melbourne, GPO 1603M, Melbourne, Victoria 3001, Australia

Jakobsen, B.F.
Rootzone Laboratories International Pty. Ltd., Kambah Heights, Canberra, ACT 2902, Australia

Kenna, M.P.
Green Section, United States Golf Association, P.O. Box 2227, Stillwater, OK 74076, United States of America

Liebao, H.
Institute of Turfgrass Science, College of Plant Science and Technology, China Agricultural University, Beijing, 100094, P.R. China

Lettner, R.G.
Agricultural Sciences, Fairview College, Box 3000, Fairview, AB, TOH IL0, Canada

McAuliffe, K.W.
New Zealand Sports Turf Institute, P.O. Box 347, Palmerston North, New Zealand

McGeary, D.J.
Turfgrass Technology, P.O. Box 1125, Sandringham, Victoria 3191, Australia

McIntyre, D.K.
Horticultural Engineering Consultancy, 5 Brimage Place, Kambah Heights, Canberra, ACT 2902, Australia

Minner, D.D.
Horticulture Department, Iowa State University, Ames, 1A 50011, United States of America

Neylan, J.J.
Turfgrass Technology, P.O. Box 1125, Sandringham, Victoria 3191, Australia

Robinson, M.R.
Turfgrass Technology, P.O. Box 1125, Sandringham, Victoria 3191, Australia

Ryan, P.
Pacific Coast Design Pty. Ltd., 2/60 Bay Road, Sandringham, Victoria 3191, Australia

Semos, P.S.
StrathAyr Pty. Ltd., Tallarook Park, School House Lane, P.O. Box 267, Seymour, Victoria 3660, Australia

Snow, J.T.
Green Section, United States Golf Association, Golf House, P.O. Box 708, Far Hills, NJ 07931-0708, United States of America

Wilson, J.R.
Tropical Agriculture, CSIRO Australia, Cunningham Laboratory, 306 Carmody Road, St. Lucia, Queensland 4067, Australia

Witherspoon, W.R.
Guelph Turfgrass Institute, University of Guelph, Guelph, Ontario N1G 2W1, Canada

PREFACE

It has been estimated that grasslands occupy about one-quarter of the world's vegetative cover. Grasses constitute a major human food source, either as grain, or as pasture for our native and domesticated animals. Grass makes life worth living, turf makes life worthwhile. Turfgrasses influence many human and environmental activities by protecting our land surface from erosion, stabilising our soils, moderating our temperatures, providing low cost safe surfacing for many sporting and leisure activities, and bringing comfort and pleasure to the landscape.

Amenity grasslands, and turf in particular, not only form a significant part of the global landscape, but are also the basis of a multi-million dollar industry. Turfgrass venues, agencies, activities and services are highly valued and varied, and require highly motivated trained personnel with enquiring minds to manage them efficiently and effectively. This handbook provides a treatise on the establishment, maintenance, and management of natural turfgrass surfaces in Australia and overseas. Developed in association with more than 20 leading turf management authorities, this manual details how the natural turfgrass surface functions and the implications for its management. For this reason it is important reading for teaching and training staff, graduate students, students undertaking undergraduate and technically oriented programs in agriculture, horticulture, and natural resource management, as well as all land managers and other allied professionals who intend to take leadership roles in the management of turf and amenity grasslands.

This handbook has been arranged into four sections. Section 1 covers Chapters 1 to 10 and systematically introduces the reader to the turfgrass industry, how turfgrasses are identified and selected, how turfgrasses grow, the significance of soils and drainage, and how grasses may be successfully established. The scientific names, and with few exceptions the common names, of the grasses in the manual closely follow Watson and Dallwitz's, *The Grass Genera of the World*, and Hanson and Juska's, *Turfgrass Science*. Subsequent chapters address irrigation and its application, nutrition and fertilisers, machinery and equipment operation, and plant health and protection. Section 2 covers chapters 11 to 13 and emphasises the management and administration of the turfgrass business, and the importance of contract establishment and management. Chapters 14 to 19 address the current performance standards in place today, and use them to effectively manage different cultural systems such as the bowling, croquet and golf green, the grass tennis court, the cricket table, the football and athletic field and arenas, the racetrack, and the golf course fairway. Both instructor and student alike will need to relate the different management strategies that will need to be adopted with other natural (amenity grasslands, airfields, road reserves, school and institutional grounds, lawn cemeteries, military installations, the private and public landscape), and sporting fields (baseball, lacrosse, polo, and field hockey). The fourth section, chapter 20, addresses the important environmental issues in turf management such as the use of scarce water resources, pesticide and fertiliser pollution, and the effects of golf course activities on people and wildlife.

Numerous publications have been consulted for information; the more significant ones are listed at the end of each chapter. Trade names found in this book serve only to identify materials or equipment and no endorsement of them is implied or intended.

Many people have contributed material and helpful critical reviews on individual chapters in the publication of this manual. Thanks are due to all contributing colleagues from Canada, the United States of America, Great Britain, China, Australia and New Zealand, who gave of their time and experience in developing their respective chapters. Thanks are also due to those who contributed by reviewing different chapters: Drs Greg Moore and Peter May, University of Melbourne-Burnley College, Melbourne; Philip Ford, Northern Institute of TAFE, Melbourne; Bruce Stephens, Chemturf Pty Ltd, Melbourne; David McGeary, Turfgrass Technology Pty. Ltd, Melbourne; Ian Chivers, Racing Solutions Pty Ltd, Melbourne; and Gary Beehag, Globe Australia Pty Ltd, Sydney, New South Wales.

For illustrations and slides, thank you to Bayer Australia Ltd, and Chipco, a Division of Rhône-Poulenc Rural Australia Pty. Ltd.

Finally an immense debt of gratitude to my wife, Kaye, who put up with my wordprocessing sessions, and my children, Matthew and Andrew, who introduced me to the workings of desktop publishing, and Janine, who ably assisted me with illustrations. Also special thanks to Rosemary Peers and Brenda Hamilton for their editing and production guidance.

David E. Aldous
Richmond, Victoria

CHAPTER 1

Introduction to turfgrass science and management

D.E. ALDOUS, Institute of Land and Food Resources, University of Melbourne, Richmond, Victoria, Australia

Introduction

Turf comes from either the ancient Sanskrit word, *darbha*, or the old English word, *torfa*, both meaning a tuft of grass. More modern definitions cite turf as a surface layer of vegetation, consisting of earth and a dense stand of grasses and roots. In profile, turf consists of verdure, the green aerial shoots remaining after mowing; thatch, the intermingled layer of dead and living stems and roots that develops between the verdure but above the soil surface; and soil, which contains underground stems and roots. Other terms used synonymously with turf include sod, which is a piece cut from this vegetative material plus its adhering soil; sward, which is the grassy vegetation often used in association with pastures; grass, which is any monocotyledonous plant belonging to the family *Poaceae*; a green, which is a smooth, grassy area used for sporting purposes; and a lawn, which had been defined as a flat and usually level area of mown grass.

Turfgrass science and management involves the art, culture and science of managing these natural grass surfaces. All turf managers aim for the best playing conditions and turf of the highest quality. Playing quality is a function of the natural grass surface and soil conditions in the field, and has a great influence on traction, impact absorbence and ball response. Measurable playing standards are now in place for many natural grass surfaces (Canaway *et al.*, 1990). Turfgrass quality involves a composite, visual assessment of the natural grass surface, and has established standards against which performance is measured (Beard, 1973).

Benefits derived from the turfgrass community

Living turf provides considerable aesthetic, ecological, functional, recreational and social benefits. Functional and ecological benefits include improved soil stabilisation at the surface by reducing the potential for erosion and wind blown soil particles, and acting as a filter for improving the quality of groundwater. In addition, turf can substantially influence heat loss, reduce noise, glare and visual pollution, and act as a safe, low cost impact surface for many sporting and recreational surfaces, highways and roadsides (Beard, 1973; Roberts, 1985; Beard and Green, 1994). Actively growing turf also maintains the fundamental abiotic components of the world's life support systems, such as air, soil and water (Carne, 1994).

In urban environments, plants generally modify temperatures by influencing the rate of energy exchange (Mastalerz and Oliver, 1974).

Grasses transpire at a rate that, in energy terms, exceeds the local radiant energy supply. If a substantial portion of this heat load is not dissipated through the processes of evapotranspiration, re-radiation, conduction and convection, temperatures at the surface of the leaf can reach lethal levels. Energy not dissipated remains to affect the specific heat balance and temperature of the leaf. Factors that influence the temperature of the leaf canopy are ambient air temperature, relative humidity, availability of soil moisture, and wind velocity. Under warm to hot conditions, the leaf surface may be up to 20°C cooler than nearby unprotected buildings or road surfaces. Finnigan *et al.* (1994) found that effective tree cover ameliorated the local temperatures and humidity by up to twice the regional average, with reductions of 4°C when compared to average temperatures of 30°C. Gibbs (1997) compared the temperatures between natural and synthetic bowling green surfaces, and found that the air temperatures were only likely to be large when the ambient air temperatures rose above 20°C. Surface temperatures of 60°C have been recorded on synthetic turf, alongside maximum temperatures for natural grass of 32°C (Mecklenburg *et al.*, 1972). Tree cover can also reduce surface temperatures by up to l5°C (Givoni, 1991), and can cool adjacent turf by conduction. Lowering the height of cut of turf will also influence surface soil temperatures. For example, soil temperature extremes are greater under a turf cut at four millimetres than at 37 millimetres. Research has shown that strategically placed vegetation can reduce noise levels by 15 to 45 per cent at distances of 9 to 21 metres (30 to 70 feet) along heavily used urban freeways, as well as reduce glare and associated eye discomfort from reflected light (Beard, 1977).

Turf provides a recreational benefit by providing a low-cost, low-impact, safe surface for many outdoor sport and leisure activities. Psychologically, aesthetically pleasing, green turf enhances the beauty and attractiveness of a landscape by improving mental health and work productivity, as well as providing an overall better quality of life. Sociological benefits can also accrue to the individual and to the general community when people interact with plants. Relf and Dorn (1995) and Kaplan and Kaplan (1989) have shown that interacting with plants can develop an improved sense of self worth, create new friendships and social placement, as well as provide feelings of freedom and being in control of one's life. Researchers have also documented that people who interact with plants can recover more quickly from everyday stresses (Bennett and Swasey, 1996) and show improved self esteem (Smith and Aldous, 1994). Nursing home clients who care for plants have shown an improvement in alertness, participation and well-being (Langer and Rodin, 1976), and display a more positive outlook on life (Ulrich, 1990).

Historical perspectives in turfgrass

References in the early scriptures make frequent mention of fields of grass and ornamental gardens set in idyllic situations. Genesis (1: 11–12) makes mention, 'And God said, let the earth bring forth grass, . . . And the earth brought forth grass . . .'. The ornamental gardens of Emperor Babar and Chosroes I of Persia (AD 531–579) have been illustrated in the weave of early Persian garden carpets (Rohde, 1927). Gardens have long been expressions of luxury, such as the gardens of one of China's early emperors, Wu Ti (157–87 BC) (Malone, 1934); for pleasure, such as the Indian Taj Mahal and its surrounding gardens (Goethe, 1955), and as an expression of affection. For example Shah Jahan, the Grand Mogul, developed a number of ornamental gardens as an expression of love for his wife Mumtaz-i-Mahal (Huffine and Grau, 1969). The natural grass surface has also played an important part in the development of many sporting and recreational pursuits. Chaugau or polo frequently occupied the day for Akbar, 1556–1605 AD, the Great Emperor of Hindustan. The Plains Indians of the United States played baggataway, an early form of lacrosse, on the grassed prairies, either on foot or on horseback. The shepherds of early England and Europe played many competitive 'ball and stick' games while tending their flocks on the lowlands. The native population of many countries threw clubs at discs bowled along the ground, and played sport with balls and stones.

However it was during the years of the Crusades (11 to 13th centuries) that we see an increasing exchange of ideas as Europe came into closer contact with the East. By the 13 to 15th centuries, grassed areas were considered an integral element in the classical gardens of medieval Europe and Britain. The English dramatist and poet, William Shakespeare (1564–1616), makes reference to grass in his play *The Tempest,* (Act ii., sc.1.); 'How lush and green the grass looks . . . Here, on the grass-plot, in this very place, to come and sport . . .'. Balthazar Nebot's series of paintings of Hartwell House, in Buckinghamshire, United Kingdom (circa. 1738), depicts gardeners with scythes and lawn rollers (Sanecki, 1997). Other literary sources include the works of Miss Eleanour Sinclair Rohde who paid tribute to grass in 'Nineteenth Century and After', 1928, CIV, 200 (Dawson, 1949). In the 1930s, Englishman John Evelyn 's instruction manual states that 'bowling greens are to be mowed and rolled every fifteen days' (Evelyn, 1932).

With the advance of the Industrial Revolution, there developed an increasing demand for goods, and gardening and sport became major leisure occupations for many of the new middle and upper class town dwellers. This stimulated the manufacture of implements and machinery for maintaining the garden. Eighteen-century lawns were originally grazed or cut with the scythe. In 1830 Englishman, Edwin Budding, an engineer at a textile mill, developed a cylinder or reel type mower, which consisted of a series of blades arranged around a cylinder with a push handle. The licence to manufacture this first lawnmower, based on Budding's design, was granted in 1832 to Ransome of Ipswich in England, although a prototype had been made as early as 1831 by another Englishman named Farrabee. In 1841 Alexander Shanks of Arbroath, Scotland, had registered a pony-drawn mower which also swept up the clippings. By 1870 Elwood McGuire of Richmond, Indiana, had also designed a machine that mowed turf. With these inventions the hand scythe and cradles were abandoned for horse-drawn machines, and the sickle-bar mower was brought from the hay field to mow larger turf areas of parkland and land put aside for golf courses.

Prior to 1700 all seed was hand sown. Around 1701 Jethro Tull (1674–1741), an English agriculturist and inventor, perfected the seed drill. The introduction of fencing wire in 1840 enabled animals to be inexpensively confined close to grazed areas, and the advent of galvanised wire in 1851 and barbed wire in 1860 improved this situation. Post-World War II saw Power Specialists of Slough, England, introduce a rotary Rotoscythe mower which offered both petrol-driven and electric models (Sanecki, 1997). In 1948 Australia entered the machinery market with the introduction of a petrol-driven Rotoscythe mower which was sold for 76 pounds, four shillings and sixpence by the Finally Brothers Pty Ltd of Melbourne. This was followed in the 1950s by the introduction of the first two-stroke rotary lawnmower by Australian designer and builder Mervyn Victor Richardson of Concord, New South Wales. By 1966 lightweight electric mowers were being developed and introduced by Flymo Ltd of Middlesbrough, England. Traction power for turf maintenance operations had progressed from draught animals (Figure 1.1(a)), or even students (Figure 1.1(b)), through steam-driven tractors to those with petrol or oil engines. Over the 1920s and 1930s there was a general tendency to abandon the horse in favor of the tractor-mounted mower, not only for constructional work, but also for regular mowing of large areas.

Figure 1.1(a) Horse-drawn cylinder mower, circa 1941; (b) Student-powered cylinder mower, circa 1890 (Photos courtesy of Burnley Archives, University of Melbourne, Burnley College, Victoria)

History of significant turfgrass sports

Lawn tennis is thought to have evolved from the indoor game of real tennis, the word being derived from a corruption of the French *tenez* meaning attention or hold. Others make mention of the term *jeu de paume* or the palm game. Lawn tennis was played in monasteries in France as early as the 11th century. The long- handled racket was not invented until about 1500. 'Field' tennis is mentioned as early as 1793 in a British magazine. The first lawn tennis club was established in Leamington Spa, Warwickshire, UK in 1872. The United States Lawn Tennis Association was formed in 1881, and the English equivalent, seven years later. The Wimbledon Championships were instituted in 1877, the US Open in 1881, and the Australian Championships in 1905. In Australia the first tennis match was played on an asphalt court laid at the Melbourne Cricket Club in 1878. The Club put down a grass court the following year. The Lawn Tennis Association of Australasia was formed in Sydney in 1904.

Lawn bowls originated with the crowned heads of Europe. During the reign of England's kings, Edward III, Richard II and Henry VII, bowls was banned as a sport because the archers were often distracted by the game, thereby endangering the sustainability of the reserve military forces. Despite these legal restrictions, the game prospered with the world's first bowling club established as the South Hampton Town Bowling Club in 1299. It is reputed that the game of bowls was made famous by the English sea captain Sir Francis Drake (1540–96) who insisted he finish his game before his ships sailed to defeat the Spanish Armada in 1588. Lawn bowls has been played in Scotland since the 16th century. The modern rules were framed in Scotland by William Mitchell in 1848–9. The game is played mostly in the United Kingdom and the other Commonwealth countries. Bowls came to Australia with the first migrants, who often established greens next to their bars and taverns, as was the custom. Australian research suggests that the earliest bowling greens were built at Sandy Bay in Tasmania, and the 'Golden Fleece' and 'Woolpack' hotels in Petersham, Sydney, as early as 1826. Victoria's first bowling green was commenced in late 1845 by a Mr W. Turner who aptly named his premises the 'Bowling Green Hotel' (Gerty, 1996). The first bowling associations were established in Victoria and New South Wales in 1880. The Australian Bowls' Council was formed on 22 September 1911. The New Zealand Bowling Association (NZBA) evolved as a national body in 1914. Bowls are now played on both grass and synthetic surfaces.

Croquet is believed to have originated in southern France during the 13th century. By the 17th century, it was a popular game with French Royalty and shortly afterwards was taken up by Royalty in England. During the 18 to 19th centuries it proved to be a favorite pastime, played mostly on the private lawns of country homes and estates in Europe and the United Kingdom. Croquet has been played in Australia since colonial times. The first croquet club in Australia was founded in Kyneton in 1866 and the next was in Kapunda, South Australia in 1869. In 1949 the Australian Croquet Council was formed. Croquet headquarters are co-located with the US Professional Golf Association (PGA) at Palm Springs, in the United States. Croquet comes in three main forms: association croquet, golf croquet, and kingball.

Hockey-like, curved sticks have been observed as part of sport in early Greek carvings (500 BC), and on Egyptian tomb paintings (2050 BC). However in its modern form, hockey developed in England around the second half of the 19th century, with Teddington Hockey Club, formed in 1871, standardising the rules. The English Hockey Association was founded in 1886. Field hockey is considered a modern version of the game played under such names as Hurley and Shinty in England, and Hoquet in France. It was thought to have been introduced into Australia by the Irish gold miners in the 1860s. Hockey clubs were formed in Adelaide in 1900, Perth in 1902, and Melbourne and Sydney in 1906.

Cricket is thought to have been brought to Britain by Flemish weavers from the Low Countries as early as the 14th century. The term cricket is suggestive of the Flemish phrase *met dekrik ketsen*, which literally means to chase with a curve stick. This was shortened to *kriket*, and finally cricket as we know it today. Other interpretations come from the Anglo-Saxon word *cricce*, meaning a crooked staff; a game that became popular among shepherds playing on the short cropped grassed downlands of southern England. Other Low Country linguistic connections have emerged. For example, the word stump does not exist in Anglo-Saxon, but did exist as the word *stomp* in Holland. Illustrations of people playing cricket can be observed on medieval French manuscripts dating from the mid-14th century. The scene depicts monks and nuns playing cricket, or a bat-and-ball game similar in style to cricket. In the background are four nuns and monks, hands held out, ready to catch the ball. Certainly by the 17th century there were numerous references to cricket being played on the commons in medieval England. As a sport, cricket must have proven popular in Ireland, for Oliver Cromwell (1599–1658), English general and statesman, ordered the destruction of all bats and balls in Dublin. Cricket even appears in a poem entitled the 'Mysteries of Love', by the nephew of John Milton (1608–1674), who wrote, 'Would my eyes had been best out of my head with a cricket ball the day before I saw thee'.

The formation of the Marylebone Cricket Club (MCC) in 1787 resulted in codified laws being developed by 1835. The International Cricket Conference (ICC), so called since 1965, allowed for membership from non-Commonwealth countries. The seven test playing full members now include Australia, New Zealand, England, India, Pakistan, Sri Lanka and The West

Indies. Cricket has been considered Australia's first organised sport, with the game organised by officers of the HMS *Calcutta* in Sydney in 1802–03. In Australia, cricket clubs flourished in the early to mid 19th century with the establishment of Military and Australian, and the Royal Victoria (1826), Melbourne (1838), Prince Albert (1840), and in 1844, the Currency and City Cricket Clubs. In Melbourne the game was played as early as 1836 on Batmans Hill, which was later levelled to become the site of Spencer Street Station.

Football, as we know it, embraces association football (soccer), American gridiron, rugby union, rugby league, Australian rules, and Gaelic football. A game resembling football, Tsu-Chu-Tsu, and meaning to kick the ball with feet (*chu* meaning leather) was played in China around 400 BC. The game calico, which is closer to the modern game, existed in Italy in 1410. Official references to football date to King Edward II's reign in England, when he banned the game in London in 1314. The first soccer rules were formulated at Cambridge University in 1846. The Football Association (FA) was founded in England on 21 October 1863. From Britain, soccer quickly spread throughout the world to countries such as Denmark (1889), and by 1904 was being played in as many as fifteen European countries. In the same year the Federation Internationale de Football Association (FIFA) was formed. In Australia the first soccer club was formed in Parramatta, NSW, in 1880 through the efforts of English schoolmaster, J. W. Fletcher. Together with J. A. Todd they formed a club known as the Wanderers, with Fletcher becoming its first secretary. The first administrative soccer body in Australia was the (NSW) English Football Association formed in 1882. Gaelic football developed from a traditional interparish 'football free for all' with no time limit, no defined playing area and no specific rules. The Gaelic Athletic Association established the game in its present form in 1884. Played throughout Ireland, the first All-Ireland Championship was held in 1887.

Gridiron, or American football, had its origins in the football played by the large ivy league colleges and universities on the east coast of the US in the second half of the 19th century. It is thought to have descended from soccer and rugby in Britain. The first recorded game, in the US, was between Princeton and Rutgers on 6 November 1869 at New Brunswick. Columbia and Yale joined this group in 1873. By 1876 the 'Boston game' based on running with the ball had gained almost universal approval. College football spread far and wide throughout the latter part of the 19th century and during this period developed in professionalism. The American Professional Football Association (APFA) was formed in 1920 and in 1922 renamed the National Football League. The American Football League was formed in 1960. These two leagues merged in 1970 and were later reorganised into the National Football Conference (NFC) and the American Football Conference (AFC). American football is played on both grass and synthetic surfaces.

Football had been played by Irish soldiers in Sydney in 1829 and by Victorian goldminers as early as the 1850s. Australian Rules was introduced through the efforts of Thomas W. Wills, the son of an eminent Victorian pioneer pastoralist. The first match was played between schoolboy teams from Scotch College and Melbourne Grammar in Melbourne on 7 August 1858. Australian football rules were codified in 1866 in Melbourne, with the use of an oval rather than round ball by 1867. Ten years later the Victorian Football Association was founded. Australian Rules football later spread to NSW (1866), South Australia (1875), Queensland and Tasmania (1879), and in 1883 to Western Australia. The VFL (now AFL) Grand Final is played annually at the Melbourne Cricket Ground. This was the site of the first commercial flood lighting using electricity in August 1879.

Rugby Union was supposedly born when William W. Ellis, while playing soccer at the Rugby School, England, in 1823, picked up the ball and ran towards the opponents' goal line. The game spread, particularly among the wealthy and privileged of Britain as well as those who attended the private school system. Certainly rugby was being played at Cambridge University by 1839. The Rugby Football Union was formed on 27 January 1871. The International Rugby Football Board was formed in 1890. International Championships between England, Ireland, Scotland and Wales were first held in 1884. Since 1888, teams representing the British Isles have toured Australia, New Zealand and South Africa. In Australia, the first rugby union game was played in Sydney in 1829, forty-five years before the formation of Australia's first administrative body, the South Union. In 1892 the Southern Union was renamed the NSW Rugby Union. Other Unions were established in 1873 (Scotland), 1875 (Ireland), 1888 (Wales), 1892 (New Zealand), and in 1889, South Africa. Rugby League developed as an extension of the game of rugby union on 29 August 1895, following concern by English players that they should receive similar payments to their soccer counterparts in the British Football Association (BFA). Rugby League has been played principally in Great Britain, France, Australia, New Zealand and Papua New Guinea. In Australia, rugby league was formed in 1907 after a dispute with rugby union representatives over the payment of out-of-pocket expenses to injured players. The Australian Rugby Football League (ARL) Board of Control was formed in 1924 and became the ARL in 1984.

Baseball appeared as a grassed sport in the English language somewhere towards the beginning of the 18th century. Early references were made to baseball in the November 1748 entries of Lady Hervey's letters, and in 1774 in a series of alphabetically arranged verses of children's sports entitled *A Pretty Little Pocket Book* (Anon, 1983). It is now generally accepted that baseball is a more sophisticated version of the English game of rounders, a game which was extensively played in England in the 18th century and taken to America in the early 1800s. In 1840, Alexander J. Cartwright, considered by many to be the real father of the modern game, codified the rules, and on 19 June 1846 the first real game of modern baseball was played between the Knickerbockers and the New Yorks in New Jersey. From this location the game spread widely throughout the US. There are two leagues: the National (NL) and the American (AL) founded in 1876 and 1901 respectively. While there is a tendency to think of baseball solely as an American sport, it is also popular in a number of Central and South American countries, Japan and Australasia where it was introduced in Sydney and Melbourne in the early 1880s. It is thought to have been introduced to Australia by the American miners who migrated during the gold rush of the mid 1850s. Australians took to the game around 1885 in Melbourne, and in 1888 the sports equipment magnate A.G. Spaulding brought two American teams to Australia. Competitive baseball in Australia began around 1899 in NSW. In 1965 the Australian Baseball Council was established.

Modern golf started with the Scottish Parliament passing a prohibiting law in 1457 declaring 'goff be utterly cryit dounce and not usit'. This is the earliest mention of golf, although games of similar style date back as far as AD 400 to Holland and other parts of Europe. The world-renowned Royal and Ancient Golf Club of St. Andrews, in Scotland, dates from around 1400 AD on linksland that is now called the Old Golf Course. In 1888, St. Andrews, in Yonkers, became the first official American golf club. Today there are in excess of 15 000 golf courses in the United States. In Australia, Alexander

Reid, a transplanted Scot, was found playing the 'feathery' ball on farmland at Ratho, Bothwell, Tasmania in the 1820s. He had just returned from a trip to Scotland complete with several wooden golf clubs and featheries, a feather stuffed golf ball, and has since been credited with establishing Australia's first ever golf course. The royal and ancient game of golf was first played in Melbourne when it was reported that the Hon. James Graham, another Scot, had laid out a course on the site of the present Flagstaff Gardens, in Melbourne, in 1847. The Australian Golf Club was formed in Sydney on 12 December 1882 through the efforts of C.E. Riddell on a primitive course at Centennial (Moore) Park. The Sydney Golf Club was established on 3 August 1893 when prominent Concord landowner, Edith Walker, gave permission for the course to be laid out on her spacious property 'Yaralla'. In Melbourne, it was not until 1891 that J.M. Bruce, Thos. Brentnall and William Knox banded together and formed the Melbourne Golf Club, later Royal Melbourne, at Caulfield and later at Sandringham. The Royal Melbourne Golf Club can claim to be the oldest golf club in Australia through the continuity of its existence, followed by The Geelong Golf Club (1892). In Adelaide, South Australia, there was an attempt to establish golf in 1869, with the course being laid out on common park land to the north-east of the present Victoria Park Royal. Other golf clubs, which also commenced prior to the turn of the century but which failed to maintain existence in their embryonic period but were subsequently revived and are now part of the establishment include The Royal Sydney (1893), North Adelaide Golf Club (1890), Adelaide Golf Club (1892), North Queensland Golf Club (1893), later to become the Townsville Golf Club in 1924, and the Newlands Golf Club, Hobart, in 1896.

The rules of golf run according to the Royal and Ancient Golf Club of St. Andrews throughout the world, except for the USA where the game has been legislated by the United States Golf Association. In Australia, golf is governed by the Australian Golf Union which originated from a meeting held at Royal Melbourne Golf Club in 1898 between interstate representatives. The Australian Ladies' Golf Union was established in 1921.

Horse racing appears to have been confined to the early chariot races of the Greeks and Romans. The first recognisable race meeting was held at Smithfield, London, in 1174. The first prize money was a purse of gold offered by English King Richard I in 1195. In Britain the Jockey Club is the governing body of flat racing, steeplechasing and hurdle racing, after merging with the National Hunt Committee in 1968. In Australia horses came in with the First Fleet. The first race meeting was held at Parramatta, NSW, in 1810. Racing during the colonial era lapsed in 1813 when the 73rd Regiment was sent to Sri Lanka and was not restarted until 1825 when a program was held at the newly established course near Bellevue Hill, Sydney. From this site the Sydney Turf Club emerged. In 1838 the Melbourne Racing Club was formed. However it was not until 1861 that the first Melbourne Cup was run. The New Zealand Racing Industry Board (NZRIB) was established in 1991, and now conducts over 600 race meetings annually (Williams, 1994).

Lacrosse is thought to take its beginnings from the early Plains Indians of the United States who played baggataway on the grassed plains and prairies. Early French settlers thought the throwing instrument used by the Iroquous Indians resembled a bishop's crozier and gave the game the French equivalent of 'La Crosse'. Another interpretation is that French settlers may have named lacrosse after their game Chouler a la crosse, played in France as early as 1381. Certainly lacrosse had reached Europe by the 1800s and England by 1867. In North America the first non-Indian club was the Montreal Lacrosse Club, formed in 1856. In 1874, a visiting Canadian to Australia, L. L. Mount demonstrated the sport to a group of boys playing in a Melbourne park. In 1879 the Victorian Lacrosse Association was formed in Australia. The International Federation of Amateur Lacrosse (IFAL) was founded in 1928. The purely Australian horse sport of colors was derived from an equestrian exercise in England, where it combined polo, lacrosse and netball, and was introduced to Australia in 1938 by Edward Hirst of Ingleburn, NSW. The early sport of polo, which also involves horse and rider, had its origins in Manipur State, India, circa. 3100 BC when it was played at Sagol Kangjei. Polo was introduced to England from India in 1869 and now has a keen following in the United States, Argentina and Australia.

Table 1.1 Turfgrass time line for Australian grass sports (1800–1900) (adapted from Blanch, 1978; Anon., 1983)

Year	Event
1802–3	First recorded game of cricket organised by officers of the HMS *Calcutta* in Sydney, NSW.
1810	First race meeting at Parramatta, Sydney on 30 April; second meeting was at Sydney's Hyde Park on 15,17 and 19 October, organised by officers of the 73rd Regiment.
1820s	First golf match in Australia at Ratho, Bothwell, Tasmania.
1826	First Australian Cricket Ground, Sydney Racecourse, now Hyde Park.
1829	First Rugby football played as 'an amusement for the military' in Sydney, NSW.
1830	First steeplechase race in Australia.
1832	First Cricket Club, Hobart, Tas. Cricket was also played in Tasmania in the 1820s.
1844	First Bowling Green, Sandy Bay, Hobart, Tasmania.
1845	First Bowling Green, 'Bowling Green Hotel', Melbourne, Victoria. First Bowling Club in NSW, 'Woolpack Inn', Petersham, Sydney. First Bowling Club, 'Boundary Road Hotel', Surry Hills, Sydney.
1856	American miners who migrated to Australia and the goldfields at Ballarat, Vic., introduce baseball.
1858	First reported Australian Rules match between Melbourne Church of England Grammar School and Scotch College, Melbourne.
1864	Melbourne Bowling Club established.
1868	First Australian cricket team to tour overseas. First Australian Croquet Club, established in Kapunda, South Australia.
1869	First golf course, Adelaide, with nine holes is constructed.
1871	First Brisbane Cup is run at Eagle Farm. Tasmanian Turf Club is established. The SA Cricket Association is established. First bowling green went down in the 1870s at the Maryborough home of James Fairlea.
1872	The Sydney Amateur Athletic Club is established.
1873	Adelaide Cricket Ground is opened.
1874	First tennis court; England's Major Wingfield applies for a patent for 'portable court for playing the ancient game of tennis'. The Southern Rugby Union Football administrative body established.
	The Tasmanian Horse Racing Club established. The South Melbourne and North Melbourne Australian Rules Clubs are founded. Lacrosse is first played in Australia. First women's cricket played at Bendigo, Victoria. First lacrosse game in Melbourne.
1875	The first official polo game is staged at Albert Park, Melbourne, between teams of service officers recruited from vice-regal staff. Morphettville Racecourse opens in SA. Australian Rules football is established in Tas. The Victorian Amateur Turf Club is established. Hobart Cup's inaugural horse race takes place in Tas. The Queensland Cricket Association is established.
1876	Sydney Bowling Club, Government House, Domain, comes into existence. Lawn bowls come to South Australia.
1877	First Test between England and Australia played in Melbourne. Victorian Football Association formed.
1878	Tennis is first played at the Melbourne Cricket Club. The Sydney Lawn Tennis Club is established. First white Australian cricket team tours England.
1879	First lawn tennis match at Melbourne Cricket Club. First lacrosse team is established in Melbourne. First lawn tennis court is built in Melbourne. Southern Tasmanian Football Association is founded in Hobart.
1880	First Victorian Tennis Championships held at Melbourne Cricket Club. First soccer club is founded in Parramatta, Sydney. First match between Clubs Wanderers and King's School, Australian Rules football begins in NSW. First golf match is played in Queensland. Formation of the New South Wales Bowling Association.
1881	Victoria wins NSW/Victoria intercolonial cricket match. NSW plays first intercolonial Rugby Union match against Queensland. NSW wins. First women's lawn bowls competition is held in Melbourne. First Queensland bowling club at Booroodabin.

continued

Table 1.1 Turfgrass time line for Australian grass sports (1800–1900) (adapted from Blanch, 1978; Anon., 1983) *continued*

Year	Event
1882	NSW Soccer Association formed. The first golf club in Australia, the Australian, formed in NSW. Victorian Lacrosse Championships are first held. South British Football Soccer Association is established.
1883	Launceston Bowling Club, Tas., established. Moonee Valley Cup is first run in Melbourne. Footscray Australian Rules Club is established in Melbourne. Queensland's first soccer team, the Rangers, establish a club.
1884	Canterbury Racecourse opens in Sydney. First Lacrosse Club established in NSW. Fitzroy Australian Rules Club is established in Melbourne.
1885	Rosehill Racecourse opens in Sydney.
1886	First women's cricket match played at Sydney Cricket Ground.
1887	NSW Amateur Athletics Association is formed. Wallsend Rovers soccer club is established in NSW.
1888	First state athletic meet is held in Sydney. The Queensland Lawn Tennis Association is established.
1889	Rugby Union Football is established in Victoria. Grace Park Lawn Tennis Club is established in Victoria.
1890	NSW Lawn Tennis Association is founded.
1891	Caulfield Golf Club is established in Melbourne.
1893	Royal Sydney Golf Club is completed.
1894	First men's golf championship is held in Melbourne. First Croquet Club formed in Melbourne (Lilydale).
1895	The first lawn bowling green opens in Perth. Perth Golf Club is founded.
1896	Newlands Golf Club is founded in Tasmania. Eight Football Association Clubs form the Victorian Football League.
1897	Adelaide Lawn Bowling Club is established.
1898	The Australian Golf Union is established. The first women's bowling club is established in Rainsford, Victoria.
1899	Albert Park Ladies Bowling Club, Melbourne, is established. The first polo club in Australia, Northern Challenge Polo Club, is established in NSW.
1900	First game of hockey played between officers and ratings of Royal Navy ships stationed in Adelaide.

Turf research

American W. J. Beal initiated the first acknowledged turfgrass evaluation trials at the Michigan Agricultural Experiment Station in the United States around 1880. These were followed in 1886 by further evaluation trials at New Haven, Connecticut. In 1890 lawn grass experiments were commenced by J. B. Olcott of South Manchester, Connecticut (Olcott, 1890). By 1890 the Rhode Island Agricultural Experiment Station had commenced some lawn experiments and had them extended in 1905. In 1894 the US Golf Association was formed, the Professional Golf Association in 1916, the Green Section of the USGA in 1920, and the Golf Course Superintendents Association of America in 1926. The USGA Green Section was also responsible for the laying down of turf experiments at Arlington, Virginia, which were directed primarily toward the requirements of golf. Today turf research is conducted at a number of universities in North America including The Guelph Turfgrass Institute and the Department of Horticultural Sciences, University of Guelph, Ontario, Canada, and in the United States, Michigan State University, East Lansing; Texas A&M University, College Station; Mississippi State University, Mississippi State; and The University of Georgia, Griffin, as well as government agencies such as the United States Department of Agriculture.

In Britain efforts were made as early as 1924 by the Royal and Ancient Golf Club to form a consultative body on greenkeeping. However it was not until 1929 that the Joint Advisory Council of the British Golf Unions established the world's first turf research station on the St. Ives estate at Bingley, Yorkshire (Dawson, 1929). In 1951 the original Board of Greenkeeping Research was re-organised into the Sports Turf Research Institute (STRI). In the United Kingdom turf research is also being undertaken at such institutions as the Institute of Biological Sciences, the University of Wales; The Institute of Grassland and Environmental Research, Plas Gogerddan, Aberystwyth; Strathclyde University, Glasgow; and the School of Biological and Earth Sciences, Liverpool John Moores University, Liverpool. The Amenity Grass Committee

(NERC, 1977) listed among their research priorities, the standards of management and measurement, establishment and renovation, species and cultivar selection, mowing and growth control, fertilising, wear, weed control, and use of semi-natural areas.

The South African effort commenced with the establishment of a few grass greens in the Durban area in 1891. C. M. Murray of Capetown pioneered much of the early turf research in the early 1900s when he established bermudagrass greens at the Royal Cape Golf Course. It is thought that bowls were first introduced to South Africa in the early 1800s with the establishment of The Port Elizabeth Club, in Port Elizabeth (Louw, 1996). It was not until 1933 that The University of Witwatersrand, Frankenwald Turf Research Section, in Johannesburg, commenced investigative work in turf.

Turf research is also being undertaken in other parts of the world. For example, in Japan, through the Chiba University and the Japan Turfgrass Inc, Chiba, the Institute of Biological Research, Yokohama City University, Yokohama-City; in the Peoples Republic of China through the Institute of Turfgrass Science, China Agricultural University, Beijing, and the Hangzhou Botanical Gardens; Germany through the Department of Crop Production and Grassland Science, University of Hohenheim, Stuttgart, and the INRA-Station d'Amelioration des Plantes Fourageres, Lusignan, France.

In Australia, early reports given by Department of Agriculture agrostologist, F. Turner (1891), made mention of the native weeping grass (*Microlaena stipoides*) as a grass having potential as 'a close turf', and couch (bermudagrass) (*Cynodon dactylon*), as the most 'valuable pasture grass on the eastern side of the Great Dividing Range' in New South Wales. King (1902) conducted research on a number of cool-season grasses in south Gippsland in south-eastern Victoria early in the 20th century. Those of turf interest were the cool-season grasses, meadow fescue (*Festuca pratense* or *F. elatoir*), tall fescue (*Festuca arundinacea*), rough-stalked meadow grass (*Poa trivialis*), crested dogstail (*Cynosurus cristatus*), Italian ryegrass (*Lolium multiflorum*), Kentucky bluegrass (*Poa pratensis*), red fescue (*Festuca rubra*), sheep's fescue (*Festuca ovina*), hard fescue (*Festuca ovina*) and chewings fescue (*Festuca rubra L. var. commutata*). In 1906 work was being carried out on the grasses, paspalum (*Paspalum dilatatum*), Kentucky bluegrass, Italian ryegrass, and perennial ryegrass (*Lolium perenne*) at Wollongbar, New South Wales (Gorman, 1906). By 1909, meadow fescue and small patches of tall fescue were sown down in the Shoalhaven and Milton areas in New South Wales, principally for pasture use for the fledgling dairy industry (Gennys, 1909). Early work on Australia's native grasses for pasture production was also carried out by Baron von Mueller in the Royal Melbourne Botanic Gardens, and a Mr Bacchus near Ballarat, Victoria.

It was not until the 1930s that three eastern Australian states were setting the agenda for addressing the need for turfgrass information, particularly for golf clubs. The Victoria Golf Association (VGA) had been investigating field trials on several Melbourne golf courses since 1938 (Beehag, 1994). In 1970 this organisation, as well as the Royal Victorian Bowling Association (RVBA), lobbied the Victorian government to establish The Turf Research and Advisory Institute (TRAI) in 1973. The Institute remained part of the Victorian Department of Agriculture until 1992 when it ceased to operate. In 1935, the NSW Golf Council, and predecessor to the NSW Golf Association, published the first issue of its turf newsletter 'The Bulletin'. In 1954 the same Association formed the Grass Research Bureau (NSW) Ltd., at Ryde, NSW, and provided funding to support the part-time teachers of greenkeeping to conduct basic research and offer advice to golf clubs. By 1955 the Bureau had established its first newsletter 'Grass Research'. One year later the Royal New South Wales Bowling Association (RNSWBA) commenced contributing to the financing and administration of the Grass Research Bureau. In 1970, the Grass Research Bureau was re-named the Australian Turfgrass Research Institute Ltd (ATRI) (Beehag, 1994). In Queensland, experimental work had commenced as early as 1934 by a small group of golf course enthusiasts. By 1936 the newsetter 'The Australian Greenkeeper' was in production. Currently little research is conducted into warm-season turfgrasses in Queensland.

In Australia, turf research is currently being undertaken by institutions such as the Cooperative Research Centre for Soil and Land Management, and The University of Adelaide, in South Australia; the CSIRO Plant Industry, Canberra; Turfgrass Technology Pty. Ltd., and The University of Melbourne-Burnley in Victoria; the Department of Crop Sciences, The University of Sydney, the Australian Turfgrass Research Institute Ltd. and the Agricultural Research Institute, Wagga Wagga, in NSW; and through CSIRO Tropical Agriculture, Brisbane, Queensland. In 1935 the New Zealand Golf Association established a Greenkeeping Research Committee. In 1949 this committee was re-organised as the New Zealand Institute for Turf Culture and has since been re-named The New Zealand Sports Turf Institute. The majority of turf research in New Zealand is carried through this organisation, often in partnership with other research organisations.

Turfgrass training and education

Degree and diploma courses in turf management are of long standing in universities, polytechnics and colleges in Canada, the United States, Britain, Australia and New Zealand (Aldous, 1997). There is also a tradition of turf management training in South Africa, Switzerland, Germany, France, Japan, Italy and Sweden, and the development of programs in other countries, such as Singapore, China and Malaysia. In the United States, turf management programs are available through a number of universities, either as a four-year program, or as two-year programs at community colleges. The university system also has provision for a comprehensive graduate education program in turfgrass science. The academic structure in Canada is similar to that of the United States with turf management available through colleges and universities. In Britain, as well as Australia and New Zealand, education and training is available at either technical and further education (vocational) level, as well as university level. Vocational training is available at some 26 colleges in Britain at N/SVQ Levels 2 to 4, and as a National Certificate and Diploma. In Australia, vocational training in turf management at National Certificate levels 1 to 6, are offered principally through the Institutes of Technical and Further Education. Britains' higher education institutions also offer turf management training at the Higher National Certificate and Diploma level, and as Bachelor of Science modules. In Australia there are no first degrees in turf management but introductory turfgrass science and management may be taken through the Universities of Sydney and Melbourne, as part of the baccalaureate degrees in agriculture and horticulture, as well as graduate diplomas and postgraduate studies. In New Zealand, turf management is available as a Diploma in Turf Culture or polytech short course level of qualification (Way, 1994), as well as a senior subject in the Bachelor of Applied Science degree and Diploma of Rural Studies at Massey University in Palmerston North.

Many institutes and associations continue to play a significant part in the training of turf personnel. For example the Canadian Golf Course Superintendents Association (CGCSA) and United

States Golf Association (USGA) provide such training, often in association with local Colleges, Institutes and Universities. In Australia these organisations include Parks and Leisure Australia (PLA), the Australian Golf Course Superintendents Association (AGCSA), and Turfgrass Technology Pty. Ltd; in New Zealand, the New Zealand Sports Turf Institute; and in Britain, the Institute of Leisure and Amenity Management (ILAM) and the Institute of Groundsmanship (IOG). Five countries offering the major turfgrass education and training programs, by population, gross domestic and national product, and estimated student and staff numbers in certificate, diploma, degree and postgraduate training in turf management, are summarised in Table 1.2.

Future growth and development in the turfgrass industry will require the implementation of a balanced set of education and training policy measures, based on industry requirements and implemented in line with labour planning requirements. It is also important that a career structure be developed that enables people in the turfgrass industry to establish and progress through career goals consistent with their aspirations and ability. To achieve this universities, community colleges, polytechnics and institutes of technical and further education must maintain strong contacts with the turfgrass industry. In turn, the industry must also be more involved with the substance and monitoring of turf management programs and the placement of graduates, diplomats and operators.

The turfgrass industry

Nutter (1965) described the turfgrass industry as comprising the production and maintenance of specialised grasses and other ground covers as required in the development, maintenance and management of facilities for utility, beautification and recreation. In 1969, Nutter and Watson analysed the industry under four sections: (1) facilities or agencies that deal with the management and maintenance of turfgrass; (2) manufacturing, or the provider of turf products to

Table 1.2 Comparision of selected socio-educational characteristics between Australia, New Zealand, Canada, the United States of America, and the United Kingdom in Turf Management (after Aldous, 1997).

Item	*Australia*	*New Zealand*	*Canada*	*United States of America*	*United Kingdom*
Population (1990) — millions	18	3.5	26.5	250	57
Values					
Gross Domestic Product ($US) 1990 — (per person)	16 050	13 490	19 650	21 360	14 960
Gross National Product ($US) 1989 — millions	242 131	39 437	500 337	5 237 707	834 166
Estimated students numbers (1996)[1]					
Postgraduate degree number per 100 000 pop'n	2.9	4.6	15.7	10.6	1.7
Diploma number per 100 000 pop'n	13.3	28.6	18.1	14.8	14.6
Certificate number per 100000 pop'n	77.8	142.5	45.3	20.0	26.5
Estimate total enrolled	1690	616	2080	11 350	2438
Academic turf management staff employed for degree and vocational programs (full-time equivalents)[2]	0.5–3.5	0.5–1.5	1.0–8.0	0.4–7.5	2.0–13.0

1 Student estimates from academic and vocational programs majoring in turf management, as well as courses having only an introductory program in turf management.
2 Staff estimates only.

the industry; (3) servicing, or those that use the products and facilities; and (4) institutions, which provide education, extension and research expertise in advancing the turf management industry. Facilities that include turf in the landscape are many and wide ranging: airports, athletic fields, bowling and croquet greens, the grounds and athletic fields of universities, community colleges and institutes, lawn cemeteries, crematoria and memorial parks, churches and synagogues, courthouses and governmental buildings, exposition and fairgrounds, garden apartments, golf courses and driving ranges, grass tennis courts, highway median strips, roadside verges, hospitals and nursing grounds, hotel, motel and caravan parks, housing projects and subdivisions, industrial parks and estates, commercial, industrial and residential lawns, military bases, parks and playgrounds, racecourse and tracks, retirement villages, schools (kindergarten, primary and secondary), zoological and botanic gardens, and the natural estate. The manufacturing sector involves machinery, equipment aides for sowing, mowing, fertilising, topdressing, irrigating, rolling, spraying, fertilisers and nutrition, growth regulator chemicals, irrigation system components, pesticides, seed and vegetative materials, instant lawn, computers, and special products such as soil components and amendments. The servicing section involves people associated with sales, distribution, retail (all products), contracting and consultancy services, architects and designers, maintenance, and soil and water testing laboratories. The institutional section includes the universities, polytechnics, colleges, trade and professional organisations, both private and publicly funded, that provide relevant teaching, research and extension for the industry.

Land area, labour and wealth

The value of the industry has been difficult to quantify, both from the value of the land, its labour requirements, as well as the benefits that turf contributes to human physical and mental health, health and safety, and improving the environment. Some attempts have been made to determine the value of the turfgrass industry in a number of countries. In 1965, American Gene Nutter estimated that turfgrass maintenance expenditure in the United States was $US4 326 546 994 to maintain more than 20 million acres of its major turfgrass facilities. This value constituted eleven categories of specialty turfgrass use, namely, airfields, lawn cemeteries, commercial, industrial and residential lawns, churches, colleges and universities, golf courses, roadsides, municipal parks, public schools, and miscellaneous areas. Labour, equipment and water were the major expenditures in the 1965 survey. In 1993, the US Environmental Protection Agency, DPRA Inc., in Manhattan, Kansas, estimated that the current turf acreage (based on the assumption that turf acreage is directly proportional to the population) was 46.5 million acres, with maintenance expenses ranging from $US58 per acre (roadsides) to $US1651 per acre (golf courses). In addition the US Department of Commence, Bureau of the Census, estimated that more than 4.3 million American households now purchase in excess of $US420M of turfgrass sod annually.

In 1895, there were 80 golf courses in the United States and by 1920, 477 member clubs. In 1997 there are now over 0.53 million hectares of high-use turf managed on 15 000-plus golf courses in the United States (Haydu *et al.* ,1997). Research studies conducted in Florida between 1974 and 1994 indicated that total maintenance costs rose from $US2818 per hectare to $US8855 per hectare for the golf course industry. However when adjusted for inflation, the differences in these costs were much smaller (Haydu *et al.*, 1997). In Florida alone, total employment in the turfgrass industry amounted to 130 000 full-time equivalent employees (FTEs), which is three times as great as all other agricultural industries in the state (Hodges *et al.*, 1997). Annual labour costs averaged $US1542 per hectare ($US624 per acre) of turfgrass maintained or $US10 825 per FTE employee.

Britian's amenity grasslands, which have been defined as grasslands having a recreational, functional or aesthetic value, and which are not primarily used for agricultural production. These grasslands constitute 8500 square kilometres in area with maintenance costs of 137M pounds (NERC, 1977). In 1973 there was an estimated 490 square kilometres of school playing fields, 871 square kilometres of golf fairway, greens, tees and roughs, and 1345 square kilometres of urban parks and open spaces in Britain (NERC, 1977). Semi-natural and amenity areas account for an estimated 586 187 hectares, ten per cent of which is golf course rough and intensively managed amenity grassland (Cobham, 1983). In 1989, the Institute of Horticulture in Britain estimated that the number of employees in the amenity horticulture area, in which turf is included, was up to 32 920 people. In the public sector 28 797 of these individuals were employed in urban parks and open spaces, 2800 in sports facilities, and 808 in public golf courses. In 1992 the British Association of Landscape Industries reported that some 13 500 individuals were employed in the field of landscaping and grounds maintenance, followed by 11 000 in golf and greenkeeping, and 4000 as curators of pitches at independent schools and universities (Anon, 1995). Bowling green curators, horse and greyhound racetrack greenkeepers only constituted 1000 and 220 employees respectively. In 1992, the European Golf Association quoted in 'Golf Enterprise Europe' that the number of golf courses in Europe was 1750, which included France (425), Germany (329), Sweden (257), Spain (131), Netherlands (119), Italy (117), Finland (71), Denmark (69), Austria (55), Belgium (49), Switzerland (40), Iceland (37), Portugal (26), Norway (19), Greece (5) and Luxembourg (1).

In Australia, some 180 000 hectares of land are dedicated to high-use turf, with established annual maintenance costs of $A1.3 billion. This is exclusive of the capital costs in establishing new facilities, as well as the annual costs of maintenance of the more naturalised amenity grasslands. In 1996, McIver estimated that there were 1500 golf courses covering 48 000 hectares and employing 6500 maintenance staff; 2000 lawn bowling clubs employing 3000 greenkeepers to maintain 3200 greens, a total of 450 hectares, and a turfgrass sod industry of 200 primary production farms of over 5000 hectares. In addition, there were 748 city and shire councils each maintaining playing fields, parks, gardens and golf courses. Valuations placed on the land used for sports turf in Australia are in the order of $A4.5 billion, or an average of $25 000 per hectare of sports turf. These land values range from $A130 000 to 140 000 per hectare for bowling and croquet clubs to golf courses at $A3700 per hectare. In addition, the sod production industry in eastern Australia has an estimated market value in excess of $A100 million per annum (Martin and Aragao, 1996), with smaller numbers of sod farms also found in Tasmania, South Australia and Western Australia. Results indicate that there is the equivalent of one hectare of sports turf to every 100 Australians, which equates favourably with 85 New Zealanders (Way, 1994), 500 Americans, and 769 Britains (McLaughlin, 1994).

Sports turf in New Zealand is managed over 40 000 hectares of sports turf and 61 000 hectares of non-sports turf, with an established total worth of $NZ86 million per annum (Dale, 1994; Way, 1997). Golf courses constitute the largest proportion (42 per cent), then schools' sporting fields, followed by councils and racing clubs. There are over 400 golf courses in New Zealand, which provides one course for every 8750 people. The total value of expenditure for New Zealand turf has been estimated at $NZ86.3

million with labour constituting just under half of this amount. New Zealand employs approximately 3200 grounds staff, with one-third of those employed on a part-time basis. New Zealand golf clubs employ 1.6 full-time workers per course with voluntary workers contributing a significant part of the workforce. Bowls in particular is the major beneficiary of voluntary workers, but also golf, racing clubs, croquet and other users, all have significant contributions from volunteers. The horse racing industry in large in New Zealand, employing some 28 000 people, and has an export income of approximately $NZ69 million (Williams, 1994).

References

Aldous, D.E. 1997. 'World trends in education and training for turfgrass science and management', in *International Turfgrass Society Research J.*, vol. 8: 1097–108. University Printing Service, University of Sydney, NSW.

Anon. 1884. 'Managing Lawns'. *The Gardeners' Chronicle*, 23 Feb., pp. 257, 260.

Anon. 1983. *The Macquarie History of Ideas*. Macquarie Library Pty. Ltd., 959 pp.

Anon. 1995. *Turfgrass Industry Pocket Book*, National Turfgrass Council, Minchinhampton, Glos., 26 pp.

Beard, J.B. 1973. *Turfgrass: science and culture*. Prentice-Hall Inc., Englewood Cliffs, N.J.

Beard, J.B. 1977. 'Turfgrasses add to the quality of life', *Texas Agricultural Progress*, vol. 23, no.4, pp. 17–18.

Beard, J.B and R.J. Green. 1994. 'The role of turfgrass in environmental protection and their benefits to humans', *J. of Environmental Quality*, (23):452–60.

Beehag, G. W. 1994. 'Four Decades On: The History of the Australian Turfgrass Research Institute Ltd', *ATRI Turf Notes*, Summer, pp. 2–6.

Bennett, E.S. and J.E. Swasey. 1996. 'Perceived stress reduction in urban public gardens'. *Hort.Technology*, April–June, 6(2):125–8.

Blanch, J. 1978.(ed.). *Ampol's sporting records*. Jack Pollard Publishing Pty. Ltd., Crow's Nest, NSW. 482 pp.

Canaway, P.M., M.J. Bell, G. Holmes and S.W. Baker. 1990. 'Standards for the playing quality of natural turf for association football'. *Natural and Artificial Playing Fields: Characteristics and Playing Features*. American Society for Testing and Materials. STP 1073, Philadelphia, pp. 29–47.

Carne, J. 1994. 'Urban vegetation: ecological and social value', *Proceedings of the 1994 National Greening Australia Conference, October 4–6, Fremantle, Western Australia*, pp. 211–26.

Cobham, R.O. 1983. 'The economics of vegetation management', in J.M. Way (ed.), *Management of Vegetation, BCPC Monograph No. 26*, British Crop Protection Council.

Dale, P. 1994. 'The importance of sport in New Zealand society', R.J. Gibbs and M.P. Wrigley (ed.), *Proceedings of the 5th NZ Sports Turf Convention, May, Massey University, New Zealand*, New Zealand Turf Culture Institute, 210 pp.

Dawson, R.B. 1929. 'St. Ives Research Station, its surroundings and historical associations', *The Journal of the Board of Greenkeeping Research* 1(1):9–11.

Dawson, R.B. 1949. *Practical Lawn Craft*, Crosby Lockwood and Sons Ltd., London, 315 pp.

Evelyn, J. 1932. *Directions for the Gardiner at Says-Court*, Geoffrey Veynes (ed), Nonesuch Press, England, pp. 1–109.

Finnagan, P., J.M. Raupach, and H. Cleugh. 1994. 'The impact of vegetation on the physical environment of cities', *Proceedings of the 1994 National Greening Australia Conference, October 4–6*, Fremantle, Western Australia, pp. 23–37.

Gennys, R.H. 1909. 'Splendid pasture grasses', *Agricultural Gazette of NSW*, September 2, p. 750, and July 2, pp. 565–6.

Gerty, P. 1996. 'Lawn bowls in Australia: The story so far', *Jack High*, March, 2 pp.

Gibbs, R. 1997. 'Further comparisons of natural and synthetic bowling greens', *The 1997 Green Pages Annual*, Strategic Publications, New Gisbourne, Victoria, pp. 43–4.

Givoni, B. 1991.' Impact of planted areas on urban environmental quality: a review', *Atmospheric Environment*, 24B-3:289–99.

Goethe, C.M. 1955. *Garden Philosopher*, The Keystone Press, Sacramento, California, 327 pp.

Gorman, C.H. 1906. 'Grass testing plants at Wollongbar', *Agricultural Gazette of NSW*, p. 273.

Haydu, J.J., A.W. Hodges, P.J. van Blokland and J.L. Coles. 1997. 'Economic and environmental adaptations in Florida's golf course industry: 1974-1994', *International Turfgrass Society Research J.*, vol. 8: 1109–16, University Printing Service, University of Sydney, NSW.

Hodges, A.W., J.J. Haydu and J.J. van Blokland. 1997. 'Employment and value-added in Florida's turfgrass industry', *International Turfgrass Society Research J.*, vol. 8:1117–25, University Printing Service, University of Sydney, NSW.

Huffine, W.W. and F.V. Grau. 1969. 'History of turf usage', A.A. Hanson and F.V. Juska (eds) in *Turfgrass Science*, American Society of Agronomy, pp. 1–8.

Kaplan, R. and S. Kaplan. 1989. *The experience of nature*, Cambridge University Press, Mass.

King, J.W. 1902. 'Experiments with grasses and fodder plants', *Agricultural J. of Victoria*, pp 535–9.

Langer, E.J. and J. Rodin. 1976. 'The effects of choice and enhanced person responsibility for the aged: a field experiment in an institutional setting', *J. of Personality and Social Psychology*, 34:191–8.

Louw, C. 1996. 'Bowls in South Africa', *ATRI Turf Notes*, Winter, pp 5–6, 10.

Malone, C.B. 1934. 'History of the Peking summer palaces under the Ching Dynasty', *Illinois Studies in the Social Sciences*, Vol. XIX Nos. 1–2. University of Illinois, 247 pp.

Mastalerz, J.W. and C.R. Oliver 1974. 'Microclimatic moderation: development and application in the urban environment', *HortScience*, Vol. 9(6):560–3.

Martin, P. and S. Aragao. 1996. 'The agronomy of turf farming systems in eastern Australia', *The 1996 Green Pages Annual*, Strategic Publications, New Gisbourne, Victoria, pp. 5–6.

McLaughlin, J. 1994. 'Council training: past, present and possible', R.J. Gibbs and M.P. Wrigley (eds). in *Proceedings of 5th Sports Turf Convention, May, Massey University, New Zealand*, pp. 21–2.

McIver, I. 1996. 'The Australian turf industry: environmental issues for turf', *Symposium, Penrith, NSW*, The Australian Turfgrass Research Institute Ltd, Concord West, NSW, pp. 1–5.

Mecklinburg, R.A., W.F. Rintelaman and D.B. Schumaier. 1972. 'The effects of plants on microclimate and noise reduction in the urban environment', *HortScience*, 7:37–9.

NERC. 1977. *Amenity grasslands-the need for research*, Natural Environment Research Council, Publication Series 'C', no. 19, London.

Nutter, G.C. 1965. 'Turf-grass is a $4 billion dollar industry', *Turf-Grass Times*, 1(1)1–22.

Nutter, G.C. and J. R. Watson, Jr. 1969. 'The turfgrass industry', A.A. Hanson and F.V. Juska (eds), *Turfgrass Science*, American Society of Agronomy Inc., Madison, Wisconsin. 2:9–26.

Olcott, J.B. 1890. 'Grass-keeping', *14th Annual Report of the Connecticut Agricultural Experiment Station for 1890*, pp. 162–74.

Relf, D. and S. Dorn. 1995. 'Horticulture: meeting the needs of special populations', *HortTechnology*, 5(2): 94–101.

Roberts, E.C. 1985. 'Lawns enhance the environment', *HortScience*, vol. 20(2):166.

Rohde, E.S. 1927. 'The garden', *II. Lawns. Nineteenth Century and After*, 104:200–209.

Sanecki, K. N. 1997. *Old Garden Tools*, Shire Publications Ltd, pp. 3, 22, 23.

Smith, D. and D.E. Aldous. 1994. 'Effect of therapeutic horticulture on the self concept of the mildly intellectually disabled student', in M. Francis, P. Lindsay, and J. Stone (eds.), *The healing dimensions of people-plant relations*, Centre for Design Research, Univ. of California, Davis, 498 pp.

Turner, F. 1891. 'The Grasses of New South Wales', *New South Wales Agricultural Gazette*, vol. 11, pp. 22, 238.

Ulrich, R.S. 1990. 'Effects of healthcare interior design on wellness: Theory and recent scientific research', *Third Symposium on Healthcare Design, San Francisco, CA*. In S.O. Marberry, (ed.). *Innovation in Healthcare Design*, New York, Van Nostrand Reinhold.

Vamplew, W., K. Moore, J. O'Hara, R. Cashman, and I. Jobling. 1997. *Oxford Companion to Australian Sport*, 2nd edn, OUP, Melbourne, 574 pp.

Way, B. 1994. 'The New Zealand sports turf industry: statistics and comparisons', R.J. Gibbs and M.P. Wrigley (eds) in *Proceedings of 5th NZ Sports Turf Convention, May, Massey University, New Zealand*, pp.15–18.

Way, B. 1997. 'New Zealand–a nation built on grass', in addendum *International Turfgrass Research Conference*, University of Sydney, NSW, pp.1–2.

Williams, T. 1994. 'The impact of substandard surfaces on the racing industry', R.J. Gibbs and M.P. Wrigley (eds), in *Proceedings of 5th NZ Sports Turf Convention, May, Massey University, New Zealand*, p. 10.

CHAPTER 2

Turfgrass identification and selection

D.E. ALDOUS, Institute of Food and Land Resources, University of Melbourne, Richmond, Victoria, and **P.S. SEMOS,** StrathAyr Pty. Ltd., Seymour, Victoria, Australia

Introduction

One of the largest angiosperm families, the *Poaceae*, contains approximately 785 genera and 10 000 species (Watson and Dallwitz, 1992). Sixteen genera are found within this family, and about 40 species in the subfamilies of *Eragrostoideae*, *Festucoideae* and *Panicoideae*, contain the world's principal turfgrasses. The *Festucoideae* contain the main cool-season turfgrasses, such as the fescues (*Festuca*) and bluegrasses (*Poa*), while the *Eragrostoideae* and *Panicoideae* contain examples of the main warm-season turfgrasses, such as the bermudagrasses (*Cynodon*), and Queensland blue couch (*Digitaria*). However less than 30 of all these grass species have the desirable characteristics to be maintained as turfgrasses worldwide (Beehag, 1995b). The taxa used in grass classification are illustrated in Table 2.1.

The need for classification has largely arisen because of the uncertainty of like grasses bearing different names, and to remove confusion in identifying other grass-like species. For example, in Australia, common couch (*Cynodon dactylon*) is better known as bermudagrass in the United States, skireek grass in South Africa, neguil in Egypt, ohoob or doob in India and serangoon in Malaysia. Similarly, nutgrass (*Cyperus rotundus*); Family *Cyperaceae*, mullumbimby couch (*Kyllinga brevifolia*); Family Cyperaceae, onion weed (*Nothoscordum inordorum*); Family *Amaryllidaceae*,

Table 2.1 Taxa used in the classification of tall fescue (*Festuca arundinacea*) cv. SR.8200 and Queensland blue couch (*Digitaria didactyla*) (adapted from Beard, 1973).

Trinomial: *Festuca arundinacea* Schreb. SR. 8200		Trinomial: *Digitaria didactyla* Willd.	
KINGDOM	Plantae, plant kingdom.	KINGDOM	Plantae, plant kingdom
DIVISION	Embrophyta, embryo plants	DIVISION	Embrophyta, embryo plants
SUBDIVISION	Phanaerogame, seed plants	SUBDIVISION	Phanaerogame, seed plants
BRANCH	Angiospermae, seeds enclosed in ovary	BRANCH	Angiospermae, seeds enclosed in ovary
CLASS	Monocotyledoneae, monocotyledons	CLASS	Monocotyledoneae, monocotyledons
SUBCLASS	Glumiflorae, having chaffy leaves	SUBCLASS	Glumiflorae, chaffy leaves
ORDER	Poales, grasses and sedges	ORDER	Poales, grasses and sedges
FAMILY	Poaceae, grass family	FAMILY	Poaceae, grass family
TRIBE	Festuceae, fescue tribe	TRIBE	Paniceae
GENUS	Festuca, fescues	GENUS	Digitaria
SPECIES	arundinacea	SPECIES	didactyla
CULTIVAR	SR.8200	CULTIVAR	none available

and rushes (*Juncus spp.*); Family *Juncaceae*, are all grass-like in appearance. The structural features that provide a basis for separating the *Poaceae* from the *Cyperaceae* and *Juncaceae* are listed in Table 2.2.

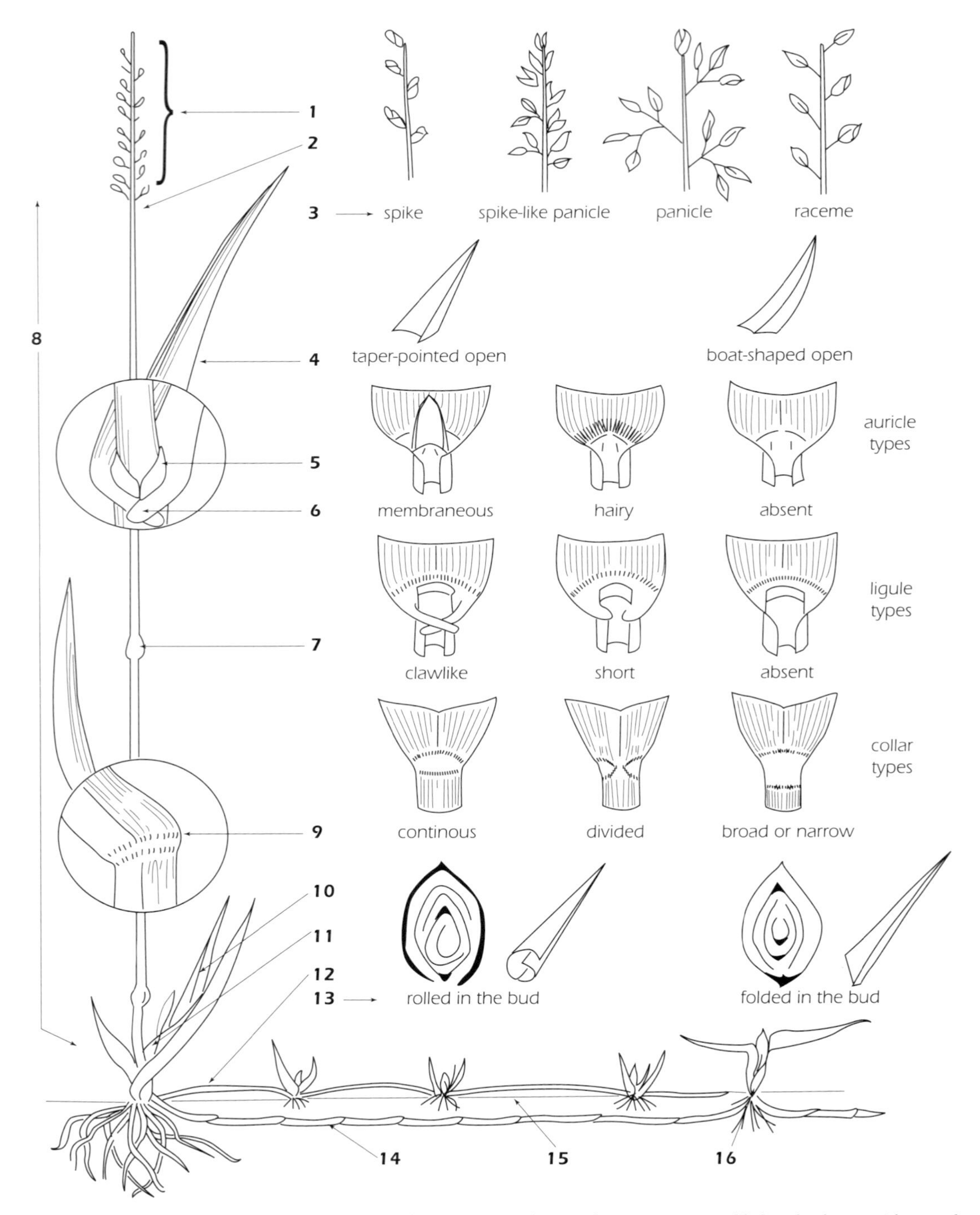

Figure 2.1 Grass plant showing typical structures: 1 inflorescence, 2 rachis, 3 inflorescence types, 4 blade, 5 ligule, 6 auricle, 7 node, 8 culm, 9 collar, 10 tiller, 11 crown, 12 stolon, 13 vernation, 14 rhizome, 15 internode, 16 new shoot

Table 2.2 Morphological and anatomical features separating the *Poaceae, Cyperaceae,* and *Juncaceae* families (adapted after Lambrechtsen, 1986).

	GRASSES *(Poaceae)*	SEDGES *(Cyperaceae)*	RUSHES *(Juncaceae)*
Leaf whorl	(Top view)–two rows	(Top view)–three rows	(Top view)–two rows
Leaf sheath	Usually split	Never split	Sometimes split, sometimes closed
Leaf blade	Flat, rolled	Folded	Usually not present expanded
Leaf margins	Rough or smooth	Rough	Smooth
Leaf bud (vernation)	Round or oval in cross section, rolled or folded in bud Hollow internodes Interrupted by nodes	Triangular in cross section Not hollow, Round nodes inconspicuous	Round in cross section Not hollow Inconspicuous nodes

The classification system is related to a pattern of nomenclature at the genus level. Further division takes place at the genus and species level. For example, bermudagrass belongs to the Family *Poaceae,* the genus *Cynodon,* and species *dactylon*. In selecting the species name, in most instances botanists have given some indication of the appearance of the plant, its geographic habitat or indicated the person who either discovered, described, or assisted with introduction of the plant. For example, South African couchgrass (*Cynodon transvaalensis*), originated in the Transvaal region of South Africa. The scientific name is then followed by the authority (or authorities) credited with naming the species. For example, Kentucky bluegrass, *Poa pratensis L.*, the *L.* referring to the Swedish botanist Carolus Linnaeus (1707–78), who identified this species and published the name in *Species Plantarum* in 1753. In other cases the botanical name has become the common name, such as with *Poa annua,* also called winter grass or annual bluegrass.

Identification of turfgrass genera and species

The typical grass structure is illustrated in Figure 2.1. The vegetative plant parts of a grass given here are essential to an understanding of the family and to the use of keys for identification.

Auricle: outgrowths of the edge of the leaf occurring in some grasses from either side of the collar at the junction of the leaf; present in varying sizes and shapes, or may be absent.

Awn: slender hairlike projection(s) arising from small flowers of grasses; variable in length and texture.

Blade: the extended upper portion of the leaf beyond the sheath; present in cross-section as flat, v-shape, thread-like; blade tip as sharply pointed, boat or canoe shape; leaf surface and margin as hairy, saw-like; and leaf colour as light to dark green.

Collar: strengthening tissue at the back of the ligule, immediately above the leaf sheath, often a different colour from the rest of the leaf blade; present as a broad, narrow or divided band.

Crown: refers to the junction of the root and stem, usually at ground level.

Culm: the stem of grasses and sedges.

Fibrous or bunch root system: a root system in which the roots are finely divided, usually in a clump.

Inflorescence: refers to the part of the turfgrass that consists of the flower bearing stalk.

Internode: refers to the stem section between the swollen nodes, often hollow.

Keel: leaf shape of lower surface of the leaf; protrudes like the keel of a boat or canoe.

Lateral shoot: Shoots originating from vegetative buds in the axils of leaves or from the nodes of stems, rhizomes or stolons.

Ligule: the small membrane or ring of hairs that occurs on the upper side of the leaves, just at the junction of the leaf blade and the sheath and wraps around the stem; reduced or absent in some species; present as fringe of hairs, acute, truncate or ciliate.

Margin: refers to the outside edges of the leaf blade.

Midrib: central vein of the blade of the leaf, often forming a pronounced ridge on the upper surface and a keel below.

Node: refers to the point of the plant stem from which the leaves or lateral stems grow.

Panicle: an open, often branched and spreading flowering structure; one type of common inflorescence.

Papery bracts: a series of small dry structures that surround or enclose the seed of almost all grasses.

Perennial: a plant that lives for an indefinite number of growing seasons; usually propagated vegetatively, perhaps seed; may or may not flower the first growing season, but continues to grow and flower thereafter.

Rhizome: refers to a horizontal, non-green, underground creeping stem. Often has short, scale-like leaves and may produce roots and/or leafy shoots from the nodes. Turfgrasses bearing them exhibit a rhizomateous habit of growth.

Rachis: the axis of the inflorescence.

Scabrous: rough and harsh to the touch; often show short stiff bristles or saw-like teeth on the part concerned.

Sheath: refers to that part of the leaf originating from the node, and surrounds part of the stem. It can be cylindrical or compressed with overlapping, open or closed margins. The leaf sheath may be split, split to near base with margins overlapping or closed.

Spike: a narrow and usually longer than wide inflorescence, the flowers borne along one stem.

Spikelet: the small flowering unit of grasses, consisting of a series of bracts placed one inside another, attached to a branch of the main flowering stem. There are many spikelets in one inflorescence.

Stolon or runner: refers to the horizontal stem, usually prostrate or trailing at ground level, always above the ground, often rooting at the nodes. Turf plants bearing them exhibit a stoloniferous habit of growth.

Tiller: a shoot or stem that arises from the base in grasses. Termed vegetative tiller if it produces leaves only, and a flowering tiller if it bears an inflorescence.

Vein: refers to the specialised organs which conduct plant foods to the sheath and leaf, and which removes from these organs, substances synthesised under the action of photosynthesis. Often prominent on the upper surface of the leaf blade.

Venation: refers to the veins running parallel to the apex of the blade. Some species have distinct mid-veins, while others have veins uniformly distributed across the leaf. The prominence of venation, on the upper and lower leaf surface, is useful in identification.

Vernation: refers to the arrangement of the youngest leaf in the bud shoot; either folded (conduplicate) or rolled (convolute) in budshoot.

Use of turfgrass key

A key is an artificial guide used to separate out unknown grasses. Keys are typically dichotomous, composed of couplets of two sentences with the same letter. Keys are intended to cover only a restricted number of grasses for which detailed descriptions have been carried out, and are not designed to provide a positive means of identification. However by using the couplets, and by the process of elimination, a point is arrived where you may be reasonably sure of the result. Confirmation requires a more detailed description.

To use this key, which has been adapted from Nowosad, Swales and Dore (1936), and Hanson, Juska and Burton (1969), begin with the first set of couplets, either Group 1 or Group 2. Group 1 describes the specimen as being leaf venation folded in the bud, whereas in Group 2 the leaves are rolled in the bud. If the specimen has a folded bud shoot, the user then proceeds to A. Auricles present, where there will again be two alternatives, i.e. Auricles present or AA. — auricles absent or small. Choose the one which applies, and proceed on as before. Eventually the specimen will align with a scientific name. If unsuccessful, study the specimen carefully, or use another representative specimen, and commence again.

Table 2.3 Vegetative indentification key for naming some common turf and amenity grasses.

Key	Species
Group 1. Leaf venation folded in bud	
A. Auricles present	
B. Blade glossy on under surface; margin of ligule not hairy; auricles blunt to claw-like	
C. Auricles with short hairs	*Festuca arundinacea* **(tall fescue)**
CC. Auricles glabrous	
D. Blade scabrous on margins, ligule generally < 0.5 mm long	*Festuca elatior* **(meadow fescue)**
DD. Blade smooth on margins near base, ligule generally 1.0 mm +	*Lolium multiflorum* **(Italian ryegrass)**
BB. Blade not glossy on under surface; margin of ligule hairy; auricles, clawlike	
C. Collar slightly pubescent; midrib not prominent on under surface of blade; upper surface not prominently ridged	*Agropyron repens* **(twitchgrass)**
D. Collar smooth; midrib conspicuous on under surface of blade; upper surface of blade prominently ridged	*Agropyron cristatum* **(fairway wheatgrass)**
AA. Auricles absent or small	
B. Ligule a fringe of hairs	*Zoysia japonica* **(Japanese lawngrass)**
BB. Ligule membranous	
C. Sheath closed	
D. Sheath and blade smooth	*Bromus inermis* **(smooth brome)**
DD. Sheath and blade hairy	
CC. Sheath split (margins generally overlapping)	
D. Hairs present on sheath, blade, or collar	
E. Sheaths compressed	
F. Sheath smooth	*Digitaria ischaemum* **(smooth crabgrass)**
FF. Sheath hairy	
G. Stolons or creeping stems present	*Digitaria sanguinalis* **(large crabgrass)**
GG. Stolons absent	
H. Blade, soft, flat, short, hairy on upper surface	*Holcus lanatus* **(Yorkshire fog)**
HH. Blades not dense, short hairy on upper surface	
I. Ligule with dense row of whitish hairs at back	*Paspalum notatum* **(bahiagrass)**
II. Ligule without whitish hairs at back	
J. Ligule rounded to acute; sheaths very hairy below, sparsely above	*Paspalum dilatatum* **(dallisgrass)**
JJ. Ligule truncate; sheath slightly hairy on margins and midrib	*Paspalum laeve* **(field paspalum)**
EE. Sheath not compressed, ligule membranous, ciliate, sometimes dense ring of hairs	*Bouteloua curtipendula* **(sideoats grama)**
DD. Lack of hairs on sheath, blade or collar	
E. Ligule >1.5 mm long	

continues

Table 2.3 Vegetative indentification key for naming some common turf and amenity grasses. *continued*

Group 1. Leaf venation folded in bud *continued*

F. Margin of collar slightly hairy; ligule with prominent notch on either side; culms with a bulbous base — *Phleum pratense* **(timothygrass)**

G. Margin of collar smooth; ligule without a notch on either side; culms lack bulbous base

G. Ligule white, papery, 2.0–8.0 mm long, acute or obtuse; leaf blade 6.0–15.0 mm wide — *Phalaris arundinacea* **(reed canarygrass)**

GG. Ligule-thin membranous, 1.5–4.0 mm long rounded or acute; blade 1.5–7.0 mm long

H. Stolons absent; rhizomes present — *Agrostis alba* **(redtop)**

HH. Stolons long, prostrate — *Agrostis stolonifera* **(creeping bentgrass)**

EE. Ligule short (<1.5 mm long), truncated

F. Blade 1mm wide, smooth upper and lower surface — *Agrostis canina* **(velvetgrass)**

G. Blade 2.0–3.0 mm wide, smooth to rough on upper and lower surface — *Agrostis tenius* **(colonial bentgrass)**

Group 2. Leaf venation rolled in bud

A. Auricles present; lower sheaths reddish at base; smooth throughout — *Lolium perenne* **(perennial ryegrass)**

AA. Auricles absent

B. Ligule a fringe of hairs

C. Sheaths greatly overlapping between nodes, rhizomes and stolons present — *Cynodon dactylon* **(bermudagrass, common couchgrass)**

CC. Sheaths not greatly overlapping between nodes, no rhizomes, stolons present

D. Blade petioled above ligule — *Stenotaphrum secundatum* **(St. Augustine grass)**

DD. Blade not petioled above ligule

E. Blade with few long hairs scattered on both surfaces — *Buchloe dactyloides* **(buffalograss)**

EE. Blade without long hairs scattered on both surfaces

F. Blade 1.0–2.5 mm wide, sharp point, scabrous or hairy on upper surface near base — *Bouteloua gracilis* **(blue grama)**

FF. Blade 4.0–8.0 mm wide, obtuse, smooth or hairy at base — *Axonopus affinis* **(narrowleaf carpetgrass)**

BB. Ligule, short-membranous; short and hairy

C. Collar continuous, broad, hairy, turfted at lower edge — *Eremochloa ophiuroides* **(centipede grass)**

CC. Collar divided, narrow, mostly hairy on margins — *Andropogon virginicus* **(bromesedge)**

BBB. Ligule membranous

C. Hairs on margins of collar, sheath and upper leaf surface — *Eleusine indica* **(crowsfoot grass)**

CC. No hairs at margins of collar, sheath smooth or very hairy

D. Blade prominately ridged on upper surface, narrow to bristle-like

E. Ligule < 0.5 mm or absent; sheath split; leaves smooth, blue-green; turfted — *Festuca ovina* **(sheep's fescue)**

EE. Ligule some 0.5 mm long; sheath closed nearly at top; leaves generally dark green; not turfted — *Festuca rubra* **(red fescue)**

DD. Blade not prominentely ridged on upper surface, flat, not bristle-like

E. Median lines present; tip of blade canoe shaped, abruptly pointed

F. Ligule truncate, <1.0 mm long.

G. Sheath keeled, ligule usually 0.5 mm long, tapering to apex, foliage blue-green, often smooth; small hairs on margin of collar absent — *Poa compressa* **(Canada bluegrass)**

GG. Sheath not keeled, ligule usually 0.5 mm long, blade parallel-sided; foliage deep green, not smooth; small hairs often present on margins of collar — *Poa pratensis* **(Kentucky bluegrass)**

FF. Ligule obtuse or acute, > 1.0 mm long

G. Blade truncate at base and tapering to a narrow canoe-shaped tip; sheath usually scabrous; perennial — *Poa trivialis* **(rough bluegrass)**

GG. Blade not tapering (parallel-sided), tip abruptly pointed and canoe-shaped; sheath smooth; annual — *Poa annua* **(winter grass)**

EE. Median lines absent; tip of blade taper-pointed; blade-broad, turfted — *Dactylis glomerata* **(cocksfoot)**

Major cool- and warm-season turf and amenity grasses

Agropyron repens (L.) Beauv. — English twitch, quackgrass, old-man twitch

Origin: Indigenous to Eurasia. Introduced into North America and Australasia during colonisation.

Description: Coarse textured, cool-season, sod-forming perennial. Leaves: rolled in the bud. Ligule: membranous (0.2–1.0 mm long), truncate to rounded, may be finely toothed, short ciliate margins. Auricles: slender, clasping the stem. Collar: medium-broad, occasionally divided, sometimes small hairs. Sheaths: not compressed, lower sheath slightly hairy, upper sheaths smooth, veins distinct, split. Blade: flat, 2.0–5.0 mm wide, rough on upper surface, smooth below, margins harsh, leaves often a bluish or grey-green. Inflorescence: long, narrow spike with spikelets facing the main stem.

Adaptation and use: Useful component in sand fairways and amenity areas. Troublesome, persistent weed in cultivated and waste areas. Species most commonly used in the United States include fairway wheatgrass (*A. cristatum*) and western wheatgrass (*A. smithii*). Valuable species for soil stabilisation in cool humid and semi-arid areas.

Agrostis stolonifera L. (syn. A. stolonifera var. palustris Huds.) — creeping bent grass, fiorin

Origin: Introduced from Europe into the United States and Australia during colonisation. Out of over 100 bentgrass species found in nature, three, creeping bentgrass (*Agrostis stolonifera*), colonial bentgrass (*A. tenius*) and velvet bentgrass (*A. canina*) have been used primarily for fine and/or golf turf.

Description: A fine-textured, cool-season, stoloniferous perennial, loosely turfed or matted (Figure 2.2). Stolons frequently red-brown colour. Leaves: rolled in the bud. Ligule: membranous, transparent or white, 0.6–2.0 mm long, acute to oblong, may be notched, slightly hairy on back. Collar: narrow to medium broad. Auricles: absent. Sheaths: round, glabrous, split with overlapping, hyaline margins and usually with a red or purplish tint almost wholly purple. Blades: flat, 2.0–3.0 mm wide, glabrous to minutely scabrous above, below, and on margins, prominent veins above, acuminate apex, rough along the edges. Inflorescence: narrow, dense, pale purple panicle, compressed.

Adaptation and use: Prefers moist localities on very diverse soil types. Well adapted to the cool, temperate regions of New Zealand, Europe, United States, and Australasia. Requires fertile, moist, fine textured, moderately acid soil pH 5.5–6.5. Very poor drought tolerance. Tolerates close mowing, 2.0–6.0 mm. Poor wear tolerance. Medium–good salt tolerance. Medium shade tolerance. Varieties vary from the Suttons mix, which is thought to comprise *A. capillaris* and *A. stolonifera* type bentgrasses, through Seaside, developed as a public variety, which originated from natural stands on low-lying tidelands of Coos County in southwestern Oregon, US, to the

Figure 2.2 Bahia grass (*Paspalum notatum*) (top), Seashore paspalum (*Paspalum vaginatum*) (right) and Creeping bentgrass (*Agrostis stolonifera*) (bottom left)

many improved varieties such as Cobra, SR-1020, Pennlinks, Penneagle, and Penncross.

Agrostis capillaris L. (syn. A. tenius Sibth.) — browntop bent, colonial bentgrass, common bent

Origin: Native to Europe and has become naturalised in New Zealand and the Pacific Northwest and New England regions of North America (Lewis, 1934).

Description: Fine-textured, cool-season, sod-forming perennial, with loosely turfted dense bottom growth of short leafy shoots. Principally spread by short rhizomes. Leaves: rolled in the bud. Ligules: membranous, short (0.3–1.2 mm long), truncate. Collar: conspicuous, narrow, may be divided, glabrous. Auricles: absent. Sheaths: smooth, not compressed. Blades: sharp pointed, rough along the edges, distinctly ridged on upper surface, 1.0–3.0 mm wide, tapered. Inflorescence: open, delicate pyramid-shaped panicle.

Adaptation and use: Cool, humid regions, where soils are moist to wet for a greater part of the year. Less aggressive than creeping bentgrass, but not as tolerant to close mowing as the creeping or velvet bentgrasses. Best adapted as a component in seed mixtures designed for fairway and tee use. Recuperative potential is only moderate, hence slow to recover from wear and injury from insects and disease. Does not tolerate heat and drought as well as creeping bentgrass. Widely used in New Zealand, northern Europe and Britain, often in mixtures with chewings fescue.

Agrostis canina ssp. canina L. — velvet bentgrass

Origin: Introduced from Europe during colonisation, but not common in temperate zones.

Description: Fine-textured, cool-season, stoloniferous perennial, with narrow-leafed shorts at the base, forming compact mats. May have short stolons. Leaves: folded well down in the sheath, but opens to a semi-fold or almost rolled on emerging. Sheath: round, hairless, split with overlapping, hyaline margins. Ligule: white or almost colourless, long (0.4–0.8 mm) tapering at the apex. Collar: medium broad. Auricles: absent. Blades (<1 mm in width) flat, glabrous above and below, margins scabrous. Inflorescence a reddish, loose, spreading panicle.

Adaptation and use: Limited to use in cool climatic regions, as it provides less heat tolerance than creeping and colonial bentgrass. Good low temperature tolerance. Performs well in low pH soils with low fertility, but does not thrive on poorly aerated and poorly drained sites. In contrast to other bentgrasses, it will make satisfactory growth in partial shade as well as in full sunlight. Where adapted, velvet bentgrass is very aggressive under close, frequent mowing. It is used primarily for turf purposes, including bowling greens, putting greens, and in seed mixtures for home lawns and parks.

Agrostis alba subsp. gigantea (Roth) Jsk. — redtop, black bent

Origin: Introduced from Europe during colonisation.

Description: A cool-season, rhizomatous perennial, turfted, with white, often long, scaly rhizomes. Spreads at the base, with erect stems. Leaves: rolled in the bud. Sheaths: round, glabrous, split with margins overlapping. Ligule: white, long, membranous, rounded to acute (1.5–5 mm long). Collar: conspicuous, divided, glabrous, may be oblique. Auricles: absent. Blades flat, stiff, 3–5 mm wide, veins prominently ridged above, margins scabrous and hyaline. Inflorescence: reddish, very loose panicle.

Adaptation and use: Usually in damp and loose open soils, tolerant of a wide soil and climate range. Becomes stemmy, coarse textured, and soon dies under close, frequent mowing. Good seedling vigor, but not as tolerant of close mowing and has coarse leaves compared to other bentgrasses. Redtop will produce a turf of low shoot density that is relatively coarse textured and stemmy. Not recommended for use on highly maintained turf, but performs well on roadsides, ditch and pond banks, and for stablisation.

Ammophila arenaria (L.) Link. — marram grass, beachgrass

Origin: Native of the Mediterranean region and coastal sands of northern Europe. Introduced from Europe into Australia.

Description: A tough, coarse, erect perennial with extensive creeping rhizomes. Leaves: rolled in bud, Ligule: membranous, 1.0 mm long, very tall, rounded. Collar: narrow, may be divided. Auricles: absent. Sheaths: not compressed, smooth. Blade: flat, stiff, wide (5.0–10.0 mm), scabrous above, glabrous and glossy under leaf, margins glabrous and hyaline. Inflorescence: pale, dense, spreading panicle, reddish colour.

Adaptation and use: Found in cooler, temperate regions. Suitable for growth on unstable beach sands of low fertility. Young plants resist scouring and the blast of sand particles. Plants grow rapidly to a height of about 150 cm (5 feet) and can emerge through heavy sand deposits. Fertile seed is produced but not in large quantities. Propagation is effected by division of the tussocks and rhizomes. Not adapted for turf purposes, but exceptional for coast protection and sand dune reclaimation. Occasionally used as a filler in turf seed mixtures.

Andropogon virginicus L. — broomsedge, whiskey grass

Origin: Native to North America.

Description: A warm-season, perennial tussock-former of low palatability. Leaves occur both at the base and on the culm and are often a purplish red colour. Leaves: folded in the bud. Ligule: rim of dense short hairs with sparser, longer hairs, membranous, acute to truncate, ciliate (0.4–2.0 mm long). Collar: small, divided, mostly hairy on margins. Auricles: absent. Sheath: much compressed, keeled, hairy along edges and near ligule. Blade: flat, compressed near base above, smooth to rough below, margins rough-scabrous and ciliate. White bulbous based hairs may occur near the bottom, sharp pointed. Inflorescence: slender, narrow and spatheate with 2–4 seeding stalks. Spikelets awned in pairs, cotton- like, with turfs of hair at base.

Adaptation and use: Not adapted for turf purposes, and of little value except to provide cover on depleted sites. In Australia occurs along roads, particularly in coastal areas, and is often associated with low fertility soils. In the United States found largely along the eastern border, where it leaves unsightly brown turfts in winter.

Anthoxanthum odoratum L. — sweet vernal, sweet-scented vernal grass

Origin: Native to Europe and the better rainfall areas of Asia and North America. Now worldwide and common throughout Australasia.

Description: A turfed short-lived, sweet scented perennial. It is often very coarse, and under close mowing becomes prostrate and smothers the finer grasses. Leaves: rolled in the bud, though sometimes compressed and rather flattened. Ligule: membranous, variable size, either tapering to a point or rounded. Auricles: absent or rudimentary. Collar: small, a tuft of hairs. Sheath: rounded, smooth or sometimes hairy. Blade: sometimes hairy, margins usually rough; colour, light to dark green. Leaves taper to a point or rounded. Inflorescence: narrow panicle with feathery branches.

Adaptation and use: Was used for its sweet smell in hay, but is now looked upon as a weed. Becomes turfted and bunchy early in the season and necessitates regular early mowing. In golf greens, its leaves become prostrate and under certain circumstances may smother finer grasses.

Not recommended for turf purposes, and of little agricultural value.

Axonopus affinis Chase — carpetgrass, narrow-leafed carpet grass (Australia), native cow grass (Malaysia)

Origin: Native to the Gulf Coast states of the United States and other tropical regions. Indigenous to Central America and West Indies. Introduced into Australia late last century (McLennan, 1936).

Description: A weakly turfted, stoloniferous, mat-forming perennial that forms a dense sward. Characterised by flat, two edged runners and by long slender seed stalks that terminate with two branches. Leaves: folded in the bud. Ligule: short, hairy fringe (less than 0.5 mm), fused at base (about 1.0 mm long). Auricles: absent. Collar: continuous, narrow, glabor occasionally with few hairs. Sheath: glabrous and strongly compressed, keeled. Blade: glabrous or ciliate at base, obtuse, margins scabrous near apex (about 4–8 mm wide). Inflorescence: slender and drooping. Branches at the apex into two slender, one-sided spikes, sometimes with a third spike below. Spikelets are oblong, acute, 20–25 mm long, pale green or tinged with purple. Seeds are yellowish brown and about 1.25 mm long.

Adaptation and use: In the United States carpetgrass is well adapted to the middle and lower southern states. In Australia it is widely naturalised throughout the humid, high rainfall coastal and subcoastal regions, on golf course fairways, parkland and airports established on low-fertility sandy country. Prolific seed production in late summer. Prefers warm, moist sandy to silty alluvial soils, where moisture is near the surface most of the year, low fertility, pH 4.5–5.5. Does not thrive either in low lying swamps under poor drainage, or where seepage is continuous. Establishment: seed or stolons. Poor wear tolerance. Carpetgrass is not drought or shade tolerant. Salt tolerance: poor. This grass is not recommended as intensive turf, as on the better class soils of coastal areas, bermudagrass and Queensland blue couch are superior species. There are no improved selections of *A. affinis* in Australia.

Axonopus compressus (Swartz) Beauv. — broad-leaved carpet grass, tropical carpet grass

Origin: Indigenous to Central America and the West Indies.

Description: A warm-season, low-growing, sod-forming perennial, which creeps by aggressive stolons (Figure 2.3). Leaves: folded in the bud. Ligule: a fringe of hairs. Collar: narrow, indistinct, hairy at edges. Auricles: absent. Sheaths: compressed, flattened. Blades: short, rounded at the tip. Inflorescence: 2–5 one-sided spikes, two at the top of the main stem, others below. Spikelets broad at the base and tapering to a point, a single seed.

Adaptation and use: In the United States *Axonopus compressus* is distributed in the southern states from Florida around the Gulf States through Texas, but seldom north of Arkansas to North Carolina. In Australia *Axonopus affinis* invades hilly as well as flat country, *A. compressus* favours low-lying areas where soil moisture is plentiful throughout the year. It also does not seed as freely as narrow-leaf carpet grass. Found as fairway turf on golf courses and parkland in sub-tropical Malaysia. Grows best on moist sandy or loamy soils, especially those rich in organic matter, and can withstand temporary flooding. Grows well in partial shade.

Bothriochloa macra (Steudel), S.T.Blake. — red-leg or red grass, B. decipiens (Hack.) C.E. Hubbard, pitted blue, B. pertusa ex. W. Scattini, creeping lawn grass cv. Dawson

Origin: The genus *Bothriochloa* contains approximately 30 species which originate from tropical and warm temperate regions (Tothil and Hacker, 1983). Indigenous to the Australian continent,

Figure 2.3 Broad-leaved carpetgrass (*Axonopus compressus*)

B. pertusa var. dawson creeping lawngrass, was introduced from South Africa by the Queensland Department of Primary Industries and released in 1994 (Beehag, 1996). Silver bluestem (*B. sacchararoides var. Torreyana*) and king ranch bluestem (*B. ischaemum var. songarica*) are perennial warm-season grasses used for the grazing of wildlife and livestock in the USA.

Description: Fine-leaf warm-season growing perennial bunchgrass with basal reddish tinge leaves and numerous wiry stems up to 75 cm high. Leaves: rolled in the bud. Ligule: short, membraneous, 1.0–2.5 mm long, sometimes hairy at the edges. Auricle: slightly hairy. Leaf blade: Rough along margins, about 3 mm, linear, usually flat, rough to touch. Leaf colour: pale to grey-green. Inflorescence: a raceme, the spikelet hairy and in pairs, characteristic three awns on seed.

Adaptation and use: Major rangeland grass found throughout Australia. Tolerant to drought and heat and low to medium frost tolerance. While not adapted to shade, performance and persistence under moderate shade is good. The variety dawson has a relatively strong stoloniferous habit of growth. Flowering time from summer to autumn. Currently used primarily for grazing and conservation purposes but has potential for non-irrigated general purpose turf in semi-arid regions. Research suggests that at high seeding rates, red-leg grass could produce a fine turf surface.

Bromus catharticus J.Vah. (syn. B. unioloides Kunth) — prairie grass, rescuegrass

Origin: Indigenous to eastern Europe and South America. Now widespread across most parts of Australia and New Zealand.

Description: Bunch-type short-lived, coarse-textured, cool-season, winter and spring-growing grass which produces annual, biennial and short-lived perennial types, all of which are free seeders. Leaves: rolled in the bud, but may appear to be folded. Ligule: membranous, fairly long (about 1.0 mm long), sometimes covered with short downy hairs. Auricles: absent. Collar: medium broad and divided. Sheath: not compressed, usually very hairy, distinctly keeled. Blade: light green, flat, width, smooth or very sparsely hairy. Inflorescence: open, nodding panicle.

Adaptation and use: Smooth brome (*B. inermis*) extensively grown in all southern states in the United States except Florida. Found sparingly in western states except Washington. Adapted to cool, moist temperate regions of Australia and New Zealand. Preference for deep, fertile, well-drained, silt loams and clay loams. On very dry soils often exhibits annual nature. Will not withstand frequent close mowing. Prairie grass sometimes establishes in newly sown turf where its coarse prostrate stems can smother other desirable grasses. Not recommended for turf purposes. Chess (*B. secalinus*) and downy chess (*B. tectorum*) are contaminants in grain fields throughout the United States.

Buchloe dactyloides (Nutt.) Engelm. — buffalograss

Origin: Indigenous to the Great Plains of North America.

Description: Warm-season, blue-gray, dioecious, sod forming perennial that reproduces by seed and creeping rhizomes. Leaves: rolled in the bud. Ligule: fringe of hairs, short in the centre, long at the edges. Collar: broad, hairy. Auricles: absent. Sheaths: short, flattened, smooth. Blades: narrow, flat twisted or curled, sparsely hairy. Inflorescence: composed of male and female flowers; male flowers in curved branches at the top of the main stem, female flowers are hard burrs found just above the leaf sheaths, which contain 1–4 fertile seeds.
Adaptation and use: Distributed in the United States from southern Canada to northern Mexico where it is used for pasture and erosion control. Buffalograss is very drought resistant. Foliage turns reddish after frost. Newer cultivars selected for density and colour are suitable for amenity areas.

Chloris truncata R. Br. — windmill grass, umbrella grass, star grass

Origin: Indigenous to Australia.
Description: Short lived perennial warm-season bunchgrass that spreads by seed. Roots at the lower nodes. Leaves: folded in the bud. Sheath: compressed and flattened. Blade: strongly folded so stems appear flattened, hairless, light-bluish green in colour. Inflorescence: digitate with 6–9 spreading spike-like branches.
Adaptation and use: Under Australian conditions, this grass makes rapid growth in early spring up to a height of 30 cm. In drier areas it may persist as an annual, regenerating from seed in spring and summer. Flowering time late winter to summer. Low frost tolerance. Moderate drought tolerance. Windmillgrass (*C. verticillata*) in the United States grows from Illinois southward through Louisiana and Texas and west through Kansas, Colorado and California. Windmill grass is an important range grass that has potential in revegetation, and for coarse low maintenance turf on non-irrigated sites.

Cynodon dactylon (L.) Pers. — couch grass, common couch (Australia and Africa), bermudagrass (USA), kweekgrass (South Africa), devil's grass, indian doub (India), gramillia (Argentina), serangoon (Malaysia)

Origin: Primary centre of origin is Africa. Introduced into the United States during the colonial period from Africa in 1751, or possibly India. Possibly introduced into Australia with the First Fleet (1778), as references record a visit to Capetown, South Africa, to collect feed and stock. First recorded in Australia by Scottish botanist, Robert Brown, during his voyage of 1802–05 (Brown, 1810). Bermudagrass was thought to be indigenous to Australia (Bacchus,1874). However Langdon (1954) has shown through mycological studies that the grass was introduced. Seed of *C. dactylon* was imported into the United States from Australia as early as the 1890s (Kneebone, 1966).
Description: Warm-season, prostrate, sward-forming perennial that spreads by wide-creeping stolons and scaly rhizomes. In common with other *Cynodon spp.*, the leaves are borne on stems which produce long internodes alternating with one or more very short internodes. Leaves: folded in the bud. Ligule: a fringe of white hairs (2.0–5.0 mm long), conspicuous. Auricles: absent, tufts of hair on collar. Collar: continuous, narrow, glabrous, hairy on margins. Sheaths: compressed to round, loose, split, a few short hairs on back. Blade: short, wide V or flat, tip involute, hairy at the base near the ligule, sharp-pointed blade tapers fairly uniformly to a point, margins scabrous. Inflorescence: 3–5 slender spikes, joining at the top of the main stem. Spikelets oval, in two rows, containing a single seed.
Adaptation and use: Bermudagrass is a warm-season perennial species adapted to tropical and subtropical regions of the world. In the United States, bermudagrass distribution extends from New Jersey and Maryland

southward to Florida and westward to Kansas and Texas. Under irrigation its distribution extends westward to southern New Mexico and California. In Australia, active bermudagrass growth is confined to most of western Queensland, New South Wales and Victoria, northern South Australia, Western Australia and the Northern Territory. In New Zealand, bermudagrass covers some areas of the North Island. Bermudagrass is of lesser importance in cooler regions, however natural selection and the development of naturally winter-hardy varieties, have extended their range. In Australia there are approximately twenty *Cyndon* selections cultivated as sportsturf (Beehag and Jacobs, 1993). Turf uses of bermudagrass are wide, with the intraspecific bermudagrass hybrids, such as Tifdwarf, Tifgreen (Figure 2.4), and Santa Ana, used for special purposes such as golf and bowling greens, cricket wickets, sports fields and tennis courts (Siviour, 1987; Robinson and Neylan, 1993; Beehag, 1997). The more successful naturalised *Cynodon* selections in Australia include Greenlees Park couch, Wintergreen, Windsorgreen and Riley's Super Sports, and are finding favour on golf course fairways, sports fields and bowling greens.

Prefers dry to very dry, sandy to slightly loamy, sometimes salty, soils. Tolerates both acidic and alkaline soil conditions. Medium to high salt tolerance. Tolerates some flooding but does not thrive on water logged soils. Breeding has predominately involved selection from common seeded sources, as well as interspecific hybridisation involving *C. dactylon* (common couch) and *C. transvaalensis* (South African couch). Descriptions of the more common seeded and hybrid bermudagrasses are available from Beard, 1973; Beehag, 1993, 1995a, 1997; Burton, 1951; Gibbs Russell *et al.*, 1991; McMaugh, 1988, 1993; and Worrad, 1987. Other Australian introductions include South African couch or Germiston grass (*Cynodon tranvaalensis*) (1930), introduced into Australia from South Africa (Whittet, 1969), *Cynodon incompletus Nees.*, (1907), African star grass (*Cynodon plectostachyus K. Schum. Pilg.*) (1920) and Cape Royal Couch, an improved strain of common couch, originally from the Frankenwald Experiment Station, Johannesburg, South Africa in 1955 and released in 1960.

Figure 2.4 Intraspecific bermudagrass hybrid Tifgreen (*Cynodon dactylon x C. transvaalensis*). (Photo courtesy of G.E.Beehag, Globe Australia Pty. Ltd, NSW)

Cynosurus cristatus L. — crested dogstail

Origin: Native to Europe and Asia, now widely distributed in temperate zones of the world.

Description: Short-lived, slender, turfted perennial, culms erect or shortly geniculate, 15–75 cm high. Persistent and free seeder. Leaf: folded in bud, but may be rolled in older tillers. Ligule: 0.5–1.5 mm long; membranous, whitish, very blunt. Auricles: absent. Sheath: somewhat flattened and yellow at the base when the dead sheathes are removed, hairless. Blade: flat and tapering to a point, distinctly veined, shiny undersurface, hairless. Inflorescence: open-shaped panicle

Adaptation and use: Introduced roadsides, landfill or waste areas, occasionally in pastures.

Often included as a component of fine lawns and mixtures in cooler, high rainfall locations. Prefers moist and moderately fertile soils. Will not take regular close mowing. Sown as a constituent of pasture and old world lawn mixtures. Rough dogstail (*C. echinatus*) is an erect annual and is considered an introduced weed of roadsides and waste areas invading declining pastures especially on light soils in areas of low rainfall. Crested dogtail is of uncertain value in fine turf mixtures.

Dactyloctenium australe (Rbr.) Beauv. — sweet smother grass

Origin: Native to South Africa (Lothian, 1974). First recorded in Queensland, Australia in the early part of 20th century.

Description: Summer growing, coarse-textured, warm-season annual and perennial that spreads by seed and stolons (Chaudhary, 1989) (Figure 2.5). Leaves: rolled in the bud. Ligule: short membrane, truncate or a fringe of hairs. Collar: narrow, smooth, with hairy edges. Auricles: absent. Sheaths: compressed, smooth, with some hairs near ligule. Blades: flat, linear, sparse hairs along edges. Inflorescence: several digitate spikes, flat, broad, short awns present, 3–5 seeds.

Adaptation and use: In Australia, parts of Northern Territory, western Queensland and north-western New South Wales. *D. australe* is relatively tolerant to shade particularly if cut relatively high (30–50 mm). Propagated vegetatively from turf or sprigs. Performs well in the Sydney environment in both full sun and under trees. Medium wear tolerance, poor summer tolerance. Close mowing: 2.5–3.0 cm (summer), 5.0–6.5 mm (winter). Durban grass (*D. aegyptium*) is an annual species which spreads by rooting at the nodes. In the United States the plant's distribution ranges from North Carolina to Florida and around the Gulf Coast to Arizona and California.

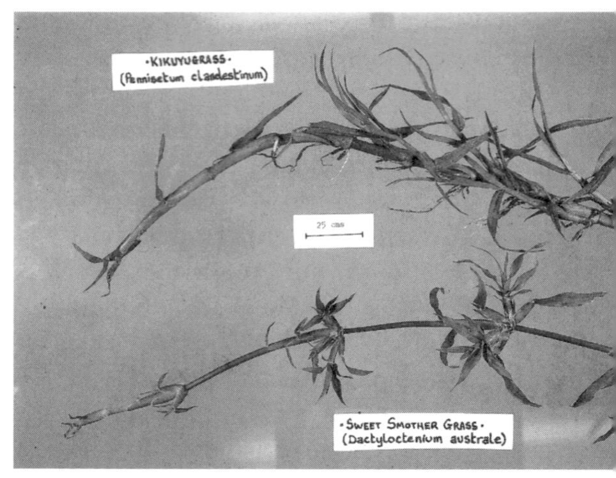

Figure 2.5 Kikuyugrass (*Pennesetium clandestinum*) (top) and Sweet Smother Grass (*Dactyloctenium australe*)

Dactylis glomerata L. — cocksfoot, orchardgrass

Origin: Indigenous to western and central Europe. Introduced to Australia during colonial times from temperate Europe.

Description: Coarse-textured, cool-season, perennial bunchgrass. Leaves: folded in the bud. Ligule: membranous, truncate, often with awn-like point at apex (2–10 mm long). Auricles: absent. Sheaths: strongly compressed, flattened, green above, white below. Collar: broad, prominent. Blade: V-shaped in cross-section at base, sharply keeled below, gradually tapering to acute point, deep furrow over midrib, margins almost smooth to scabrous. Inflorescence: panicle, stiff branches terminating with fan-shaped spikelets.

Adaptation and use: Prefers moist, loamy soils but occurs on a wide range of soil types except sandy soils, and stands drought well. Found in coastal, tablelands and slopes of eastern Australia, and waste places in North and South Island of New Zealand where it is used primarily for hay and pasture purposes. It is a troublesome weed in turf, rather than having actual turf value.

Danthonia caespitosa Gaud. — common wallaby grass, satin top, white top

Origin: Native to Australia. Common throughout temperate New Zealand.

Description: Fine leafed, turfted, year-long growing perennial, leafy at the base with stems erect. *Danthonia spp.* are cold tolerant, persistent during drought, respond well to nutrients and can withstand heavy grazing. These grasses can flower over a long period peaking in both spring and autumn. Freshly harvested seeds may be dormant for up to four months. Leaf: folded in the bud. Ligule: a ring of silky hairs, hairs up to 5 mm long. Collar: medium, continuous, hairy along the edges. Auricles: absent. Sheath: narrow, grooved, usually hairy. Blade: narrow, usually flat but often needle-like, smooth to hairy, rough on the lower surface and edges, pale or blue green. Sheath: not compressed, smooth. Inflorescence: open panicle with single spikelets on stiff branches. Distinctive white to straw coloured seed head.

Adaptation and use: One of the more important of the native grasses for pastoral use in the higher rainfall and temperate areas of Australia (Mitchell, 1994). Good for fairways on light dry soils. Can make a dense mat of wear-resisting turf for fairways. High frost and drought tolerance. Extremely slow to establish from seed but once established is encouraged by close mowing. In Australia, Hume wallaby grass (*Danthonia richardsonii*) is currently being evaluated for various dryland situations (Groves and Lodder, 1989). Another selection, *D. setacea,* makes a rough grassland if left unmown, or a reasonable lawn for low wear areas that require occasional mowing. Povertygrass (*D. spicata*) is a fine bladed, turfed perennial found on the poorer soils in the United States.

Digitaria didactyla Willd. — blue couch, Queensland blue couch

Origin: Thought to be native to Madagascar and neighboring islands, this grass is also found in Mauritius and Indo-China. Indigenous to Australia, particularly Western Australia, Queensland, and northern New South Wales, and has been reviewed by Webster (1983). The grass was originally described in 1809, with specimens recorded in 1823 (Stapf, 1911) and 1903 (Henrard, 1950) in NSW and Nudgee, Queensland, about 1906 (Black, 1939; Smith, 1941).

Description: Medium-textured, warm-season non-rhizomatous perennial that spreads by stolons, rooting at the nodes (Figure 2.6). Leaves: rolled in the bud. Ligule: membranous (0.5 to 3 mm long), rounded to acute, sometimes undulate or toothed, often reddish. Auricles: absent. Collar: broad, mostly divided by midrib, hairy at least on margins. Sheaths: compressed, split, long hairy, green but sometimes purplish-veined. Blade: pilose on both surfaces with a few longer hairs at base on upper surtace, bluish-green colour, sharply pointed, margins scabrous and occasionally hairy (4.0–18.0 mm wide). Inflorescence: 2–3 digitate racemes with spikelets 2.0–2.75 mm long.

Adaptation and use: Queensland blue couch possesses shorter, broader leaves of a distinctive bluish colour. The first descriptive reference to blue couch in was made by Maiden (1910), who described its use on golf courses, croquet greens, tennis courts, bowling greens and home lawns around the Rushcutters Bay and Longueville suburbs of Sydney, as well as the Government House Grounds. Blue couch remains an important lawn grass in south-east Queensland, northern New South Wales and Western Australia. This grass demonstrates medium to high temperature tolerance and is slightly more cold tolerant than bermudagrass. However it does not tolerate frost. Prefers moist sandy soils to heavy clay loams of pH 5.5–6.0. Medium–good wear and salt tolerance. Poor shade tolerance. PS 21 swazi grass (*Digitaria didactyla (syn. D. swazilandensis)*) is a fine leafed stoloniferous turfgrass best suited to warm areas with high rainfall as in south-east Queensland, the north coast of New

South Wales and the southern part of Western Australia and in the south eastern states of the US (Scattini, personal communication, 1998). PS 21 is expected to become an important turfgrass destined to substitute for the current major warm-season turfgrass, Queensland blue couch. It is more vigorous than Queensland blue couch. It produces a small quantity of viable seed, but can also be propagated from stolons or turf.

Figure 2.6 Queensland blue couch (*Digitaria didactyla*)

Digitaria sanguinalis (L.) Scop. — summergrass, crabgrass

Origin: Introduced from Europe and Asia. Widespread in all temperate and warm-season climates of the world.

Description: A coarse-textured, semi-tufted, warm-season weedy annual that spreads by decumbent, branching stems, which may root at the nodes. Leaves: rolled in the bud. Ligule: membranous (0.5–3.0 mm long), short, serrated. Auricles: absent, sometimes small turfts of hair at junction of leaf and sheath. Collar: broad, mostly divided by midrib, hairy at least on margins. Sheath: rounded, split, long hairy, green but sometimes red-brown. Blade: long, flattened or slightly rolled, margins often scarbrous, tinged red and crimped, and occasionally hairy (4.0–18.0 mm wide). Inflorescence: several digitate racemes at or near summit of culm.

Adaptation and use: Smooth crabgrass (*D. ischaemum*) is distributed in most northern states of the United States and from southern California eastward through Tennessee and northern Georgia. In Australia found more commonly in the warmer northern and eastern states. A serious weed in lawns and cultivated areas, as it is a prolific and early seeder, and difficult to eradicate.

Eleusine indica (L.) Gaertn. — crowsfoot, goosegrass, finger grass

Origin: Native of tropical regions.

Description: Coarse-textured, warm-season, short-lived perennial, often regarded as an annual bunchgrass. Flat stemmed with prostrate growth habit. Leaves: folded in the bud. Ligule: membranous (0.6–1.0 mm long), toothed, divided at the centre, pubescent on back, margin sometimes short ciliate. Auricles: absent. Collar: broad, sparely hairy at the edges. Sheaths: compressed, flattened, hairy at top, split overlapping margins sometimes hairy. Blade: sparsely long, hairy near base above, smooth and keeled below, margins smooth to rough or sometimes long hairy. Inflorescence: two to several narrow spikes at or near summit of the main stem. Spikelets flattened, containing 3 to 6 seeds.

Adaptation and use: Widely distributed summer weed of roadsides, waste areas and cropping lands. In the United States and Australia it is mainly a weed of coastal areas but it can also grow in the better rainfall zones and in areas that can be irrigated. A common weed, especially troublesome in turf, as the leaves and stems are tough and wiry and are difficult to cut with the mower. Although very hard wearing it is of little use as a turfgrass surface.

Eragrostis cilianensis (All.) Link ex Vign. — stinkgrass, lovegrass

Origin: Native to the Mediterranean area, has been introduced to South Africa, North and

South America, and all Australian states.

Description: Warm-season, free-seeding, annual bunchgrass. Leaves: rolled in bud. Ligule: replaced by ring of short stiff hairs. Collar: continuous, narrow, long hairs at the edges. Auricles: absent. Sheaths: square leaf sheaths are crowded at the base, have prominent veins, and are often yellow near the ground. Blades: flat, hairless, often yellowish green, dull above, glossy below. Inflorescence: spreading gray–green open panicle.

Adaptation and use: Main areas in which stinkgrass has become naturalised are in the higher rainfall areas of eastern Australia. Develops coarse dense tufts. Despite its unattractive odour, it is palatable to stock. Not suitable for use as a turfgrass. Other species include African lovegrass (*E. curvula*), a declared noxious weed in Victoria, and Brown's lovegrass (*E. brownii*), a low, golden green coloured grass which grows to a maximum of 15 cm. Annual lovegrass (*E. pilosa*) is found from Texas into Oklahoma and across the eastern states of the United States.

Eremochloa ophiuroides (Munro) Hack. — centipedegrass, chinese lawngrass

Origin: Native of southern China and Southeast Asia. USDA plant explorer, Frank N. Meyer, introduced seed into the United States in 1916. Centipede grass was introduced into Australia from the US (Whittet, 1969) and has naturalised along the northern coastal regions, often in association with *A. affinis*.

Description: Warm-season, coarse-textured, sod-forming perennial that spreads by thick, short-noded, leafy stolons. Leaves: folded in the bud. Ligule: short, membranous with cilia, purplish membrane to 0.5 mm in length. Collar: broad, continuous, constricted by fused keel, pubescent, ciliate, tufted at lower edge. Auricles: absent. Sheath: glabrous with grayish tufts at throat, very compressed, margins overlapping. Blade: compressed or flattened (3.0 to 5.0 mm wide), short, keeled, ciliate with margins papillose toward base, otherwise smooth. Inflorescence: solitary spike-like racemes. Spikelets broad at the base tapering to a rounded tip, and a single seed.

Adaptation and use: Distribution is in the warm, humid areas of the southeastern United States, ranging from South Carolina to Florida and westward along the Gulf Coast states to Texas. The grass is also found throughout the West Indies, South America and along some areas of the west coast of Africa. In Australia it can be successfully grown in any of the areas where St. Augustine grass has adapted. Centipede grass has a high temperature tolerance, very low temperature hardiness and low temperature colour retention. Well adapted to soils and climatic conditions in the warm-season growing areas of Australia where rainfall is in excess of 100 cm (40 ins). The grass tolerates very low soil fertility levels and thrives on moderately fertile soils. pH preference 5.0–6.0. Centipede is also inferior to bermudagrass and St. Augustine grass in drought tolerance, largely put down to its limited root system. Wear tolerance: does not tolerate heavy traffic. Shade tolerance: moderate but grows best in full sunlight. Used principally for lawns, parks, golf course roughs, roadside and amenity turf. In the US improved varieties with superior drought and cold tolerance are found.

Festuca arundinacea Schreb. (syn. *F. elatior var. arundinacea*) — tall fescue

Origin: Native to Europe and temperate Asia.

Description: Coarse-textured, cool-season, turfted perennial, although occasional plants develop a few short, thick rhizomes. Leaves: rolled in the bud; sheaths round, glabrous, split with overlapping margins, reddish at base. Ligule: short membranous, light green, truncate, (0.4–1.2 mm long). Collar: broad, divided, usually pubescent on margins, continuous. Sheaths: not compressed, reddish-pink below ground. Auricles: absent. Blades: flat, wide (5.0–10.0 mm long), harsh to the touch, dull with veins prominent above, keeled and glabrous below at least at base, midrib prominent, margins scabrous and hyaline. Inflorescence: an erect, compressed nodding panicle, with spikelets containing 6 to 8 seeds.

Adaptation and use: Major pasture and forage grass in cool temperate regions of Australasia, mid-south and across the northern United States and other cool temperate regions. Turf type tall fescues have been released as well as dwarf types which can be maintained at a closer mowing height. The grass does not tolerate high or low temperatures. Prefers wet, heavy soils, and can tolerate a wide range in pH level. Medium wear tolerance. Good drought, salt and shade tolerance. Useful general-purpose turfgrass for home lawns, sports and athletic fields, and institutional areas.

Festuca longifolia (L.), Koch — hard fescue

Origin: Indigenous to northern hemisphere.

Description: Fine-leafed, cool-season, small, turfted perennial. Leaves: folded in the bud. Ligule: membranous (about 0.3 mm long), rounded. Auricles: absent. Collar: medium broad, divided, indistinct. Sheath: flattened, smooth or short hairy, split. Blade: bristle-like (1–2 mm wide), smooth to rough above, deeply ridged on upper (inner) surface, pale bluish-green, glaucous. Inflorescence: narrow panicle.

Adaptation and use: Adapted to Europe, North America and the cooler half of Australia. Useful in erosion control and soil improvement. Leaf blades are tougher and less drought tolerant than sheep's fescue.

Festuca ovina L. — sheep's fescue, sheep fescue

Origin: Indigenous to northern hemisphere.

Description: Fine-leafed, cool-season, small, turfted perennial. Leaves: folded in the bud. Sheath: round, short, hairy, split with overlapping margins. Ligule: membranous (0.3–0.5 mm long), rounded, truncate; Collar: narrow, distinct, glabrous. Auricles: absent, slightly enlarged at margins. Blade: bristle-like, (0.5–2.0 mm wide), smooth to rough above, deeply ridged above, glabrous below, pale bluish-green, margins smooth. Inflorescence: narrow contracted panicle.

Adaptation and use: Adapted to cool temperate Europe, North America and Australasia, where it can make satisfactory growth on dry, sandy, gravelly or rocky soils. It grows in shade and in poor acidic soils. Under high maintenance situations sheep fescue is often too wiry or grainy to blend well with other turfgrasses (Pepin, 1994). Used to limited extent as a durable turfgrass on poorer, sandy soils, and for erosion control. Fine-leafed sheep's fescue (*F. tenuifolia*) is better adapted to shade, poor, acidic, infertile or very dry conditions.

Festuca pratensis Huds. (F. elatior L.) — meadow fescue

Origin: Introduced from Europe during colonisation.

Description: Cool-season, short-lived, perennial bunchgrass. Leaves: rolled in the bud. Sheath: not compressed, not keeled, glabrous, split with overlapping hyaline margins, coloured reddish to purple at base. Ligule: membranous, short (0.2–0.5 mm long), truncate to obtuse. Collar: broad, distinct, glabrous, pale yellow. Auricles: small, blunt, cilia present. Blade: bright green (3–8 mm wide), glabrous and dull above, glabrous and shiny beneath, veins prominent above. Inflorescence: erect or nodding panicle.

Adaptation and use: Well adapted throughout cooler parts of Europe, North America and the southern temperate Australian states. Prefers moist, fertile soils in humid localities; not drought tolerant. It is susceptible to rust and when sown is slow to establish and usually does not persist. Seldom planted for turf except as a component in poor quality lawn seed mixtures.

Festuca rubra L. subsp. rubra (F. rubra L. var. genuina Hack.) — red fescue

Origin: Introduced from Europe during colonisation. Now in most temperate countries of the world.

Description: Coarse-textured, cool-season, sod-forming perennial, with slender rhizomes up to 90 mm high. Leaves: folded in the bud. Sheath: smooth and flattened glabrous to finely hairy, lower sheaths reddish, split part way; ligule membranous, very short (0.2–0.5 mm) truncate. Collar: narrow, continuous, glabrous. Auricles: absent or slightly enlarged margins. Blades: folded or involute, (0.5–1.5 mm wide), deeply ridged above, bluntly keeled, lower smooth and shiny, margins glabrous. Inflorescence: narrow, contracted panicle, reddish when ripe.

Adaptation and use: Well adapted to cool, humid regions of Europe, North America and Australia where it was introduced into all southern states as a component of fine lawns. Requires moderate to good drainage but grows on poor, light-textured soils, moderately moist to dry, low to medium fertility soils. Suitable species for open sun and moderate shade. Forms dense wear-resistant turf when seeded heavily. Australia's first golf greens, established on Alexander Reid's family property 'Ratho' in Tasmania in 1839, were originally put down to fescues but have since been oversown with creeping bentgrass cv. Penncross.

Festuca rubra L. var. commutata Gaud. (or var. fallax Hack.) — chewings fescue

Origin: Introduced from Europe during colonisation. Now widely distributed in most temperate countries.

Description: Fine-textured, cool-season, perennial bunchgrass, similar to red fescue in appearance except it has a more erect growth habit, turfted appearance and lacks rhizomes. Leaves: folded in the bud. Sheath: distinctly compressed, glabrous, lower sheaths red at base. Ligule: membranous (0.2–0.4 mm long). Auricle: absent, but swellings present at the edge of the collar. Collar: narrow, continuous, glabrous, hairless. Sheaths: slightly rough, not compressed, wider than the blades. Blades: bristle-like, (1–2 mm long), bluntly keeled, glabrous margins. Inflorescence: a closed, narrow, contracted panicle.

Adaptation and use: Similar to red fescue but preferred by some for use in shaded areas. Warm to cool temperatures, moderately dry to moist, low to medium fertility soils. Prefers light-textured and fairly acidic soils. A variety of red fescue which was first found in the South Island of New Zealand where it grows abundantly on the lighter and poorer soils. First cultivated and harvested in New Zealand, this species received its name from the person who first sold the seed in New Zealand.

Holcus lanatus L. — Yorkshire fog, velvetgrass

Origin: Introduced from Europe during colonisation. Now widespread in the United States, United Kingdom, Australia and New Zealand.

Description: Coarse-textured, grayish-green, cool-season, turfted perennial. Leaves: rolled in the bud. Ligule: membranous, medium tall, densely hairy, soft, wide. Auricles: absent. Collar: continuous, hairy, narrow. Sheaths: compressed, greyish green and bright green with distinct pink veins, densely hairy. Blade: flat, soft, toothed, sometimes covered with soft hairs above and below leaf surface, margins, short hairy. Inflorescence: long contracted plume-like, purplish, panicle. Spikelets contain a short, soft awn, and two seeds.

Adaptation and use: In North America *Holcus lanatus* grows throughout the northern states and is a particular problem in the Pacific Northwest.

Common throughout temperate Australia and New Zealand in pastures and waste places on moist soils of moderate fertility. Considered too coarse for turf purposes.

Lolium multiflorum Lam. — Italian, common ryegrass

Origin: Native to Europe, Asia and North America. Now distributed in all cool temperate countries.

Description: Cool-season, annual or short-lived perennial bunchgrass. Coarse-textured in thin stands but fine-textured at heavy seeding rates. Leaves: rolled in the bud. Ligule: membranous (0.5–2.0 mm long), short, blunt end. Collar-broad, distinct, smooth, pale to yellowish-green. Auricles: narrow, long, clawlike. Sheath: round, not compressed or keeled, hairless, yellow-green at the base, split overlapping, hyaline margins. Blades: long (3.0–7.0 mm wide), tapered, upper surface dull. and prominently ridged, the lower surface smooth, glossy and slightly keeled, prominate veins, margins smooth. Inflorescence: long, narrow, flat spike with awned spikelets placed edgewise along the main stem.

Adaptation and use: Introduced to Europe, North America, New Zealand, and all Australian States and Territories except the Northern Territory (Curtis *et al.*, 1994). Adapted to cool, moist conditions and fertile soils. Widespread on roadsides and waste areas. Italian ryegrass dies during the summer in southern Australia. Seed is used in lawn seed mixtures, and in overseeding warm-season perennial turfgrasses. In Australia, annual or Wimmera ryegrass (*L. rigidium Gaud.*) is a serious weed of cropping lands and often requires control to allow establishment of pasture species.

Lolium perenne L. — perennial ryegrass

Origin: Indigenous to Europe, Asia and northern Africa. Introduced to Australia and New Zealand from Europe post-colonisation.

Description: Cool-season bunchgrass that can behave as an annual, short-lived perennial or erect, decument perennial, depending on environmental conditions (Figure 2.7). Leaves: folded in the bud, but may be rolled in older tillers. Ligule: membranous (0.5–2.0 mm long), long, may be toothed near apex. Collar: narrow, divided, glabrous. Auricles: small, soft. Sheath: rough or smooth but hairless, loose at the top but close at the base, where it is reddish in colour below ground, split with overlapping margins. Blade: bright green and glossy, (2.0–5.0 mm wide), keeled, prominently ridged above, glossy below, margins slightly rough, hairless. Inflorescence: narrow erect, flat spikes with unawned spikelets placed edgewise along the main stem.

Adaptation and use: Introduced to all North American states and provinces, Europe, Britain, the southern states of Australian, and throughout the temperate regions of New Zealand. Perennial ryegrass does not tolerate either high or low temperatures, and is better adapted to moist, cool environments that exhibit mild winters (Aldous and Neylan, 1997). Prefers moderately moist, high fertility soils for satisfactory growth, pH 5.5–7.0. Wear tolerance: medium to good, recovers quickly. Close mowing; does not tolerate heights less than 35 mm, although new turf type ryegrasses can be maintained at lower heights of cut.

Figure 2.7 Perennial ryegrass (*Lolium perenne*)

Drought tolerance: medium to poor. Salt tolerance: medium. Shade tolerance: medium–good. Useful on athletic fields, racecourses, fairways, inexpensive lawn seed mixtures. Categories of perennial ryegrass now available in Australia include the high quality fine leaf 'turf-type' varieties, the 'sports turf type' varieties, the pasture type varieties, and the annual and tetraploid varieties of ryegrass (Coles, 1996).

Microlaena stipoides (Labill.) R. Br. — weeping grass, meadow ricegrass, New Zealand ricegrass

Origin: Native of New Zealand and Australia. Approximately ten species of Microlaena are found in a region encompassing South-east Asia, the Pacific Islands, Australia and New Zealand (Jones and Whalley, 1994). Two species of Microlaena native to Australia, *M. stipoides* (*var. stipoides and var. breviseta*) and *M. tasmanica*, as well as a number of different ecotypes (Aldous and Chivers, 1996).

Description: *M. stipoides* (*var. stipoides*) is a year-long, green, low growing, turfted perennial with slender culms arising from a scaly rhizome. Leaves: rolled in the bud. Ligule: very short and blunt, membranous, often almost inconspicuous, fringed with a few hairs. Auricles: small, occasionally hairy, but collar margins may overlap to suggest short blunt auricles. Collar: often with tuft of silky hairs. Sheath: round, often with short bristly hairs near collar. Blade: flat, light or yellow green, fine leaf (2–5 mm wide), often outwardly curved and drooping, spearshape leaf blade. Leaf usually rough to touch or with soft hairs. Leaf colour usually lime green. Each side of midrib there are two well defined veins from which there may be distinct rows of short hairs. Flowering time summer to autumn. Inflorescence: slender drooping panicle of delicate appearance bearing between 10–30 spikelets, green or purplish in colour.

Adaptation and use: Weeping grass is widespread throughout eastern and southern Australia. High frost and drought tolerance (Mitchell, 1994). Common in shaded woodlands and open forests from sea level to 670 m. Prefers wide range of soil types, including those of very low pH. Wear tolerance: medium. Drought tolerance: will survive relatively long periods of dry weather. Salt tolerance: medium. Shade tolerance: medium to good. Good turf former where it is established on fairways and other low maintenance areas. Weeping grass is a highly competitive species which responds well to increased fertility and moderately to intensive mowing while it is actively growing. A number of varieties available for Australian use include Micro L50 and Runolaena, both of which have a finer leaf than *M. stipoides*, as well as holding their colour over winter.

Panicum laxum — panic grass

Origin: Tropical and sub-tropical America.

Description: Slow growing, short, persistent, warm-season turfed perennial grass. Leaves: rolled in the bud. Ligule: a fringe of hairs. Collar: continuous, broad, hairy. Auricles: absent. Sheath: slightly compressed, hairy. Blade: flat, hairy, sharply pointed. Inflorescence: spreading seeded panicle, with spikelets containing a single seed.

Adaptation and use: Adapted to shaded to very shaded situations in coastal subtropic and wet tropic climates. Grows well under both high and low light regimes and does not response to the red and far red light regimes. The seed has a clear requirement for light. The plant is sensitive to low water status during germination (Paterson and Wilson, 1997). In 1997, *Panicum laxum* cv. Shadegro, was released cooperatively from the University of Queensland and CSIRO Tropical Agriculture in Brisbane, Australia. Hairy panic (*P. effusum*) can also be commonly found on poorer or shallower soils, with growth commencing in spring, and species provide useful herbage even under dry conditions, In the United States, witchgrass (*P. capillare*), a turfed, coarse, hairy annual that seeds heavily in the autumn, and *P. dichotomoflorum* can be found in amenity areas.

Paspalum dilatatum Poir. in Lamk. — paspalum, dallisgrass, caterpillar grass

Origin: Native of South America; now widely distributed throughout the warmer regions of the world. Possibly introduced into Australia from Uruguay or Argentina during mid-19th century.

Description: Robust turfted perennial grass with short creeping rootstock. Leaves: rolled in the bud, but compressed and often almost flattened. Ligule: membranous, tall (up to 2 mm long, rounded, and frequently fringed with hairs. Auricles: absent. Collar: glabrous, broad, continuous, often hairy on margins. Sheath: compressed, mid-vein prominent, hairy towards the base but otherwise smooth, split. Blade: long, wide, flat, keeled, tapering to rounded narrow base, sparse hairs near the ligule, no distinct veining on the upper surface, margins often tinged with red, sometimes crimped or wavy. Inflorescence: 3 to 5 one-sided racemes (spike-like branches) pointed upwards along the main stem, with flat, round to oval seed.

Adaptation and use: There are 200 to 400 species of paspalum distributed throughout the warmer regions of the world. The paspalums are found between 30°N–S near sea-level. This species has been widely cultivated and does well on soils of moderate to high fertility in tropical to warm temperate climates. Temperature tolerance: high and low temperature hardiness, good. Soil fertility: best growth on moist, fertile, land. Wear tolerance: good. Drought tolerance: medium to good. Salt tolerance: good. Shade tolerance: poor to medium. Used as pasture species in coastal districts of mainland Australia, United States and S.E. Asia. Flowers sporadically throughout the year. Mat-former on stream banks and can form useful turf in restricted locations. This species prefers moist ditches, disturbed areas and is a common invader of garden beds. Limited use as a rough, low maintenance turf. Selected forage types of paspalum are available.

Paspalum distichum L. (P. paspalodes (Michx.) Scribn.) — water couch, mercer grass, knotgrass

Origin: Native to warm temperate, sub-tropical and tropical regions.

Description: Perennial grass, which spreads by both stolons and rhizomes, to form an extensive matted culm. Leaves: rolled and usually somewhat flattened. Ligule: tall, rounded, membranous. Auricles: absent but small turfts of hairs at collar. Sheath: compressed, smooth, hairy edges. Blade: wide V and open almost flat but margins somewhat inrolled, keeled especially near base, wide but tapering uniformly to tip, narrower and shorter than *P. dilatatum.* Usually hairless on upper surface with few fine hairs beneath and on margins and near ligule. Stem nodes often hairy. Inflorescence: Mostly two spikes at the top of the main stem, occasionally one below. Spikelets elliptical, broad at the base, pointed at the top, with a single seed.

Adaptation and use: *Paspalum distichum* is a particular problem in the United States from South Carolina into Florida and westward through Texas and Oklahoma, but it also thrives in states west of the Rockies. The grass inhabits the littoral, warm-season growing areas in Australia. Principal use as pasture, however finding more use in soil stabilisation and beach protection. Suitable as an amenity grass. *Paspalum distichum* can be found in freshwater habitats, whereas *P. vaginatum* inherently colonises the saline flats and swamps.

Paspalum notatum Fluegge — bahiagrass

Origin: Introduced into the United States from Brazil in 1913; found extensively around the Gulf

Coast. Entry of species into Australia and other warm-season growing countries unknown.

Description: Coarse-textured, warm-season perennial that spreads slowly by short, strong, woody rhizomes (Figure 2.2). Leaves: folded in the bud. Ligule: membranous, short, with dense row of whitish hairs at back (1.0 mm long). Auricles: absent. Sheath: compressed, keeled, rather glossy, glabrous or ciliate toward summit or rarely pubescent throughout. Blade: flat or folded at base, usually sparsely ciliate toward base, sometimes almost to summit, otherwise glabrous. Inflorescence: 2 or 3, rarely 5, slender one-sided racemes.

Adaptation and use: Both *P. notatum* (bahiagrass) and *P. vaginatum* (salt water couch or seashore paspalum) are used in Australia as sportsturf. Adapted to a wide range of soil conditions, bahiagrass is found growing in poorly drained heavy soils as well as sandy soils that are comparatively infertile. Bahiagrass has some measure of drought tolerance, but grows best in areas with high or well distributed rainfall. Fair shade tolerance. In Australia bahiagrass is limited to the coasts of far north Queensland and the Northern Territory, whereas in the United States, *P. notatum* is distributed all around the Gulf Coast from the coastal areas to central North Carolina to eastern Texas. It is used for pasture, erosion control of salt-effected areas, and areas to be reclaimed from tidal influences.

Seashore paspalum is a warm-season, salt tolerant, prostrate growing perennial grass that spreads by rhizomes and stolons. *P. vaginatum* was introduced into Western Australia from Africa (Burvill and Marshall, 1951). The grass produces a soft textured couch type surface and can withstand very high salt levels (Duncan, 1997). Dwarf varieties of seashore paspalum Saltene/Salpas have also been selected and sold in Adelaide and Perth.

P. vaginatum has been used on lawns and bowling greens in South Australia. Other examples of paspalum include Bull paspalum (*P. boscianum*) and field paspalum (*P. laeve*), both of which form clumps of turf in many of the southern states of the US.

Pennisetum clandestinum Hochst. ex Chiov. — kikuyu grass, giguyu, gekoyo, agekoyo

Origin: The genus *Pennisetum* contains about 130 species scattered across the warmer regions of the world and has been reviewed by Mears (1970) and Skermos and Riveros (1990). Kikuyu grass is endemic to the Kenyan highlands in South Africa and the common name is derived from a native tribe. Introduced into Australia from Africa in 1908 and into the United States in 1920.

Description: Relatively coarse-textured, summer growing dioecious perennial spreading by rhizomes and stolons, rooting freely at the nodes and forming a dense turf (Figure 2.5). Stems are horizontal to weakly upright and produce numerous leaves. Leaves: folded in the bud, covered with soft, short hairs, and somewhat flattened near tips. Ligule: a ring of hairs, collar often milky-white. Auricles-absent, tufts of hairs at sheath and blade collar. Blade: flat or wide V, slightly hairy, with a waxy appearance, light green colour, tapering to a point. Sheath: flat and usually hairy, pale green, often almost white. Inflorescence: flowering stems are topped with 2 to 4 flowers which bloom on short side shoots. Fertile and male-sterile strains are recognised. Anthers and stigmas are exerted on the fertile strains but only stigmas on the male-sterile.

Adaptation and use: In the United States it is found in the warmer season states such as California. Similarly it is now found in most Australian states and territories. Kikuyu grass is adapted to a temperate/sub-tropical climate that receives regular rainfall and where winter temperatures do not fall below 8 to 10°C for significant periods of the year (Aldous, 1998).

Also adapted to the higher elevation (2000 to 3500 m) in the moist tropics. *Pennisetum clandestinum* will not stand severe frost, although it recovers readily in the following warm season. Good high temperature and drought tolerance. Prefers medium textured, fertile, imperfectly drained soils. Good wear tolerance. Medium salt and shade tolerance. Principally grown as a pasture, kikuyu grass is used for sports fields, recreation areas and golf course fairways (Beehag, 1997), as well as for waterways and general-purpose turf (Curtis *et al.*, 1994). Often regarded as an objectionable plant and a weed. Not recommended as a turfgrass in continental US.

Phalaris aquatica L. (syn. P. tuberosa L.) — canary grass, phalaris

Origin: Native of the Mediterranean region. Introduced into Australia under the name of *Phalaris commutata* by R.R. Harding, Curator, Botanic Gardens, Toowoomba, Queensland in 1883 from Italy.

Description: Coarse-textured, cool-season, deep-rooted perennial. Leaves: rolled in the bud. Ligule: long (3–5 mm), membranous, translucent, smooth or serrated top, often overlapping in front of stem, minutely hairy on back (2.0 to 5.0 mm long). Auricles: absent. Collar: distinct, glabrous, continuous, oblique. Sheath: long, round, split, edges membranous, hairless. Blade: long, wide and flat, upper side indistinctly ridged, lower side slightly so, margins scabrous, hairless. Inflorescence: densely flowered, narrow spike-like panicle. Seeds are very shiny, and readily shed when the seedhead is ripe.

Adaptation and use: Used primarily for soil conservation purposes and in some areas for pasture and forage. Not suitable as a turfgrass. *P. arundinaceae* (reed canarygrass) is also a perennial, which has a use in swampy areas growing over mid-spring to mid-autumn, but is winter dormant.

Phleum pratense L. — timothy, large-leafed timothy

Origin: Introduced from Europe during colonisation.

Description: Coarse-textured, cool-season, bluish-green turfy perennial. Lower internodes on stems enlarge to form haplocorms. Leaves: rolled in the bud. Sheaths: hairless, pale green, nearly white at base, hyaline margins. Ligule: membranous, medium long and white, distinct notch on either side (1.0–4.5 mm long). Collar: broad, distinct, continuous, usually divided; Auricles: absent. Blade: flat, light-greyish green, often with a heavy margin, hairless, mid vein distinct on under surface, (4–8 mm wide), margins scabrous. Inflorescence: densely flowered, cylindrical, spike-like panicle.

Adaptation and use: Winter-hardy species, being resistant to low temperatures as would be experienced in the cold, wet soils of cool temperate Australia. Prefers clay, loams or heavy soils, long-lived in cool, humid regions, fairly shade tolerant. Not resistant to close, continuous mowing. Poor recovery for limited moisture; does not tolerate drought, high temperatures or alkaline soils. An important hay grass that is also used in pasture mixtures, for silage, and for conservation purposes. It is not adapted for turf use. Where established on fairways it can be a useful species, however deliberate sowing not advocated. The small-leafed timothy (*P. bertolonii*) is also hard wearing, has no rhizomes, and does not tolerate wet heavy soils.

Poa annua L. — winter grass, annual bluegrass, annual poa

Origin: Native of the Mediterranean region. Introduced into North America and Australia during colonisation. Now distributed throughout most temperate countries. Believed to have originated from a cross between *Poa infirma H.B.K.*, an annual species, and *Poa supina Scrad*, a creeping perennial.

Description: Cool-season, loosely to compactly turfted annual or short-lived perennial. Leaves: folded in the bud. Sheaths: compressed, slightly keeled, glabrous, whitish at base, split with overlapping, hyaline margins. Ligule: long thin membranous, 0.8–3.0 mm, rounded at the apex. Collar: distinct, narrow, glabrous, V-shaped. Auricles: absent. Blades: V-shaped, 2–3 mm wide, usually pale green, glabrous, soft, canoe-shaped apex, transparent lines on each side of midrib, parallel sided, flexuous transversely, margins slightly scabrous and hyaline. Inflorescence: small, open greenish-white panicle, branches few and spreading.

Adaptation and use: Common weed in pastures and lawn areas, roadsides and waste areas over a wide range of cool temperate habitats. Adaptable as a turfgrass surface in cool temperate regions. Prefers low to medium fertility soils of wide moisture range. Growth limited by acid soil conditions. Desirable turf characteristics include toleration of low mowing, moist soil conditions, and compacted soils. It is dense and fine textured, and exhibits moderate shade and wear tolerance. Undesirable turf characteristics include its high temperature susceptibility, low drought tolerance, susceptibility to numerous diseases, unslightly seed heads, poor traction, and the fact that it re-establishes well through self-seeding. Principally infests bowling greens, golf greens and lawns during late autumn, winter and early spring months. Because of its turfted nature, it creates trouble in hybrid bermudagrass bowling greens by deflecting the bowl. More akin to *Poa annua* than *Poa pratensis* are the newly released varieties of *Poa supina*, which have proven low light and wear tolerance (McMaugh, 1998).

Poa pratensis L. — Kentucky bluegrass, English meadow grass, smooth-stalked meadow grass

Origin: Native to Europe, north Africa and Asia. Now widespread in Europe and North America. Fairly common in Australia and New Zealand.

Description: Fine-textured, loosely turfted or turf-forming rhizomatous perennial. Leaves: folded in the bud. Ligule: membranous, very short 0.2–1.0 mm long, slightly pointed. Collar: medium keeled, divided, glabrous, closed when young but later split, yellowish green. Auricles: absent. Sheaths: slightly compressed, glabrous, somewhat keeled, split with overlapping, hyaline margins. Blades: V-shaped to flat, 2–4 mm wide, soft, usually glabrous, keeled below, double veins on upper surface, parallel sided, abrupt canoe-shaped tip, transparent lines on each side of midrib, finely scabrous margins. Inflorescence: an erect, open, greenish white pyramidal panicle with slender, spreading lower branches usually occurring in whorls of five.

Adaptation and use: Wide distribution throughout temperate North America and Europe. Distributed in all Australian states except the Northern Territory. Good tolerance to high and low temperatures. Prefers regions with higher rainfall, cool temperate climate and mild summers. Soil fertility: moderate to low, moist to dry soils. Drought tolerance: poor but can survive long, dry periods in dormant state. Wear tolerance: medium to good. Salt tolerance: poor. Shade tolerance: poor. In the US and Australia, a valuable grass for turf and widespread on roadsides, waste areas in gardens and occasionally in pastures from sea-level to about 800 m altitude. A number of varieties and cultivars exist for turf, lawns, pastures and other uncultivated areas.

Poa trivialis L. — rough bluegrass, rough-stalked meadow grass

Origin: Native to Europe, North Africa and Asia. Now widespread and common in many moist and wet areas of the world.

Description: Fine-textured, loosely turfted stoloniferous perennial which spreads by stolons. Tufted while young but later, under moist to wet conditions, forms dense mats or patches. Leaves: folded in the bud. Ligule: membranous, 2.0–6.0 mm long, greenish in colour, tapering and glossy, toothed at the tip. Collar: distinct, broad, glabrous, divided by ridrib. Auricles: absent. Sheaths: compressed, rough to the touch. Blades: flat or V-shaped, 1.0–4.0 mm wide, soft, rough on the edges, terminating in canoe-shaped tip, transparent lines on each side of midrib, margins scabrous. Inflorescence: an oblong panicle.

Adaptation and use: Favours damp, cool temperate regions. Prefers fertile, moist soils; requires shade. Grows well in wetlands and along ditch banks. Not drought tolerant. Included in some lawn and turf seed mixtures developed for shady areas. Often mixed with perennial ryegrass for overseeding because it does not compete with warm-season turfgrasses during the spring transition. Other species include Canada bluegrass (*Poa compressa*) a wiry, tough, blue-green perennial, which spreads by short rhizomes and forms a very open sod, and fowl bluegrass (*P. palustris*) a wiry perennial, spreading by short rhizomes, requires frequent moist, open ground in cooler growing climes.

Stenotaphrum secundatum (Walt.) Kuntze — buffalo grass (Australia, New Zealand and South Pacific), St. Augustine grass (USA), crabgrass (Bermuda and the West Indies)

Origin: Native to Gulf of Mexico region, the West Indies and Western Africa, St. Augustine grass has naturalised along the Gulf Coast in the U.S, in southern Mexico, throughout the Caribbean region, South America, South Africa, Western Africa, Australia and the South Pacific, and the Hawaiian Islands. Prior to 1800, the species was reported in Uruguay, Brazil, Nigeria, Sierra Leone, the West Indies, Bermuda and South Carolina. Since its introduction in 1840 to Australia on the ship S.S. *Buffalo, Stenotaphrum secundatum* has been called buffalograss. The grass should not be confused with *Buchloe dactyloides (Nutt.), Engelm.* from the American mid-western prairies. The most common strain, a fertile diploid with a white stigma colour, is native to the Gulf Coast–Caribbean and West Africa.

Description: A coarse-textured, warm-season, robust, sod-forming perennial, that spreads by long, tough, woody stolons to form a spongy mat. Leaves: folded in the bud. Ligule: a ring of inconspicuous fringe of hairs (approx. 0.3 mm long). Auricles: absent. Collar: continuous, glabrous, broad, much constricted to form short stalk or petiole for blade. Sheath: compressed, broad and flattened, membranous on margins, sometimes sparsely hairy on the edges near the blade. Blade: very short, smooth, erect, with a blunt or boat-shaped point, and often split. The upper surface gray-green and dull, the lower paler and shiny. Usually hairless. Leaves and stolons tinged reddish-brown. Inflorescence: a thick spike with few spikelets imbedded along the sides.

Adaptation and use: In the U.S. St. Augustine grass is found from the Carolinas to Florida and westward along the Gulf Coast to Texas and in southern and central California. In Australia St. Augustine grass is adapted to the moist coastal areas of Queensland, New South Wales and northern Victoria that display mild winter temperatures. It is known to tolerate high summer temperatures, and can retain its colour at temperatures lower than those that discolour couchgrass. Shade tolerance is as good or better than other warm-season grasses. Under heavy shade develops thin, spindly turf. Soil fertility: tolerates a wide range of soil types. Prefers fertile, well-drained soils with a pH of 5.5–8.5, but develops a chlorotic appearance in highly alkaline soils (above pH 7.5). Does not tolerate compacted or waterlogged conditions. Salt tolerance: very good, producing satisfactory growth at salt levels as high as 16 mmhos. Principal use has been as a pasture or lawn grass. St. Augustine grass may cause a skin irritation to some people. Several selections of St. Augustine grass, such as Floratine, Bitter Blue and Floratam, are available from the US. In Australia, the following soft St. Augustine varieties are available: ST85, Shademaster, Velvet, Sir Walter and Shadow.

Themeda triandra Forsk. (syn. T. australis (R.Br.) Stapf) — kangaroo grass

Origin: Native to Australia.

Description: Densely turfted, warm-season perennial with short leaves concentrated at the base (Lamp *et al.*, 1990). Leaves: folded in the bud. Ligule membranous, not conspicuous, few loose spreading hairs on the edges. Collar: brown hairs. Auricles: absent. Sheath: absent. Blade: light green, sometimes bluish often with a crimped edge and may have long sparely distributed hairs near the base. Inflorescence: borne on loose panicle on culms with dark nodes and consist of a number of spathes, leafy looking structures with long dark awns protruding from the spikelets.

Adaptation and use: Useful pasture and range plant. Leaf turns green to blue-green to purple and later to golden brown from spring through to winter. Extremely tough plant that will grow in most conditions, and is relatively maintenance free once established. Annual slashing is usually adequate in retaining plant vigour. Its coarse texture limits its use as a turfgrass surface.

Vulpia bromoides (L.) S.F. Gray — rat's tail fescue, silver grass, squirrel tail grass, hair grass

Origin: Native of Europe. Introduced into Australia during colonisation.

Description: Small, slender, turfty annual, leaves fine and not numerous. Leaf: rolled and not flattened. Ligule: very short, rarely 1 mm, almost inconspicuous. Auricles: absent, but slightly swollen glands in their place. Blade: tightly involute when dry, tip slightly rough, fine and needlelike. Sheath: hairless, straw coloured when older, rounded or slightly compressed and keeled. Inflorescence: oblong, open panicle.

Adaptation and use: Adaptable to cool temperate regions and poor, open and infertile soils. Has been used to give cover for winter use of fairways. Prolific seeder on unmown areas. Not recommended for turf use.

Zoysia japonica Steud. — Japanese lawn grass, Korean lawn grass; *Zoysia matrella*, Manila grass; *Zoysia tenufolia*, Mascarene grass or Korean velvet grass

Origin: Indigenous to southern China, Korea, and Japan. *Zoysia japonica* was introduced into the US in 1895 from the Manchurian Province of China. In 1911, *Zoysia matrella* was introduced into the US from Manila by USDA botanist C.V. Piper. *Zoysia japonica* was introduced into Sydney, Australia in 1947.

Description: Warm-season, medium to fine sod-forming perennial that spreads by both stolons and rhizomes. Leaves: rolled in the bud. Ligule: a fringe of hairs (up to 0.2 mm long). Collar: medium, covered with hairs. Auricles: absent. Sheath: absent. Blade smooth or occasionally short, hairy above with at least a few long hairs near base, smooth below, margins smooth to scabrous and occasionally long, hairy near base. Inflorescence: short, terminal, spike-like raceme with spikelets alternate on two sides. Spikelets blunt at the base, round and tapering to a point.

Adaptation and use: Zoysiagrasses are major warm-season turfgrasses suited to the warm temperate and subtropical areas of the world. They have been used as turf in Japan since 750 and farmed commercially as turf since 1700 (McMaugh, 1991). In Japan it grows as far north as 40° 10' along the coastal areas (Kitamura, 1989). In the United States, zoysiagrasses have adapted along the Atlantic coast from Florida to Connecticut and along the Gulf Coast to Texas. In Australia, zoysiagrass is limited to the north-eastern coastline. Three species are utilised for turfgrass purposes: Japanese or Korean lawngrass (*Z. japonica*), Manilagrass (*Z. matrella*), and mascarenegrass or Korean velvetgrass (*Z. tenufolia*). All are distinguished primarily on the basis of size, vigour, coarseness and winter hardiness. Degree of low temperature hardiness: Japanese lawngrass > Manilagrass > marcarenegrass. Zoysia discolours with 10–12°C temperatures and remains in a state of dormancy throughout the cool to cold winters. Shade tolerance: more shade tolerant than couchgrass. Unattractive to insect pests, such as army worms, because of their high fibre, but susceptible to nematode attack. A native species, *Z. macrantha* called spiky couch is available along the eastern coast of Australia, but is not cultivated (Wheeler *et al.*, 1982).

References

Aldous, D.E. 1998. 'Management of kikuyugrass (*Pennisetum clandestinum*) as a recreational turf', *Australian Parks and Recreation*, vol. 34, no. 2, June, pp. 28–31.

Aldous, D.E. and I. Chivers. 1996. 'Agronomic and quality factors of weeping grass (*Microlaeana stipoides*)', *New Zealand Turf Management Journal*, February, Vol. 1: 31–2.

Aldous, D.E. and J.J. Neylan. 1997. 'Performance of cool-season perennial turfgrasses on the Australian continent', in addendum *International Turfgrass Society Research Conference*, University of Sydney, Sydney, NSW, pp. 11–18.

Bacchus, W.H. 1874. 'Some Victorian and other Australian grasses', Description of *2nd Annual Report of Secretary of Agriculture, Victoria*. p. 143.

Beard, J.B. 1973. *Turfgrass: science and culture*, Prentice-Hall, Englewood Cliffs, N.J., 658 pp.

Beehag, G. 1993. 'Grass profile: Tifgreen (328) couch', *Turfgrass Notes*, Autumn, Australian Turfgrass Research Institute, p. 8.

Beehag, G.W. and S.R.L. Jacobs. 1993. 'Couchgrass culture in Australia' *Australian Turfgrass Research Institute Turf Notes*, vol. 11, no. 3, pp. 10–11.

Beehag, G.W. 1995a. 'Grass profile: windsor green couchgrass', *Australian Turfgrass Research Institute Notes*, Summer, p. 2.

Beehag, G.W. 1995b. *Turfgrass identification manual—an introduction about basic identification of turfgrass species*, Australian Turfgrass Research Institute, Concord West, NSW, 32 pp.

Beehag, G.W. 1996. 'Grass profile: dawson creeping lawngrass', *Turfgrass Notes*, Spring, Australian Turfgrass Research Institute, p. 6.

Beehag, G.W. 1997. 'An historical review of the utilisation of the warm-season turfgrasses in Australia', in addendum *International Turfgrass Society Research Conference, University of Sydney*, Sydney, NSW, pp. 7–10.

Black, R.S. 1939. 'Australian turf grasses—some notes on blue couch', *The Australian Greenkeeper*, March, pp.18–19.

Brown, R. 1810. *Prodromus florae*, Novae Hollandiae et Insulae Van Diemen,189–93.

Burvill, G.H. and A.J.T. Marshall. 1951. 'Paspalum vaginatum or sea shore paspalum', *J. of Ag. W.A.*, vol. 28, pp. 10–11.

Burton, G.W. 1951. 'The adaptability and breeding of suitable grasses for the southeastern states', *Advances in Agronomy*, 3:197–241.

Chaudhary, A. 1989. *Grasses of Saudi Arabia*, Ministry of Agriculture and Water, Safir Press, 465 pp.

Coles, D. 1996.' The two kinds of ryegrass', *TurfCraft*, September/October, no. 50, p. 65.

Curtis, W.M. and D.I. Morris. 1994. *The Student's Flora of Tasmania. Part 4B, Angiospermae: Alismataceae to Busmanniaceae*, St. David's Park Publishing, Hobart, pp. 194–355.

Duncan, R.R. 1997. 'Environmental compatibility of seashore paspalum for golf courses and other recreational uses', *International Turf Research Conference*, July, University of Sydney Press, NSW, 22 pp. (in press).

Gibbs Russell, G.E., L. Watson, M. Loekemoer, L. Smook, N.P. Barker, H.M. Anderson and M.J. Dallwitz. 1991. *Grasses of Southern Africa—memoirs of the botanical survey of South Africa*, No. 58, National Botanic Gardens, South Africa, 437 pp.

Groves, R.H. and M.S. Lodder. 1989. 'Native grasses—do they have potential as turfgrasses?', *TurfCraft*, March/April, p. 11.

Jones, C.E. and R.D.B. Whalley 1994. 'Microlaena

(Microlaena stipoides (Labill.) R.Br.)—a native turf for native gardens', *A.T.R.I. Turf Notes*, Autumn, 2 pp.

Hanson, A.A., F.V. Juska, and G.W. Burton. 1969. 'Species and varieties', in Hanson, A.A. and F.V. Juska, (ed.), *Turfgrass Science*, American Society of Agronomy, 13: 370–409.

Henrard, J.T.H. 1950. *Monograph of the Genus* Digitaria, Universitaire Pers. Leiden, 999 pp.

Kneebone, W.R. 1966. 'Bermuda grass—worldly, wily, wonderful weed', *Econ. Botany*, vol. 20, p. 94–7.

Kitumura, F. 1989. 'The climate of Japan and its surrounding areas and the distribution and classification of zoysiagrass', in H. Takatok (ed.), *Proceedings of the 6th International Turfgrass Research Conference, Tokyo, Japan*, pp. 423–6.

Lambrechtsen, N.C. 1986. *What grass is that? A guide to identification of some introduced grasses in New Zealand by vegetative characters*, New Zealand Department of Scientific and Industrial Research, Government Printing Office, Wellington, Information Series no. 87, 151 pp.

Lamp, C.A., S.J. Forbes and J.W. Cade. 1990. *Grasses of Temperate Australia —a field guide*. Inkata Press, 310 pp.

Langdon, R.F.N. 1954. 'The origin and distribution of *Cynodon dactylon* (L.) Pers.', Dept. of Botany, Uni. of Qld, vol. IV, no. 5, pp. 41–4.

Lewis, I.G. 1934. 'A greenkeeper's guide to the grasses. 3. The genus *Agnostis*', *J. of the Board of Greenkeeping Research*, 3(11): 200–206.

Lothian, N. 1974. 'Durban grass—*Dactyloctenium australe*', *Aust. Parks*, Aug., pp. 10–11.

Maiden, J.H. 1910. 'Blue couch as a lawn grass', *Agricultural Gazette of NSW*, September 2:789–91.

Mitchell, M. 1994. *Indentification handbook for native grasses in Victoria*. Speciality Press, June, 34 pp.

McLennan, L.W. 1936. 'Carpet grass. Has it a place in north coast pastures?' *The Ag. Gazette*, vol. VII, part 1, pp. 555–8.

McMaugh, P. 1988. 'Australian couch', *TurfCraft*, May/June, pp. 21, 36.

McMaugh, P. 1991. 'Zoysia—golden hope or great hoax?', *TurfCraft*, May, p. 20.

McMaugh, P. 1993. 'Couchgrasses—facts and fairy tales', *Landscape Australia*, pp.42–3.

McMaugh, P. 1998. 'Grasses—Grass selection and use in parks', Australian Parks and Recreation, vol. 33, no. 4, December, pp. 32–4.

Mears, P.T. 1970. 'Kikuyu (*Pennisetum clandestinum*) as a pasture grass—a review', *Trop. Grasslands*, vol. 4, no. 2, p.139–53.

Nowosad, F.S., D.E. Newton Swales and W.G. Dore. 1936. 'The identification of certain native and naturalised hay and pasture grasses by their vegetative characters', *Tech. Bull. 16*, Macdonald Coll., McGill Univ.

Paterson, M.F. and Wilson, J.R. 1997. 'Germination characteristics of *Panicum laxum* cv. shadegro, a new lawngrass for shaded areas, Poster presentation for the *International Turfgrass Research Conference, University of Sydney, Sydney, Australia*, July.

Pepin, G. 1994. 'Fine-leaf fescues', *Grounds Maintenance*, May, pp. 34–5.

Robinson, M. and J. Neylan. 1993. 'National couchgrass evaluation', *TurfCraft*, August, pp.17–20.

Scattini, W.J. 1998. Personal communication.

Siviour, T. 1987. 'Santa Ana—answering the criticisms', *TurfCraft*, August/Sept, October, pp. 26.

Skermos, P.J. and F. Riveros. 1990. 'Tropical grasses', *FAO Plant Production and Protection Series. No. 23*, Food and Agriculture Organisation of the United Nations, Rome.

Smith, L.S. 1941. *Some Queensland couch grasses*, Queensland Department of Primary Industries, Brisbane, unpublished, 6 pp.

Stapf, O. 1911. 'Blue couch: a new lawn grass', *Kew Bulletin*, pp. 256–61.

Tothil, J.C. and J.B. Hacker. 1983. *The grasses of southern Queensland*, University of Queensland Press, St. Lucia, Qld.

Watson, L. and M.J. Dallwitz. 1992. 'The grass genera of the world', *CAB International*, 1038 pp.

Webster, R.D. 1983. 'A revision of the genus *Digitaria* Haller (Paniceae:Poaceae) in Australia', *Brunonia*, vol. 6, pp.131–216.

Wheeler, D.J.B., S.W. Jacobs and B.E. Norton. 1982. *Grasses of New South Wales*, University of New England Publishing Unit, NSW.

Whittet, J.N. 1969. 'Turf grasses and their management', in *Pastures*, 2nd edn, Department of Agriculture, NSW, pp. 605–16.

Worrad, D. 1987. 'Couch grass—a national review', *TurfCraft*, September/October, pp.17–18.

CHAPTER 3

Turfgrass growth and physiology

D.E. ALDOUS, Institute of Food and Land Resources, University of Melbourne, Richmond, Victoria, and **J.R. WILSON,** C.S.I.R.O., Tropical Agriculture, St. Lucia, Queensland, Australia.

Introduction

Within the broad geographical distributions of Europe, North America, Latin America, the Middle East, South East Asia, Africa and Oceania, native and naturalised grasslands occur over many climatic regions, although there are chief discernible differences between wet and dry seasons (Langer, 1972). In the wet and dry tropical climatic zone, where the mean temperatures are not less than 18°C, are found the savannah grasslands of South and Central America, parts of Africa and northern Australia. The desert and steppe regions, characterised by dry climates with low rainfall, include the grass steppes of the Argentine pampas, the prairie lands of North America and the shrub steppes of Patagonia and parts of Australia. In the humid mesothermal climatic regions, which include Europe, North America, Australia, New Zealand and parts of Asia, the mean monthly temperature does not fall below 3°C in winter, or become excessively high in the summer months. The fourth climatic zone in which grasslands occur is the humid microthermal, such as the steppes, prairies, tundras and alpine grasslands, which experience severely limiting winter growing temperatures (Moore, 1964).

Regions of turfgrass adaptation

The landscapes of the world are a reflection of topography, climate and vegetation. Many of the cool-season turfgrasses are widely distributed throughout the cool humid, cool subhumid, and cool semiarid climatic regions of the world, as well as their adjacent transitional zones. Warm-season grasses extend from these transitional zones to include the warm humid, warm subhumid and warm semiarid climates of the world. For example, in the United States the cool, humid regions include the states in the north-east, north central, Appalachia, and the Pacific Coastal north-west (Juska *et al.*, 1969), whereas sections of Tennessee, Kentucky, Arkansas and Missouri are in the transitional zone and exhibit warm, humid and cool, humid conditions (Holt, 1969). The United Kingdom and northern Europe also represent the cool humid, cool subhumid climatic regions of the world, with southern Europe, the Indian sub-continent, Japan and parts of China, within the transitional zone. Much of southern Asia and Australia are better suited to the growth of warm-season turfgrasses.

Australia's specific cool-season zones include Tasmania, except in the extreme north-east corner, regions on or near the Great Dividing Range in Victoria and New South Wales, small areas in Western Australia and the southern-most extension of the Stirling Range in South Australia (Lush, 1991). The northern-most cool-season region includes the high altitude areas of the New England ranges. Transitional growing grasses are confined to a narrow coastal strip in Victoria and southern NSW, as well on the northern and western sides of the cool season zone. The Atherton Tablelands in Queensland is considered transitional, as is a patch of the Gregory Range. Most of the south-west corner of Western Australia is marked as transition, and so is a strip just inland from the Great Australian Bight. Warm-season growing grasses are best suited to the rest of Australia, and include much of north Queensland, Western Australia and the Northern Territory. The cities of Darwin, Brisbane and Perth are classified as warm-season, Sydney, Adelaide and Canberra as transitional, Melbourne on the boundary of the cool and transition zones, and Hobart as cool season. In New Zealand, the upper half of the North Island is considered transitional, with the remaining area better suited to the growing of cool-season turfgrasses.

Optimum air temperatures for shoot growth for cool season grasses, such as Kentucky bluegrass (*Poa pratensis*), are 17 to 24°C and optimum soil temperatures for root growth 10 to 18°C. Shoot growth ceases when maximum daily air temperatures are above 32°C, whereas root growth ceases when soil temperatures are above 25°C (Beard, 1973). These temperatures are often exceeded on cool-season grassed areas in Australia over summer. Warm-season turfgrasses, such as zoysiagrass (*Zoysia japonica*), have an optimum temperature above 35°C and a minimum of approximately 10°C (Youngner, 1961). Winter dormancy restricts the use of warm-season grasses in southern Australia (Beehag, 1997). Many temperate grasses originated in areas from east Europe to central Asia, while the warm-season grass species have come from diverse centres of origin such as Africa, South and Central America, and tropical Asia. A number of morphology, ecology and physiology comparisons have been made between cool- and warm-season grasses (Hirose, 1973) (Table 3.1 and Figure 3.1).

Turfgrass life cycle

Turfgrass plants are normally perennial, and grow and increase in size through tillering, stolon and rhizomatous growth, and ideally, develop new vegetative shoots continually throughout the year. Cool-season turfgrass species are adapted to rapid periods of growth during the cool, moist periods of the year; and usually are dormant during the short warm to hot summers. They commence growth in autumn, grow strongly in winter, flower and set seed in spring. Warm-season perennial grasses are always summer to autumn growing, often sensitive to frost, with many species flowering throughout summer and autumn (indeterminant flowering), while others have only one flowering period (determinant flowering) (Lodge *et al.*, 1990). Warm-season growing grasses provide a good turf in summer but are inadequate in winter, whereas cool season growing grasses provide a good surface over autumn and winter but are not ideal in summer. The turfgrass root, stem, leaf or plant increases in size, and follows a distinct pattern from the time of initiation to senescence thoughout its life cycle. The rate and order of these tissues are partially influenced by plant growth hormones (Evans, 1969). The potential for growth is initially contained within the grass 'seed'.

Grass 'seed'

The 'seed' of a grass is shed as a structure known as a caryopsis. In most turfgrasses the caryopsis is enclosed by a tightly adhering lemma and palea.

Table 3.1 Morphological, ecological and physiological characteristics of temperate and tropical grasses (adapted from Hirose, 1973; Wilson, 1993).

	Temperate grasses	*Tropical grasses*
Morphological features:		
Seedling	narrow leaf blade, upright leaf arrangement	wide leaf blade, spreading leaf arrangement.
Auricle	general absence of hairs	hairs often present
Root	epidermal cells of the root hairs alternately long and short, with only short cells giving rise to root hairs	epidermal cells in the region of the root hairs similar in nature, all capable of giving rise to root hairs
Seed	small embryo, narrow in shape	larger embryo, roundish in shape
Spikelet	spikelet of one to several florets, terminal one often small and usually sterile	spikelets of the *Eragrostoideae* similiar to temperate grasses; those of the *Panicoideae* usually of two flowers, terminal one fertile, the other sterile
Leaf anatomy	epidermal longitudinal walls straight (Figure 3.4a), some species sinuous, low vein frequency across width of leaf, 4 to 12 mesophyll cells between adjacent veins, thin-walled indistinct parenchyma, chloroplast-containing bundle sheath around veins	epidermal walls sinuous, high vein frequency, two mesophyll cells between veins, thick-walled distinct chloroplast-containing bundle sheath around vein (Wilson, 1993)
Cytology	chromosomes size large, basic chromosome number 7 with most genera	basic chromosome number 9 to 10 with most genera
Ecological features:		
Photoperiodic response	Long-day plant; flowering is generally in spring and occasionally autumn and responds to long days; vernalised under cool temperatures, internode elongating under long-day conditions	short to middle-day plant; flowering usually summer to autumn and occasionally spring; cool temperatures are unnecessary for vernalisation
Habitat	subfrigid to temperate regions; under tropical and subtropical conditions at higher elevations	subtropical to tropical regions; often dormant in winter in temperate regions
Physiological features:		
Optimal growth temperature	range of 15 to 22°C	25 to 35°C
Growth limiting temperature	5°C	10°C
Germination temperature	slow but achievable at 0°C. maximum upper temperature is approximately 35°C	not achieveable at 0°C, very good at 35°C
Photosynthetic capacity	20 to 30 mg CO_2/sq.decimetre/h	40 to 60 mg CO_2/sq. decimeter/h
Saturating light intensity	116 to 233 W m^2 (20 to 30 Klux @ 300 ppm CO_2 conc.)	390 to 465 W m^2 (60 Klux or more)
Optimal CO_2 concentration	300 ppm +	300 ppm
Carbohydrate reserves	fructosan	starch, sucrose
Carbon pathway in photosynthesis	C-3 cycle	C-4 cycle
Photorespiration	present	absent
Carbon dioxide compensation point	50 to 60 ppm	0 ppm

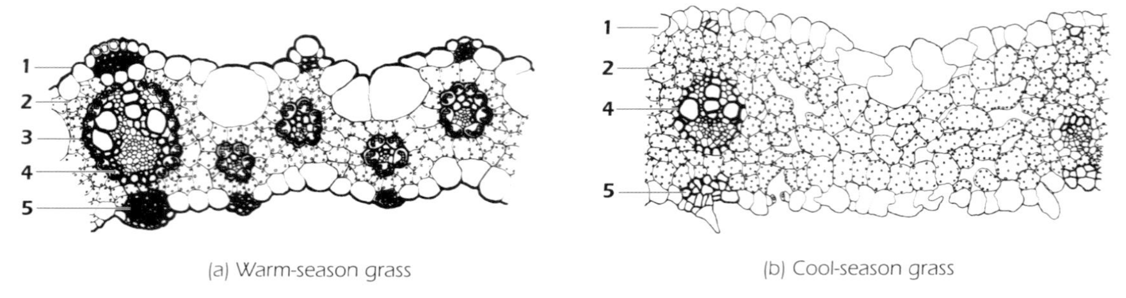

Figure 3.1 Cross-section of grass leaf structure for: (a) a warm-season grass (*Cenchrus ciliaris*), (x150) and (b) a cool-season grass (*Phalaris aquatica*) (x100). (1 = epidermis, 2 = mesophyll, 3 = parenchyma bundle sheath, 4 = vascular tissue, 5 = sclerenchyma strand (adapted from Minson and Wilson, 1980).

In certain species, such as bermudagrass (*Cynodon dactylon*), these tissues may be separated from the caryopsis at threshing. The caryopsis contains the true seed, which consists of an embryo and endosperm, the latter being responsible for the carbohydrate supply for the developing embryo. The embryo can be formed from the fertilised egg, or through 'apomixis' or self reproduction that substitutes for sexual reproduction in higher plants (Berg, 1972). The structure and homologies of tissues of the embryo are discussed more fully by Barnard (1964). The embryo contains a shield-like structure, the scutellum or single cotyledon, which functions in enzyme secretion and the transport of nutrients from the endosperm to the developing seedling during germination. Central to the scutellum are the embryonic shoot and root axes. The radicle or primary root and the root cap are surrounded by parenchymatous tissue called coleorhiza and are found at the basal end of the embryo. At the opposite end is the plumule or shoot, surrounded by similar tissue called the coleoptile. In most Festucoid grasses, coleoptile attachment is direct to the scutellar node, while in warm-season grasses, such as the Panicoid and Chloridoid, it is attached above the scutellar node. Just inside the seed coat is the aleurone layer, a shallow proteinaceous tissue, which is important in the biochemical processes of germination (Figure 3.2).

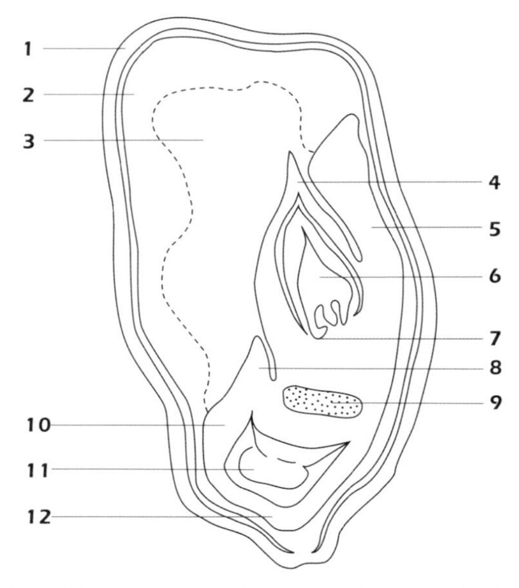

Figure 3.2 Median longitudinal section of a grass seed embryo showing typical structures: 1 fused testa and pericarp, 2 aleurone layer, 3 endosperm, 4 coleoptile, 5 scutellum, 6 plumule, 7 growing point, 8 epiblast, 9 scutellar node, 10 radicle, 11 root cap, 12 coleorhiza

Universal requirements for germination of nondormant seeds include adequate moisture, temperature within a suitable range, and oxygen. Optimum temperatures for germination vary with grass species but range from 15.5 to 32°C. In general, warm-season grasses have a higher optimum than do the cool-season grasses. Alternating temperatures are often beneficial to germination of grass seeds and may be a requirement for improved germination. Some species, such as Kentucky (*Poa pratensis*) and annual bluegrass (*P. annua*), and the Panicums, e.g. *P. maxium* and *P. laxum* (Wilson, 1997a), seem to benefit from a light requirement. During germination the caryopsis absorbs water and swells. If moisture levels are high, saturation of the seed may occur in a few hours. The rate of uptake depends not only on the moisture level, but also on the nature of the seed coats. Floral parts that surround the seed may retard water uptake, especially if moisture levels are low. As water is absorbed, the cells become turgid but cell elongation may not occur for many hours. The first physical indication of growth is the appearance of the radicle, which breaks through the coleorhiza and extends two pairs of lateral rootlets. The first true leaf, or primary shoot, breaks through the coleoptile at approximately the same time as the radicle emerges from the coleorhiza. At this point germination is complete and subsequent development must be considered seedling growth.

Seedling growth

Whalley *et al.* (1966) recognised three phases in the growth of seedling grasses: the heterotrophic stage, which begins with the imbibition of water and germination, and is concluded with the emergence of first leaves above the soil surface; the transition stage when the seedling manufactures the necessary organic compounds for growth from both the photosynthetic process and the remainder of the endosperm; and lastly the autotrophic stage, in which the endosperm reserves are exhausted and the seedling is required to obtain all its complex organic compounds as products of photosynthesis. Following emergence of the radicle from the coleorhiza, additional seminal or primary roots develop. These vary in number depending upon the grass species, and are generally well developed by the time the coleoptile appears above the soil surface.

The grass coleoptile and the mesocotyl, or first internode between the coleoptile and scutellum, may be forced up from a considerable depth by elongation of this structure. Coleoptile growth is promoted by light while that of the mesocotyl is promoted by darkness or inhibited by light. At the growing point of the seedling enclosed by the coleoptile, subsequent foliage leaves also arise through the coleoptile. The next seedling structures to develop are the secondary or adventitious roots, which form at the nodes of the new shoot. Timing of secondary root formation varies and may be as late as initiation of the first tiller (Troughton, 1957). Secondary root primordia may also be present in the embryo of some grass species.

Shoot growth

Leaf initiation and development

The stem apex consists of a dome-shaped growing point, comprising a central core of cells (corpus) and surrounding tissue (tunica). The growing point consists of crescentric ridges, termed leaf primordia, which arise alternately on each side of the meristem. As these young leaves develop from the lower ridges, they form a hood which encloses the younger leaf primordia and the growing point. Sharman (1947) has classified grass species into three groups according to the length of the vegetative stem apices. These range from Italian ryegrass (*Lolium multiflorum*) and timothygrass (*Phleum pratense*) (Figure 3.3(a)), with fifteen to twenty leaf primordia, through winter grass, bentgrass (*Agrostis stolonifera*) and perennial ryegrass (*Lolium perenne*) with approximately five to ten leaf primordia, to bermudagrass (*Cynodon dactylon*), and many other tropical species, with two to three leaf primordia. The vegetative apical stem apex in most cool-season turfgrasses has been estimated to be between 0.5 to 1.0 millimetres in length and is usually much shorter in most warm-season grasses.

As the turfgrass plant develops the shoot, or culm, consists of a series of successive leaves, which are initiated by the stem apex, but often remain short during vegetative development. The individual turfgrass leaf is borne on a stem and consists of a flattened upper lamina and a lower basal sheath that encompasses the stem (Beard, 1973). Both the lamina (leaf blade) and sheath function in photosynthesis and respiration, and vary in length and width between and within grass species. Formation of the leaf primordium is initiated by periclinal divisions in the tunica resulting in a microscopically visible protuberance or ridge (Figure 3.3(a)). This is followed by rapid, lateral cell divisions along the ridge margin to form the leaf tip which are nearly always initiated on opposite sides of the apical meristem. The whole leaf primordium is meristematic, but soon cell division activity becomes confined to an intercalary meristem at its base. This region divides into two zones through the formation of a band of parenchyma cells, and this coincides with the appearance of the ligule, which is located on the inner (adaxial) side of the leaf at the sheath and lamina interface. These series of events mark the beginning of separate development within the foliar organ, for the upper portion of the meristem is associated with growth of the lamina, while activity in the lower portion leads to growth of the sheath. The same processes are repeated continuously provided the shoot apex remains in a vegetative state. Emergence of the lamina from the enclosing sheath of lower leaves is accompanied by several changes, for not only do the cells of the exposed portion cease expansion but they also encounter a new environment in which they undergo photosynthesis and transpiration. For many grasses the ligule becomes just visible from within the sheath of the previous leaf, after the

Figure 3.3 Development of the inflorescence of Timothy (*Phleum pratense*): (a) vegetative apex consisting of ten alternating crescentric ridges or leaf primordia (x16); (b) early differentiation of the inflorescence (x16) (adapted from Wilson, 1959)

lamina of that leaf has just reached full cell expansion, whereas the sheath is still elongating at this stage.

Leaf area and rate of leaf emergence above the cutting height is a function of genetic and environmental influences. Climatic and seasonal conditions include temperature (Mitchell, 1956), light intensity and photoperiod (Youngner, 1959; 1961), and mineral nutrition. These characteristics vary inversely, with the leaves being smaller with an increased rate of leaf emergence. The rate of leaf emergence is most rapid in the warmer temperatures of spring and summer in both cool and warm season grasses, but may decline during periods of heat, low temperature or water stress.

In practice, as the emerging leaf develops within the sheath of the next oldest leaf, the newly formed leaves eventually emerge, and are removed in the mowing process. However while the shoot remains vegetative the growing point remains near the soil surface and is not removed by mowing. Leaf area is also influenced by environment. Reduced light intensity, except when severely limiting, usually causes grass leaves to become larger, longer and narrower; whereas in shade situations, grass leaves may also be quite large in area but low in weight. Leaves, having reached their final size, remain on the plant for a certain period and then commence to senesce basipetally (Soper and Mitchell, 1956). Cell constituents are mobilised and redistributed, and the leaf loses weight. Photosynthesis begins to fall, at first slowly for the initial third of each leaf's life, but then more rapidly as senescence approaches. The leaf eventually becomes an additional respiration load on the turfgrass plant.

Three distinct types of tissues comprise and traverse the grass leaf: the epidermis, mesophyll and vascular bundles. The epidermis is a surface protective layer of cells located on both the upper and lower leaf surface. With the exception of the stomata, the epidermis may have an especially thick wax-like layer or cuticle, which provides support and assists in the prevention of water loss. Stomata are specialised cells found primarily on the upper surfaces of most turfgrass species, which provide for exchange of water, oxygen and carbon dioxide through the stomatal pore. Other types of epidermal cells that have been modified include the cells of *Zoysia* and *Paspalum spp.*, which contain silica, and bulliform cells, which allow for the rolling in of turfgrass leaves, to expose only the lower (abaxial) epidermis, and hence reduce transpiration during drought. Bulliform cells are found on each side of the main midrib in Kentucky bluegrass, and between the veins across the leaves of warm-season grasses.

Between the epidermis and surrounding the vascular bundles of cool-season grasses is specialised photosynthetic ground tissue, the mesophyll, which is seldom differentiated into palisade and spongy ground tissue as in many broad-leaf plants, but rather is composed of thin-walled chlorenchyma containing chloroplasts utilised in photosynthesis. Vascular bundles serve the same function as the phloem and xylem in the stem, which is the conduction of water and nutrient elements from the soil to the leaves (xylem), and in the translocation of photosynthetic products from the source of manufacture (leaf surface) to other non-photosynthetic organs of the plant (phloem) (Figure 3.4).

Tillering

Buds that develop in the axils of the leaf sheaths on the grass shoot, and develop into new shoots, are called tillers. These tiller buds arise from sub-hypodermal cell layers of the meristem and are usually found in the axils of all leaves, the number and vigor of which will control the density of leaves in the sward (Adams and Gibbs, 1994). Tillers, as distinct from stolons and rhizomes,

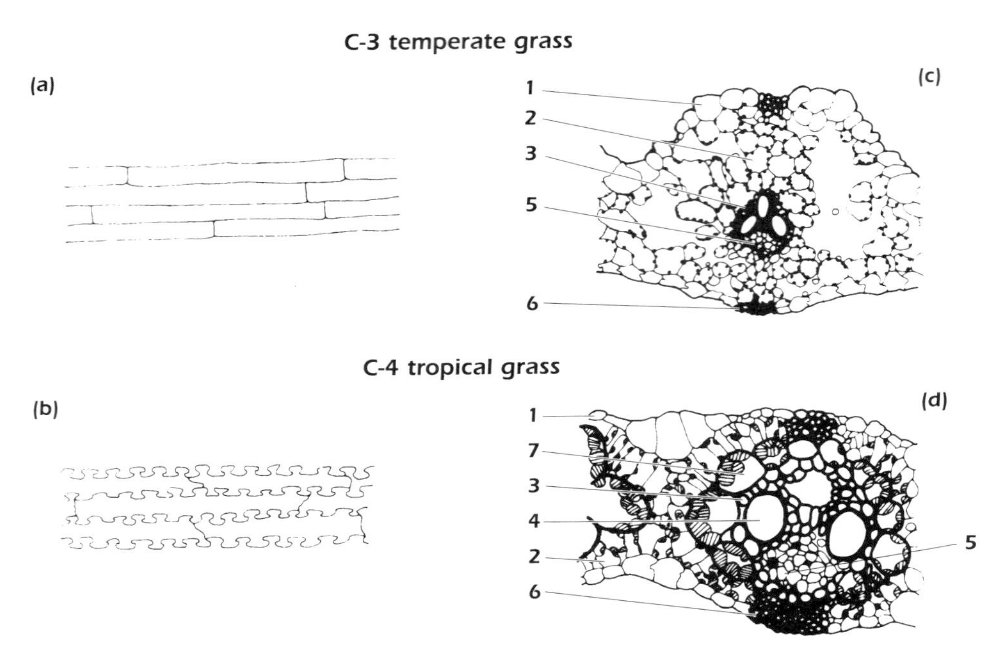

Figure 3.4 Paradermal view of epidermal cells (a) and (b), and cross-section of a main vein (c) and (d) of a temperate (*Lolium multiflorum L.*) and tropical grass (*Panicum maximum var. trichoglume*)-PCK type: 1 epidermis, 2 mesophyll, 3 mestome sheath, 4 xylem, 5 phloem, 6 sclerenchyma and 7 parenchyma bundle sheath (adapted from Wilson, 1993)

develop from the base upwards (acropetally) with the young vegetative shoot growing upwards within the basal leaf sheath (intravaginal growth).

Although complete in all respects, tillers remain in vascular connection with the parent shoot. Little is known of tiller development prior to their emergence externally, although under constant conditions of leaf appearance, tillers are known to appear at a linear rate (Youngner and Nudge, 1976). Tiller development is influenced in cool-season grasses by short day length, low temperature, high light intensities, nutrient levels, and differences in mowing height and frequency (Peterson and Loomis, 1949). Tillering in cool-season grasses occurs throughout the season at temperatures above 0°C (32°F). Optimum temperature for tillering of warm-season paspalum (dallisgrass), was found to be 27°C (80°F) (Mitchell, 1956), although Langer (1972) found little decline in the tillering of tropical grasses up to 35°C. Maximum tillering in cool-season turf occurs during early spring and late autumn if temperatures and other conditions are favourable, and is minimal during periods of extremely high or low temperatures, or during water stress. With warm-season grasses, tillering occurs vigorously from late spring to mid-autumn.

Tiller control is a physiological process and involves the concentration of auxin produced by the apical meristem (Yeh *et al.*, 1976) and carbohydrate level (Youngner and Nudge, 1976). The age of a tiller generally does not exceed twelve months. Little is known about the length of life of a tiller under turf conditions of frequent mowing. Frequent defoliation of both perennial cool- and warm-season grasses usually results in a higher density of smaller tillers. Similarly with perennial grasses, tillers undergo a cyclic pattern of production and can be produced over prolonged periods. With annual grasses, tillers either form inflorescences or cease production. Beard (1965) and Beard and Daniel (1965) demonstrated that tiller numbers per unit area can differ greatly among species and strains of turfgrasses under identical shade conditions. Reduced tiller density grown under shade may result from the reduced tillering rate under low light intensities (Mitchell, 1953). Increasing light intensity has been shown to increase tiller development as tiller buds compete with other parts of the plant for carbohydrates (Evans *et al.*, 1964). Under a wet tropic and sub-tropic environment incident PI (photon irradiance) is relatively high at 8.5 to 10.8 megajoules per square metre per day. Under these conditions, and with adequate nutrition, shading can reduce input to 70 per cent of incident radiation, but often does not reduce grass yield. However shading to 20 per cent of incident radiation can reduce relative yield of most warm-season grasses to 12 to 20 per cent of full sun yield (Wilson, 1997b). Similarly, shade from trees or artificial structures usually has little effect on the grass canopy air temperatures, generally causing decreases of only 1 to 2°C. However shade can cause marked decreases in surface soil temperatures, averaging at least 10°C.

Stem growth

Grass stems fall into three general categories: aerial culms, underground rhizomes and surface stolons (Gibbs Russell *et al.*, 1991). Aerial culms bear the leaves and inflorescence that will determine overall appearance. The initiation of upright stem, rhizome or stolon primordia appears to be similar to that of the tiller. In the case of intervaginal growth, the vegetative stem grows up inside the sheath of the subtending leaf, and the grass as a whole forms a dense turf. However, if the young vegetative stem penetrates the basal leaf sheath (extravaginal growth), a spreading or creeping habit of growth often results. Such secondary lateral stems that arise extravaginally with above ground stem elongation are called stolons, whereas similar lateral stems that elongate underground are called rhizomes. Both stoloniferous and rhizomatous growth can either be indeterminate (bermudagrass) or determinate (Kentucky bluegrass and creeping red fescue), the latter forming new leaf and roots at the nodes as they grow. Some grasses, such as summer grass (*Digitaria sanguinalis*), siro shadegro® (*Panicum laxum*) and paspalum (*Paspalum dilatatum*), have stolon-like procumbent stems, which may root at several nodes before the stem apex turns upward. In other grasses, such as bermudagrass, Queensland blue couch (*Digitaria didactyla*) and kikuyu grass (*Pennisetum clandestinum*), there is an intergradation of rhizomes and stolons, with both stem types capable of initiating shoot and root growth from the meristematic regions of each node. Branching of lateral stolons and rhizomes can also occur at some of the nodes which bear reduced colorless leaves without blades, called scale leaves. Scale leaves are always found on rhizomes where they function in protecting the growing points of the young shoots.

Rhizome and stolon growth occurs throughout the year provided the soil temperatures are above 0°C (Troughton, 1957). Conditions that favour the emergence of the rhizome apex above soil level include short day lengths, excessively high temperatures and levels of nitrogen nutrition (Brown, 1939). Long days, high light energy levels and lower levels of nitrogen nutrition favour rhizome and stolon growth in both cool- and warm-season turfgrasses (Evans and Watkins, 1939). Species and cultivar variation in the degree of stolon and rhizome development can be exploited for turfgrass breeding.

Grass roots

Grasses have two root systems: the seminal roots, and the secondary, nodal or adventitious roots

(Gibbs Russell *et al.*, 1991). Both types provide for water and nutrient absorption and anchorage. Seminal roots arise from the germinating embryo, and seldom live longer than twelve months in annual grasses. They also include the primary root which results from the growth of the radicle of the embryo. Mature grasses have a fibrous root system consisting of nodal roots which form at the lower nodes of the grass stem, two to three weeks after germination, and can be observed as whorls of roots at several successive nodes just below the internodal intercalary meristem. Such roots may also develop from the nodal points of stolons and rhizomes of perennial turfgrass species. Nodal roots may last one to several years, their age being influenced by the grass species, the season of development (roots produced in autumn or winter live longer than roots produced in spring or summer) and environmental conditions. Root initiation and growth of cool-season grasses principally occurs during the spring and early summer, with root senescence more likely to correlate with summer heat stress periods. Root growth is influenced by similar environmental conditions that affect shoot growth; temperature, light and nutrients. Bermudagrass roots have been shown to be adversely affected by relatively low soil temperatures of 21°C (70°F) even if the air temperatures are higher (Langer, 1972). Supraoptimal soil and air temperatures (>35°C) has been shown to reduce root activity in Kentucky bluegrass cultivars (Aldous *et al.*, 1977; Aldous and Kaufmann, 1979). Root growth is also greatly reduced under shade conditions (Beard and Daniel, 1965; Wilson, 1997a, Wilson, 1997b). Optimum temperatures for cool-season root growth are several degrees lower than the optimums for shoot growth (Brown, 1939). Maximum root growth of both warm-season and cool-season turfgrasses appears to take place at moderate soil temperatures, but with a slightly higher optimum for warm-season than for cool-season grasses.

A longitudinal section through a young root will show, at the distal end of the root, a root cap of relatively loose parenchyma cells. The root cap protects the meristematic zone directly behind it, and its cells are regularly removed, and replaced, by abrasive action. Beyond the meristematic region is the zone of cell elongation in which there is little cell division but where there is provision for cell elongation. Still further back from the root tip lies the zone of cell differentiation in which vascular tissue reaches maturity. Root hairs form part of the epidermis. In cool-season (festucoid) turfgrasses root hairs originate from specialised epidermal cells called tricoblasts. Root hairs greatly increase the absorption surface of roots, and persist for a longer time in grasses than in many other families (Barnard, 1964).

Root–shoot relationships

A high root–shoot ratio (ratio of the root weight to shoot weight) is preferred for a well managed turfgrass. In general, environmental conditions such as temperatures above the optimum for root growth and development, close mowing, excessive nitrogen and irrigation, and low light intensities favour top growth of cool and warm-season grasses and therefore reduce the root–shoot ratio. An association between onset of flowering and the decline in the root–shoot ratio has also been noted (Troughton, 1956). Root and shoot growth of bermudagrass, as evidenced by growth and rooting of stolons, occurs simultaneously throughout the warm season. However little specific information is available about the seasonal response, and the subsequent root–shoot ratio, for most warm-season turfgrasses.

Sources of carbohydrate reserves

Carbohydrates are non-structural materials that are stored in various plant tissues of perennial grasses and later utilised in respiration and for the maintenance and growth of plant tissues (Smith, 1967). Carbohydrates are stored primarily in the root, rhizomes, stolons and crowns of the grass plant. Tropical and temperate grasses differ in the form of their soluble carbohydrate (Weinmann and Reinhold, 1946). Grasses of tropical and subtropical origin accumulate starch and sucrose, and grasses of temperate origin accumulate fructosans and oligosaccharides (DeCugnac, 1931). The primary carbohydrate reserve in turfgrass seeds is starch. Reserve carbohydrate levels in temperate grasses are usually much higher than warm season grasses, especially under conditions of cool temperatures and low nitrogen (Wilson and Ford, 1973; Wilson, 1975). The percentages of non-structural carbohydrates in the stem bases of perennial grasses decline with the commencement of growth and following mowing, and increase with an advancement in maturity to flowering and seed formation (Rochecouste, 1968; Weinmann and Goldsmith, 1948). The climatic, edaphic and cultural factors that influence the rate and level of carbohydrate accumulation have been reviewed by Troughton (1957) and May (1960). Carbohydrate accumulation is greatest during periods of minimal shoot growth and high light intensity, and in the case of cool-season grasses this occurs in late autumn just prior to the onset of winter dormancy. By contrast optimum growing temperatures, close and frequent mowing, and high nitrogen fertility levels stimulate shoot growth and cause a decline of carbohydrate reserves. The more severe and frequent the mowing, the lower will be the carbohydrate reserves.

Inflorescence development and reproduction

Floral initiation and the subsequent morphological development of the growing point has been previously described for herbage grasses, with specific studies carried out on ryegrass, cocksfoot and timothy (Evans and Grover, 1940; Wilson, 1959). The developing floral apex is first at the base of the stem and emerges inside the flag leaf, the latter protecting the developing inflorescence during the period of extension. During inflorescence development the initiation of new leaf buds is inhibited and the apical meristems develop into the type of inflorescence inherently associated with that grass genera. Following floral induction, double ridges are formed on the vegetative apex (Figure 3.3(b)) upon which the spikelets and branches arise in the axils of each leaf primordia. Subsequently, each primordium differentiates further to form florets with their floral organs. With these changes to the vegetative apex, internode extension occurs via an intercalary meristem at the base of each internode, and the developing inflorescence moves up within the encircling leaf sheath and emerges from within the flag leaf.

Three stages in the flowering process are generally recognised: induction, initiation, and development and differentiation. Induction involves a physiological and hormonal change in the plant in response to specific environmental conditions. Although no morphological changes are evident, the plant is induced to flower. In the initiation phase, the shoot apices change from a vegetative (Figure3.3(a)) to a floral primordia (Figure 3.3(b)). In the development and differentiation phase, growth of the floral primordia continues, inflorescence structures and reproductive organs are differentiated, and stem internodes commence to elongate. Floral induction of turfgrasses is controlled principally by day length and exposure to low temperatures (Peterson and Loomis, 1949). Many cool-season perennial turfgrasses must undergo a period of chilling as the first requisite for flowering, although this will vary between genotypes and among varieties (Johnson and White, 1997). This chilling process, termed vernalisation, may also be effective when applied to seeds. Vernalisation is thought to have a direct effect on the shoot apex. The floral stimulus, resulting from the chilling, may be transmitted to the apices of tillers arising from the treated tiller. Other perennial grasses may not be vernalisable until they have entered a juvenile stage of seedling development (Calder, 1966). The temperature range for effective vernalisation is from 0 to 10°C (32 to 50°F), although chilling can also occur at a reduced rate at both warmer and colder temperatures. The effect of vernalisation is quantitative, the longer the chilling period, the more prolific the flowering up to a maximum length of time beyond which there is no further increase in flowering. Exposure to a period of high temperature during or following chilling, as may occur in late spring and summer, will have a devernalising effect, returning the plant to a state in which it is incapable of flowering. However, this stimulus is not indefinitely transferable or autocatalytic since late-formed tillers do not necessarily flower (Chouard, 1960). Perennial grasses, therefore, require revernalisation each year. Short days may substitute for chilling to produce the vernalisation effect in some grasses (Calder, 1966) such as perennial ryegrass and several bentgrasses. Floral growth and development of most grasses occurs in late spring or early summer while floral induction occurs in the autumn. Warm season turfgrasses have no chilling or short day requirement for floral induction, and flowering and stem elongation occur in response to increased day length in spring and continue throughout late spring to autumn as successive tillers develop.

Response to turfgrass management

Interrelationships with mowing

Mowing removes newly emerged leaves, leaving only the older leaves and sheaths (verdure). Only leaves that are not fully extended continue to grow after mowing (Beard, 1992). Grass growth habit and the location of leaf primordia can influence tolerance to close mowing. Differences in growth habit, leaf angle and location of meristems, are the primary reasons cited in cultivar differences. Cultivars with more horizontal leaves (lower angle) are more tolerant of close mowing. Rhizomes and stolon growth are little affected by mowing (Weinmann and Goldsmith, 1948). Moderate defoliation actually enhances their production and is used to increase stolon growth during establishment of creeping bentgrasses. However, frequent close mowing can reduce rhizome and stolon growth. As the mowing height is lowered, low growing grass surfaces demonstrate the following responses: decreased carbohydrate synthesis and storage, leaf width, root growth rate and total root production; and increased shoot growth per unit area, shoot density, succulence of shoot tissues, and quantity of chlorophyll per unit area (Beard and Daniel 1965; Beard, 1973). Madison (1962) demonstrated that the yield of shoot growth (clippings) is decreased either by reducing the mowing height or increasing mowing frequency, and that closer mowing stimulated tillering and increased shoot density per unit area, provided the mowing height was within the tolerance range of the particular species. For example, in a polystand of Kentucky bluegrass and annual bluegrass (winter grass), mowing below the tolerance level of the Kentucky bluegrass will favour annual bluegrass production. Mowing frequency is influenced by shoot growth rate, cutting height, purpose, and the environmental conditions that will affect growth (Beard, 1973). Mowing frequency is primarily dictated by shoot growth rate, with Madison (1962) showing that tall fescue has a rate nearly three times more rapid than bentgrass. Mowing height and frequency are also interrelated, that is as height is lowered, frequency must be increased to maintain turfgrass quality. No more than 40 per cent of the existing leaf tissue should be removed at any single mowing.

Interrelationship with fertilisation

Fertiliser can affect leaf colour, and the level of tolerance from pest injury and climatic stress, as well as the general growth parameters of shoot density and the extent of rooting (Madison, 1982). Application times and rates will depend on the nutrient status of the soil, grass species, the purpose and the level of turf maintenance. Nitrogen, phosphorus and potassium are the major minerals that generally produce a yield response. Excessive nitrogen use in the spring to cool-season turfs may overstimulate growth, increase thatch and disease incidence, and cause adverse physiological effects, and it is interrelated with temperature, moisture, light and other factors (Wilson, 1975). Tiller growth and root development can be stimulated during the autumn to winter period with judicious late season nitrogen applications. Similarly bermudagrass and other warm-season grasses should receive nitrogen in spring to stimulate growth for summer production. Winter nitrogen applications can sometimes cause winter injury of warm-season species in transition zones, and late autumn nitrogen applications may encourage the encroachment of foreign grasses and weeds; all factors that will compete with turfgrass growth. Reduced production of turfgrasses in many warm-season climates exhibiting heavy summer rainfall may result from nutrient loss through excessive run-off and leaching.

Interrelationship with aeration

Thatch is an intermingled, undecomposed, organic residue layer of dying and dead shoots, stems and roots that accumulate between the leaf and soil surface. Thatch has a high lignin content and resists microbial breakdown. Mat is thatch intermixed with soil mineral matter. Thatch and mat accumulation is a direct result of management practices that relate to abundant vegetative growth, vigorous growing species, high rates of nitrogen fertilisation, infrequent mowing, and excessive use of plant pesticides (Madison, 1962; 1982). Thatch can minimise the movement of air and fertilisers into the soil layer weakening the turf, create localised dry spots, render certain pesticides ineffective, and provide an environment for disease-causing organisms: all factors that will reduce active growth and development of the turfgrass sward. Thatch control is achieveable through dethatching and appropriate cultural practices (Beard, 1973).

Conclusion

This chapter has reviewed a number of aspects of grass growth and physiology that contribute to the special ability this botanical group provides as a turf surface for a wide variety of cool and warm season habitats. Particularly relevant to the success of turfgrass is the continuous nature of leaf, tiller, rhizome and stolon development, and their ability to adapt both morphologically and physiologically to close and frequent defoliation and to maintain a low, high density, green sward under such conditions. Important physiological attributes such as the grasses' relationship to mowing, fertilising and aeration were also discussed.

References

Adams W.A. and R.J. Gibbs. 1994. 'Natural turf for sport and amenity—science and practice', *CAB International*, 404 pp.

Aldous, D.E., R. Spauling and J.E. Kaufmann. 1977. 'Method of soil temperature control for plant research', *Agron. J.*, 69:325–6.

Aldous, D.E. and J.F. Kaufmann. 1979. 'Role of root temperature on shoot growth of two Kentucky bluegrass cultivars', *Agron. J.*, 71:545–7.

Barnard, C. 1964. 'Form and structure', in C. Barnard (ed.), *Grass and Grasslands*, MacMillan, London.

Beard, J. B. 1965. 'Factors in the adaptation of turfgrasses to shade', *Agron. J.*, 57: 457–9.

Beard, J.B. and W.H. Daniel. 1965. 'Effect of temperature and cutting on the growth of creeping bentgrass roots', *Agron. J.*, 57:249–50.

Beard, J.B. 1973. *Turfgrass: Science and Culture*, Prentice-Hall Inc., Englewood Cliffs, 658 pp.

Beard, J.B. 1992. 'What happens after you mow? Turf regrowth', *Grounds Maintenance*, May, pp.36, 38.

Berg, A.R. 1972. 'Grass reproduction', in V.B. Youngner and C.M. McKell (eds.), *The Biology and Utilization of Grasses*, Academic Press, New York.

Beehag, G.W. 1997. 'An historical review of the utilisation of the warm season turfgrasses in Australia', in addendum *International Turfgrass Society Meetings*, University of Sydney, Sydney, pp. 7–10.

Brown, E.M. 1939. 'Some effects of temperature on the growth and chemical composition of certain pasture grasses', *Missouri Agr. Exp. Sta., Res. Bull.*, 299: 1–76.

Calder, D.M. 1966. 'Inflorescence induction and initiation in the Gramineae', in F.L. Milthorpe and J.D. Ivins (eds), *The Growth of Cereals and Grasses*, Butterworths, London.

Chouard, P. 1960. 'Vernalisation and its relations to dormancy', *Annu. Rev. Plant Physiol.* 1: 191–238.

DeCugnac, A. 1931. 'Recherches sur les glucides des Gramineae', *Ann. Sci. Naturelles*, 13 :1–129.

Evans, L.T., I.F. Wardlaw and C.N. Williams. 1964. 'Environmental control of growth', in C. Barnard (ed.) in *Grasses and grasslands*, MacMillan, London.

Evans, L.T. 1969. '*Lolium temulentum L.*', in L.T. Evans (ed.), *Induction of flowering—some case histories*, Macmillian, New York.

Evans, M.W. and J.M. Watkins. 1939. 'The growth of Kentucky bluegrass and Canada bluegrass in late spring and in autumn as affected by length of day', *J. Amer. Soc. Agron.*, 31:764–74.

Evans, M.W. and F.O. Grover. 1940. 'Developmental morphology of the growing point of the shoot and the inflorescence in grasses', *Journal of Agricultural Research*, 61:481–520.

Gibbs Russell, G.E., L. Watson, M. Koekemoer, L. Smook, N.P. Barker, H.M. Anderson and M.J. Dallwitz. 1991. 'Grasses of southern Africa', *Memoirs of the Botanical Survey of South Africa No. 58.*, National Botanic Gardens, South Africa, 437 pp.

Hirose, M. 1973. 'Comparision of physiological and ecological characteristics between tropical and temperate grass species', ASPAC, Food and Fertiliser Technology Center, *Extension Bulletin No. 26*, January, Orient Printing, Taiwan, 20 pp.

Holt, E.C. 1969. 'Turfgrasses under warm, humid conditions', in A.A. Hanson and F.V. Juska (eds), *Turfgrass Science*, American Society of Agronomy, Madison, Wisconsin, pp. 513–27.

Johnson, P.G. and D.B. White. 1997. 'Vernalisation requirements among selected genotypes of annual bluegrass (*Poa annua* L.)', *Crop Sci.*, 37:1538–42.

Juska, F.V., J.F. Corman and A.W. Hovin. 1969. 'Turfgrasses under cool, humid conditions', in A.A. Hanson and F.V. Juska (eds.), *Turfgrass Science*, American Society of Agronomy, Madison, Wisconsin, pp. 491–508.

Langer, R.H.M. 1972. *How grasses grow*, Edward Arnold, The Institute of Biology's Studies in Biology, 60 pp.

Lodge, G.M., G.G. Robinson and P.C. Simpson. 1990. 'Grasses—native and naturalised—recognition, value, distribution', *Agfact P25.32*, 1st Edition, N.S.W. Department of Agriculture, 28 pp.

Lush, M. 1991. 'The greenhouse effect—the climate zones of Australia and the grasses which grow there', *TurfCraft*, September, pp.16,18.

Madison, J.H., 1962. 'Turfgrass ecology—effects of mowing, irrigation, and nitrogen treatments of *Agrostis palustris* Huds., 'Seaside' and *Agrostis tenuis* Sibth., 'Highland' on population, yield, rooting, and cover', *Agron. J.*, 54:157–60.

Madison, J.H. 1982. *Principles of Turfgrass Culture*, Robert E. Krieger Publishing Company, Florida, 431 pp.

May, L.H. 1960. 'The utilization of carbohydrate reserves in pasture plants after defoliation', *Herb. Abstr.*, 30:239–45.

Mitchell, K.J. 1953. 'Influence of light and temperature on the growth of ryegrass (*Lolium* spp.) 2. The control of lateral bud development', *Physiol. Plant.* 6:425–43.

Mitchell, K.J. 1956. 'Growth of pasture species under controlled environments. 1. Growth at various levels of constant temperature', *New Zealand J. Sci. and Tech*, 38A:203–15.

Minson, D.J. and J.R. Wilson. 1980. 'Comparative digestibility of tropical and temperate foliage—a contrast between grasses and legumes', *J. Aust. Inst. Agric. Sci.*, 46:247–9.

Moore, C.W.E. 1964. 'Distribution of grasslands', in *Grasses and Grasslands*, Barnard, C. (ed.), Macmillan, London, pp.182–205.

Peterson, M.L. and W.E. Loomis. 1949. 'Effects of photoperiod and temperature on growth and flowering of Kentucky bluegrass', *Plant Physiol.*, 24:31–43.

Rochecouste, E. 1968. 'Considerations on the role of phasic growth in the control of perennial grasses with special reference to couch (*Cynodon dactylon* (L). Pers), *Proc. of the 1st Victorian Weeds Conference, Melbourne*, November, The Weed Society of Victoria, 2-62-13.

Sharman, B.C. 1947. 'The biology and developmental morphology of the shoot apex in the Gramineae', *The New Phytol.*, 46:20–38.

Smith, D. 1967. 'Carbohydrates in grasses. II. Sugar and fructosan composition of the stem bases of bromegrass and timothy at several growth stages and in different plant parts at anthesis', *Crop Sci.*, 7:62–7.

Soper, K. and K.J. Mitchell. 1956. 'The developmental anatomy of perennial ryegrass (*Lolium perenne* L.)', *New Zealand J. Sci. and Tech.*, 37A:484–504.

Troughton, A. 1956. 'Studies on the growth of young grass plants with special reference to the relationship between root and shoot systems', *J. Brit. Grassland Soc.*, 11 :56–65.

Troughton, A. 1957. *The underground organs of herbage grasses*, Commonwealth Bur. of Pastures and Field Crops, Hurley, Berkshire, England.

Weinmann, H. and L. Reinhold. 1946. 'Reserve carbohydrates in South African grasses', *J. South Afr. Bot.*, 12:57–73.

Weinmann, H. and E.P. Goldsmith. 1948. 'Underground reserves of *Cynodon dactylon*', in *Better Turf Through Research*, African Explosives and Chemical Ind. Ltd.

Whalley, R.D.B., C.M. McKell and L.R. Green. 1966. 'Seedling vigor and early nonphotosynthetic stage of seedling growth in grasses', *Crop Sci.*, 6: 147–50.

Wilson, J.R. 1959. The influence of time of tiller origin and nitrogen level on the floral initiation and ear emergence of four pasture grasses, M.Sc.Agr. Thesis, University of Sydney, Australia.

Wilson, J.R. and C.W. Ford. 1973. 'Temperature influences on the in vitro digestibility and soluble carbohydrate accumulation of tropical and temperate grasses', *Aust. J. Agric. Res.*, 24,187–98.

Wilson, J.R. 1975. 'Comparative response to nitrogen deficiency of a tropical and temperate grass in the interrelation between photosynthesis, growth, and the accumulation of non-structural carbohydrate', *Neth. J. Agric. Sci.*, 23:104–12.

Wilson, J.R. 1993. 'Organization of forage plant tissues', in *Forage Cell Wall Structure and Digestibility*, ASA-CSSA-SSSA, Madison, W1, USA.

Wilson, J.R. 1997a. 'Adaptive responses of grasses to shade: Relevance to turfgrasses for low light environments', in *Proceedings of the International Turfgrass Society Research Journal, University of Sydney, NSW*, vol. 8:575–91.

Wilson, J.R. 1997b. 'Adaptive responses of turfgrasses to shade', in M. Agnew (ed.), *The 1997 Green Pages Annual*, Strategic Publications, New Gisbourne, Victoria, pp. 76–8.

Yeh, R.Y., A.G. Matches and R.L. Larson. 1976. 'Endogenous growth regulators and summer tillering of tall fescue', *Crop Sci.*, vol. 16:409–13.

Youngner, V.B. 1959. 'Growth of U-3 bermudagrass under various day and night temperatures and light intensities', *Agron. J.*, 51:557-559.

Youngner, V.B. 1961. 'Growth and flowering of Zoysia species in response to temperatures, photoperiods and light intensities', *Crop Sci.*, I :91–3.

Youngner, V.B. and F.J. Nudge. 1976. 'Soil temperature, air temperature, and defoliation effects on growth and nonstructural carbohydrates of Kentucky bluegrass', *Agron. J.*, 68:257–60.

CHAPTER 4

Turfgrass soils and drainage

B.F. JAKOBSEN, Rootzone Laboratories International Pty. Ltd, Canberra, ACT, and
D.K. McINTYRE, Horticultural Engineering Consultancy, Canberra, ACT, Australia

Introduction

Soils consist of mineral particles that are structured by physical and living processes into a pore system that constitutes the space below ground level (Stewart, 1980). In addition to the mineral particles, which are derived from the decomposition of rocks, calcium carbonates and phosphates, soils contain three other fractions: organic material, derived from ancient organisms; residues of plants and microorganisms that have recently been incorporated into the soil; and soil water, which is a solution of soluble and partially soluble salts (Russell, 1962). Soils also contain a network of channels, filled with air and water, which are bounded by the soil surfaces. Properties fundamental to soil depend on the geometry of this interconnected network of pore spaces. In this chapter discussion will focus on the soil properties that are relevant to the drainage and construction of sporting facilities, and how they perform under intensive use.

Soil structure and texture

Because soil is a three-phase material of solid, liquid and gas, we can describe soil structure in terms of either the solid phase or as pore spaces. Here we shall refer to soil structure as the size, shape and arrangement of the soil particles. The movement of water through soils is determined by the arrangement of these various particles, and how these particles respond when pressure is applied, particularly under wet or intense playing conditions. Playing pressure causes rearrangement of these particles and can increase the soil density (the mass of soil per unit volume including the solids and the pore space), reduce porosity (the percentage of total volume filled with air), as well as restricting water movement. Poorly drained and compacted soils lead to shallow rooted, poor quality growing turf. Soil texture refers to the proportion of these individual mineral particles in a soil, and is often expressed as various proportions of sand, silt and clay. The size differences between the various soil particles is large, and can differ by a factor of 1000 (Table 4.1). Two particular groups of particles have a major impact on the behaviour of soils; the clay and the silt particles. These particles are often classified as 'fines' and have the ability to move within the soil when it is saturated and pressure is applied. Clay particles have a special significance as they act as a cementing agent between larger particles, such as silt and the finer of the sand grains, causing them to cling together. Large numbers of particles can be joined to form aggregates the size of sand grains or larger.

Table 4.1 Different soil particles and their particle diameter compared with objects of similar relative size (after McIntyre and Jakobsen, 1998).

Soil particles	*Diameter (mm)*		*Objects of similar relative size*
Gravel	>2		
Very coarse sand	1–2		Soccer ball
Coarse sand	0.5–1		Tennis ball
Medium sand	0.25–0.5		Golf ball
Fine sand	0.1–0.25		Marble
Very fine sand	0.05–0.1	Fines	Match head
Silt	0.002–0.05	Fines	Sesame seed
Clay	<0.002	Fines	Table salt

When these soil materials aggregate they behave as larger particles, not separately as fine silt or clay. This can often mean that a well structured and aggregated clay soil will drain much better than a soil which has far less clay but may have a higher silt content. This process of forming aggregates can also occur with organic material cementing the small particles together, so that they also behave as larger particles. For this reason, soils with a high organic content will often drain very much better than a soil with the same inorganic particle size distribution, but with a low organic matter content. Some soils are known to have a very low stability and are readily dispersed when immersed in water. The particles break away from each other and 'float' in solution. Other soils can be very stable because the clay particles have cemented into larger aggregates, which do not disperse into primary particles when immersed in water. Such soils may require intensive mechanical treatment for them to be dispersed, even when undergoing particle size analysis in a laboratory. Therefore soils that have their origins as lake silt or flood plain soils are undesirable for sports fields because they are usually quite unstable, particularly when they are worked or played on under wet conditions. The degree of soil stability can be determined by using the dispersion test, or determined from knowledge gained from particle size analysis.

In field soils, particles are linked by physical and chemical bonds. To disperse these particles the bonding agents, such as polymerised hydroxides of iron and aluminium in various stages of hydration, have to be eliminated (Blake, 1967). A simple test can be carried out to test for soil stability. Take five to ten grams of soil and carefully place it in a small beaker of distilled or deionised water. If the water immediately becomes cloudy around the soil, then it is considered unstable. If the water remains clear and the aggregate remains intact, then the soil is considered stable. Soils which show instability will almost certainly compact badly when used in a sportsturf, and should not be used in construction.

Three other different laboratory tests can be used to predict soil behaviour under intensive use. The first of these is the water holding capacity of the soil at a standard of one metre suction. This reading predicts how droughty the soil may become in the summer, and how much water may be held at any time for grass growth. A minimum of 12 per cent by weight is considered necessary for loam soils. The second test, that of compacted hydraulic conductivity of the soil, is determined by the drop method (McIntyre and Jakobsen, 1998), and consists of dropping the saturated sample in a 250 mm diameter tube a distance of 150 mm. The value obtained after sixteen drops is approximately equivalent to heavy football use, and 32 drops equates to heavy horse racing. These values allow valid predictions to be made about the future behaviour of the soil under use, particularly under wet conditions. Hydraulic conductivity values above five millimetres per hour are essential for these soils to drain, and for good grass growth to occur. The third test is determining the particle size distribution using a wet sieve analysis. This involves breaking up the soil sample with a mortar and pestle into small aggregates, placing the sample into distilled water, and adding an agent such as Calgon® to disperse the clay particles. By vigorous stirring the material is broken down into primary particles, then washed through a nest of sieves to determine its particle size distribution. The various sieves and all the material caught on them are dried in an oven at 105°C to a constant weight (at least four hours), and weighed. This produces a particle size distribution based on dry weight. Silt and clay content can be determined separately by using a hydrometer, which measures the density of the water–silt–clay suspension. After five hours, the silt particles will have settled to the bottom and the colloidal clay will stay in suspension. It is important to have an accurate knowledge of the amount and type of fines in any soil, as these are the particles that have the potential to migrate in the soil and influence its behaviour.

Soil–water relations

Pore spaces

Soil consists of a complex arrangement of individual particles and aggregates of differing shapes and sizes, within a series of channels and spaces between these particles. The spaces between the larger particles are not completely filled by smaller particles, and these spaces, channels and voids are collectively called pore spaces. Although soil particles and pore spaces are usually depicted in the literature as two-dimensional, they are actually three-dimensional. Their arrangement will depend on the degree of soil compaction, and on the stability of the soil aggregates.

Pore spaces are very important in the behaviour of soils because they provide pathways for the downward movement of water, spaces in which roots grow, and air space which is essential for good turfgrass growth. Water is stored in the smaller pores, and these are important in determining the water holding capacity of soils. Plant roots require pores in excess of 0.1 millimetre to be able to easily force their way through the soil, as well as requiring sufficient oxygen for growth. This is because grass roots range in diameter from 0.06 to 0.25 millimetres (Adams and Gibbs, 1994). If the soil is of a dense, silty nature with a large number of very fine pores, then not only will water move down through it very slowly, but it will not be very suitable for good root growth. Sometimes these pores are inter-connected and sometimes they are not, and the movement of water in soils is greatly dependent on the frequency and orientation of these pores. The size of individual pores depends on the size of the soil particles and on how densely these particles are packed together. When the fines are all packed in between the larger sand grains, the total pore space becomes small and the size of the pores is also very small. If the clay and fines are assembled in aggregates, then there will be a system of large-sized pores between the aggregates, plus a system of smaller pores inside the aggregates. Such soils will drain well and are ideal for turfgrass growth because there is much more air space and more room for root growth.

Behaviour of water in soils

Water movement in soils is influenced by three forces: gravity, the surface tension of the water, and water adhesion to particles (McIntyre and Jakobsen, 1991b). Gravity is the driving force for most of the water movement in the soil. It pulls water downwards through the soil: the greater the vertical distance, the stronger the pull. This is fundamental in the understanding of how water moves in soil. Even when water is moving sideways in soil, except for capillary movement, it is being moved by gravity.

Where water meets air, surface tension exists because all the water molecules are attracted to each other much more strongly than they are to the molecules in the air. This process 'shrinks' the surface of the water which is in contact with the air, acting like a 'drawstring' pulling the surface of a water drop together. This makes a water drop round, as the surface tension shrinks it and holds it together. The smaller the drop, the stronger the surface becomes, and the more difficult it is to break (McIntyre and Jakobsen, 1998). It also makes the surface of a water meniscus strong to the touch, without 'breaking' this meniscus. The surface of water can be bent by gently touching it with an object, but if the force applied by the object is too strong the surface will break.

Water can also adhere to soil particles even more strongly than to other water molecules. In some instances, soil particles may become covered with a wax, which makes them water repellent, and then the attraction between water molecules is stronger than to the wax. This is the common cause of water repellence in sands. This waxy material is thought to be of plant origin. The strong attraction of the water molecules to solid surfaces bends a meniscus where water is in contact with both air and solids.

The combined forces of surface tension and adhesion can hold water back in small pores against the pull of gravity. The particle size and the pore space in a soil will determine how much water can be held in these pores, or how quickly it drains. When the soil is made up of predominantly large particles, for example in a coarse sand, the pores will be large; if the soil particles are very small, such as in a silt, the pores will be very much finer. Water adheres to soil particles and a film of water is always held around each particle. Where the particles touch, or where there are fine pores and channels, water is also held. At the ends of the pores which are filled with water there are menisci where the water comes in contact with the atmosphere. These menisci are held together by surface tension. The closer the particles, the finer the pores, the more tightly the water is held because there is a greater component of adhesive forces than surface tension holding the water in the pore. Adhesive forces are much stronger than surface tension.

Water entry and movement down through soils

Water in soil is held in two principal forms: by adhesion as a surface film around the particles, and by surface tension or water held in the pore spaces. When gravity pulls water downwards in the soil, these two forces are acting against it and attempting to hold the water molecules near to the soil particles or in the pore spaces. Gravity stretches the menisci and bends them inwards towards the surfaces of the particles. The more curvature there is on the menisci the stronger they become, which means that more force is required to remove further water from that pore. When water falls on the surface of the soil, it rapidly enters the large pores near the surface. As these become filled, the rate at which further water enters will slow down. As all the pores at the surface become full, the soil becomes saturated at the surface and run-off or ponding will occur. When a soil becomes saturated, all the pore spaces are filled with water and the menisci become flat surfaces.

Gravity then begins to pull water downwards and the soil begins to drain. Gravity is pulling the water down through the profile, and if it can remove the water faster than it is reaching the soil surface, then no surface ponding or run-off will occur. Surface tension is the weakest of these forces and is the first to break under the pull of gravity, making it the limiting factor in the amount of water a pore can hold. When the large pores are full, most of this water is only held by surface tension as it is too far away from the edges of the pores to be influenced by the adhesive forces. Hence it is easily pulled down by gravity, causing most of the water in these large pores to be drawn down quite quickly. Once the soil is no longer saturated, water is prevented from moving downwards through the soil at the rate of the saturated hydraulic conductivity. This is because even though there are large pores in the soil below the wetting front, water is prevented from filling them because it is being held in the smaller pores above by the forces of adhesion, and by the surface tension of the menisci. If the soil in the immediate vicinity is saturated, water will rapidly enter and fill these large pores, and move downwards faster. 'Gravitational water' is that water in the soil, which after rain or irrigation, moves down through the profile by gravity. When field capacity has been reached, gravity will have removed all the gravitational water and an equilibrium has been reached. This equilibrium is usually reached in 24 to 48 hours in the upper part of a soil profile.

When all the gravitational water has been removed from the larger pores, they are filled with air. This air space is necessary in the soil to allow oxygen to reach the surface of roots through the soil. Ten per cent (by volume) of air space is considered a minimum for plant growth, and as grass roots range in diameter from 0.06 to 0.25 millimetre, then a good volume of air-filled porosity is required to sustain unrestricted grass growth (Adams and Gibbs, 1994).

When rainfall or irrigation ceases, water is pulled down through the profile by the force of gravity. As more water is slowly drawn down out of the larger pores, water is held in the smaller pores and around particles by adhesive forces and by the surface tension of the menisci. A point is reached where there is no further downward movement of water. This is a state of field capacity (McIntyre and Jakobsen, 1991a). The amount of water remaining in the soil at field capacity will vary quite markedly, depending on the soil's particle size distribution. For example, a loamy soil may have 25 per cent water remaining, a USGA sand about 5 to 7 per cent, and a gravel only 2 per cent, yet they have all drained to field capacity. The amount of water retained by a soil at field capacity is also determined by the number of large and small pore spaces. Soils vary markedly in their pore space and size, hence their behaviour in relation to water movement varies; for example a heavy clay soil has very small pores and water moves very slowly through it, whereas water moves very rapidly through a coarse sandy soil with very few fines and a large number of large pores. Once field capacity has been reached, further water removal from the soil is through the plant system via the transpiration process. Plants cease extracting water at wilting point. At this point it should be noted that there is still water in the soil, but it is held very strongly on the particles and in the small pores by the adhesive forces. When the wilting point is reached, the conductivity of this water is 10 000 times smaller than when the soil was at field capacity.

A soil or part of its profile may become saturated when all pores have become filled with water. This situation rarely occurs because there is almost always air trapped in some pores, however for our purposes the above definition is adequate. Saturation may occur throughout the whole profile or it may occur as a band at varying depths as water moves down through the profile. When water reaches the soil surface, it enters it; the rate at which it enters the soil is called the infiltration rate. The infiltration rate will vary according to soil structure, and the moisture status of the soil at the time irrigation commences. The rate will be slow on a heavy clay soil, and much quicker on a very sandy soil. If the soil is very dry, the water enters quickly as all the large pores near the surface are empty. The remaining water in these pores, and the water in the smaller pores, is attracted much more strongly to the edges of the particles by the stronger adhesive forces, and requires greater force to remove it. The water in the smallest of these pores will not be drawn down by gravity at all. The rate at which water moves downward through the soil is even more affected by the pore size through which it has to move. This rate is approximately proportional to the square of the pore diameter, for instance water will move at least a hundred times faster through a USGA sand than a silt–clay loam. From laboratory tests carried out on these types of materials at the Technical Services Unit's laboratory in Canberra, Australia, it was found that the silt–clay soils used on local sportsfields had a hydraulic conductivity between 3 to 10 millimetres per hour, whereas USGA sands have a hydraulic conductivity between 600 to 1500 millimetres per hour. Thus as the water content of a soil decreases due to the pull of gravity, the rate of drainage decreases at an exponential rate because the remaining water is held in more narrow pores.

When the soil is dry and rainfall or irrigation has commenced then water will enter the soil very quickly, and the rate of entry will gradually slow down until the top of the profile is saturated. The rate at which water then enters the soil becomes constant, and this constant rate is called the saturated hydraulic conductivity. This rate is reached because as the large pores fill up at the top of the profile, additional water enters more slowly; and the rate of entry continues to slow down as saturation is approached. This is a very important parameter which is used to judge the rate at which soils drain. Sports turf is almost always different to that of a pasture sward, because the sports turf profile often ranges in depths from 75 to 200 millimetres, and is laid over a clay subsoil base which is quite often highly compacted. Under these conditions, turfgrass root zones rarely exceed 150 millimetres and on shallow topsoil profiles are as shallow as 30 millimetres. The shallow nature of topsoil–base combinations has to be considered differently when compared to agricultural profiles, as when it becomes saturated, further water entry into the soil will only be at the drainage rate of the clay subsoil base. This rate could be as low as one to five millimetres per day. Topsoil can become saturated to its entire depth by as little as 25 millimetres of rain, and will always be saturated during prolonged rain. Table 4.2 shows the differences in infiltration rate between measurements taken at wilting point, visible grass stress, and field capacity of four soil types taken from sportsfields. When irrigation began, the soil moisture content was at permanent wilting point; the point where there is visible grass stress (McIntyre and Jakobsen, 1998).

Table 4.2 Rates of infiltration between wilting point and saturation in four different soil types in Canberra, Australia (from McIntyre and Jakobsen, 1998).

Soil type \ *Soil moisture content at time of irrigation*	*Permanent wilting point (mm/h)*	*Visible grass stress (mm/h)*	*Field capacity (mm/h)*	*Top of profile saturated (mm/h)*
Sandy loam	122	114	107	60
Loam	35	33	31	20
Silt/loam	17	15	12.5	5
Silt/clay	7	5.3	3.6	1

Lateral movement of water through soils

Water moves laterally through soil at a rate that may be in excess of a hundred times slower than it moves vertically down through the soil profile. Four factors influence lateral movement of water in topsoil: (a) 'free water' at the bottom of the saturated zone; (b) the position of this saturated free water zone; (c) the slope on the subgrade (which determines the component of gravitational force contributing to the lateral movement of this water); and (d) an adequate exit point (for water to move significant distances laterally from the zone, it must have the ability to be removed from the profile by some means such as a subsoil drain), otherwise it will pond. The rate at which water drains is constrained by the cross-sectional area of the saturated free water zone through which it is moving.

Where a section of the topsoil becomes saturated, but does not extend right down to the subgrade, the rate at which the water moves down through the soil is similar to the saturated hydraulic conductivity of the topsoil. In this situation there is no free water zone, hence there is no lateral movement of water other than the very small amounts due to capillary action. This only occurs if there is an area of adjacent soil which is much drier than the saturated area. In the situation where the whole surface of the soil is wet by rain, there will be virtually no sideways movement of water because there will be a section of the profile which will be saturated to approximately the same depth right across the whole area. The major difference in the variation of this saturated zone which has not yet reached the subgrade, will be due to the soil texture or pore space distribution between adjacent areas of topsoil. Once the saturated zone reaches the base a new set of conditions apply, and the lateral movement of water is now much quicker. The depth of this topsoil layer is important, as it will determine how quickly it becomes saturated, and how quickly a saturated zone reaches the very slow draining base. The rate at which the subgrade drains is also important, as even a drainage rate of one millimetre per hour can be significant in removing water from the surface of the soil. In shaping the subgrade traffic from earth moving machines can cause compaction rendering it almost impermeable, and adversely influencing the drainage profile.

Water only moves sideways if there is a component of gravity pulling it, and this is usually achieved by the slope on the subgrade. The steeper the slope, the higher the component of gravity. Water will move laterally towards a drain where it is being removed, even though the base may be flat. Under these circumstances the gradient on the depth of the saturated free water zone, as it is being depleted, provides the gravitational component. For example, if the slope on the subgrade is 1:100, then the water is only being pulled sideways across that slope at one-hundredth of the rate that it can move downwards, or one-hundredth of its saturated hydraulic conductivity. If the slope is 1:5, then it will move sideways at one-fifth of this rate. If the hydraulic conductivity of the soil was 20 millimetres per hour, then the maximum rate that water could move sideways, even on this very steep slope (of 1:5) would be 4 millimetres per hour. If the slope on the base is only 1:100, then the maximum rate at which the above soil can move sideways is 0.2 millimetre per hour ($20 \times 1/100$). Therefore even on a steep slope water cannot move very quickly sideways, but on the more common gentle slopes, water moves sideways very slowly indeed.

When the saturated zone reaches the subgrade and free water begins to move laterally down the slope, water must be removed from the system. If this does not occur an equilibrium will be reached, often resulting in ponding and damp conditions at the end of a slope, or against an impermeable obstruction such as a wall. This can result in a build-up of hydrostatic pressure which can cause the collapse of walls and land slips. There will be free surface water at the end of the slope if there is no drainage source to remove it. The rate at which water 'flows' laterally is influenced and restricted by three factors: (a) the depth of the saturated free water zone in the topsoil; (b) the hydraulic conductivity of the topsoil; and (c) the slope on the subgrade. This is explained diagramatically in Figure 4.1. For example, when the saturated zone of a topsoil of 300 millimetres in depth, exhibiting a hydraulic conductivity of 20 millimetres per hour has reached the subgrade, it has built up to 200 millimetres of free water. If we consider that the surface area over which the water was collected was one square metre, and that this water is moving laterally towards a drain which will remove it, then it will be only moving through a 'pipe' with a cross-sectional area of 1 metre × 0.2 metre or 0.2 square metres. The maximum rate that water could move laterally is 20 millimetres per hour × 0.2 or 4 millimetres per hour. This rate will be further reduced by the gradient on the subgrade.

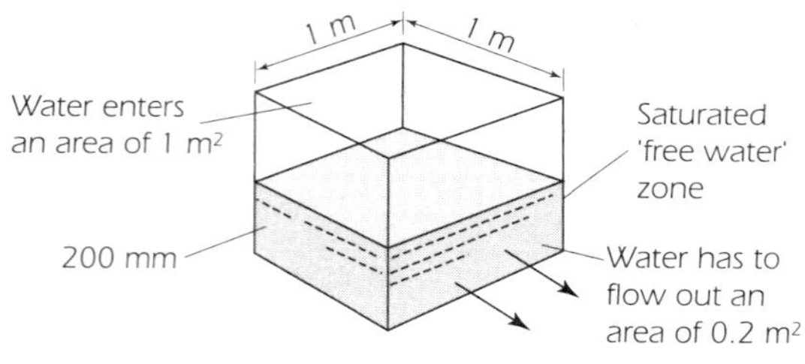

Figure 4.1 Water collected from an area with a surface area of 1 square metre has to move laterally through a 'pipe' with a cross-sectional area of 0.2 square metres, which will constrict the flow rate to one-fifth of its hydraulic conductivity (from McIntyre and Jakobsen, 1998)

In this scenario, the water is reaching the subgrade at a rate of 20 millimetres per hour, and moving sideways at a maximum rate of 4 millimetres per hour. The depth of the topsoil layer itself is a limiting factor in this process. If the depth of the topsoil is only 100 millimetres, then even when the soil has free water to the surface, water can only move sideways at a rate one-tenth the rate it can move downwards. If the topsoil depth was 300 millimetres, then there is the potential for water to move towards drains much quicker because the saturated free water zone can be deeper. Therefore the depth of the saturated free water zone of this topsoil is important in determining how quickly water will travel laterally in soil towards the drains. The depth of the free water zone (in metres) as a fraction, in the above case 1/5, will determine the rate at which water can move sideways in saturated soil. The rate at which water moves sideways through saturated soil is directly proportional to the rate at which it moves downwards through saturated soil. Water will move much faster through a coarse sandy material than it would in a silty clay soil. When the saturated zone reaches the base and a zone of free water builds up, it begins to flow towards a drain through the force of gravity. Using the example above, the potential rate for the lateral movement of water was only 4 millimetres per hour, even though the hydraulic conductivity of the soil was 20 millimetres per hour. Further if the slope on the subgrade was 1:100, then the maximum rate at which water could move sideways would be one-hundredth of the 4 millimetres per hour or 0.04 millimetres per hour. Even if the slope was very steep at 1:5, then the rate would only be 4/5 or 0.8 millimetres per hour. This is extremely slow, particularly as this soil is a sandy loam with a good drainage rate of 20 millimetres per hour, and is 300 millimetres deep with a free water zone 200 millimetres deep. Even under these very good conditions water is only able to move sideways at the rate of 0.04 millimetres per hour on a 1:100 slope, and 0.8 millimetres per hour on a steep slope of 1:5.

Water entry and movement into drains

How water enters sub-soil drains can be explained by establishing a moisture release or water retention curve. We can take a plastic tube 200 millimetres deep, cover the bottom of the tube with an open gauze to prevent the soil from coming out and allow free drainage, and then almost fill it with topsoil. The tube is then filled with water until water begins to drain freely out the bottom. Once all the water is allowed to drain from the tube, the moisture content of the soil in the tube is taken at various levels down the tube. A graph of moisture content against depth in Figure 4.2 can be plotted.

At the lower end of the tube there is a zone where the soil is still saturated, but no water can move sideways from it because the force of gravity needed to pull it sideways is not as strong as

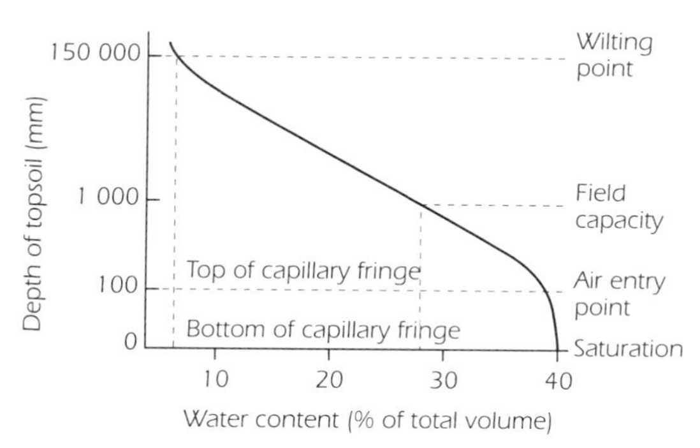

Figure 4.2 Moisture retention curve showing the percentage of soil moisture at various depths in the column after free drainage has ceased (after McIntyre and Jakobsen, 1993)

the meniscal forces at the top of the large pores, which are holding this water in place. Sometimes this water is referred to as 'suspended water', but it is the capillary fringe. The capillary fringe is of great significance as the water in this zone is not able to move sideways, and therefore it cannot move into drains. When water ceases draining from the tube an equilibrium is reached, and there is considered to be 'nil suction' at the very bottom of the column. If at this stage the bottom of the tube is sealed and more water is added from above, an extra zone of free water develops below the capillary fringe. This water has the ability to move sideways and can enter drains (McIntyre and Jakobsen, 1998). As more water is added to the top of the column, the depth of this free water zone increases. The point where there is nil suction also rises, and this is at the interface with the capillary fringe and the free water zone (Figure 4.3).

In the field situation where, as the zone immediately above the base becomes saturated, more rain will cause the formation of a free water zone directly above the subgrade. If there is a gradient on the subgrade, or if there is a drain that is taking water away which in effect creates a gradient, then water can move sideways from this free water zone. There is a point on the moisture release curve called the air entry point. This is the point above which the pull of gravity is strong enough to break the surface tension on the top of the menisci of the large pores, and the water in these large pores drains down the profile. The nearer the surface, the stronger the pull of gravity, so there is more water being pulled down out of ever increasingly small pores. This continues until the soil reaches field capacity, where no more water can be pulled down through the profile because the combined forces of adhesion and surface tension are equal to the force of gravity. This is why the water content decreases from the

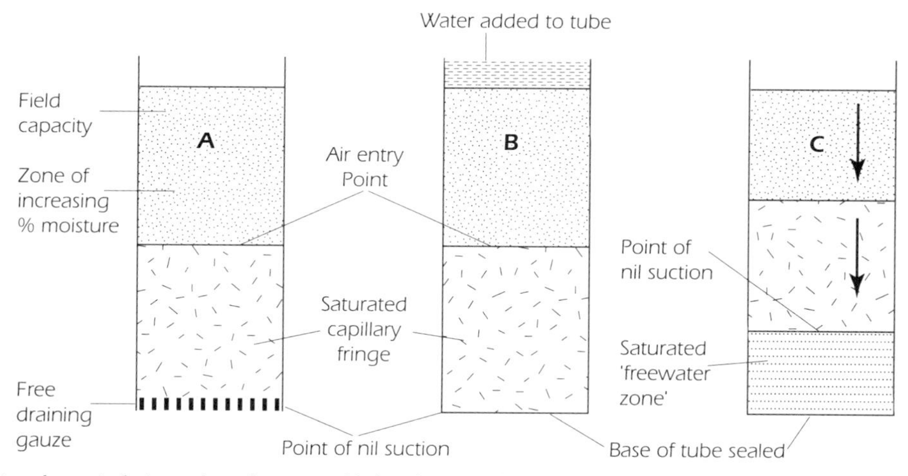

Figure 4.3 In column A, drainage through a permeable base has ceased. In column B, the base is sealed and extra water is added to the top of the column. In column C, there is a rise in the point of nil suction and a zone of free water developing at the bottom of the column.

air entry point to field capacity, as shown by the moisture content in column A.

Provided the soil profile is deep enough, the surface can drain back to field capacity while there is still free water at the top of the base. The air entry point must be below the soil surface, otherwise the soil profile can remain completely saturated for long periods. As the free water is moved down into the base, the top of the capillary fringe moves down at the same rate. If a hole is dug into the soil in this situation, when the hole reaches the saturated free water zone, water moves sideways and enters the hole because there has been sufficient gradient created by the depth of the hole for water to move sideways into it. Even though the soil immediately above the free water saturated zone is also saturated, no water will move sideways into the hole. Similarly, when the soil is in this situation (capillary fringe), no water can move sideways to enter drains. This is because the water is held in the pores by the combined forces of the surface tension of the menisci and the adhesive forces of the soil particles, which are stronger than the component of gravity that is trying to pull it sideways. The height of the capillary fringe increases as the average pore size of the soil decreases. This means the height of the capillary fringe in loam soils may be as high as 200 millimetres. In most sports turf situations, this will mean that the topsoil will remain saturated to the surface while ever there is any free water at the top of the base. Similiarly, areas of compacted soil will have less pore space, hence a higher capillary fringe with the result that they will stay wetter longer than the adjacent less compacted topsoil.

In situations where the soil profile is deep enough, there will be a saturated free water zone above the base, a capillary fringe, and the soil above the top of the capillary fringe will be nearing field capacity. Only water that can move into the sub-soil drains from the free water zone; and no water from the capillary fringe can move sideways towards the pipe, even though this zone is saturated. As water moves into the drain the height of the free water zone diminishes, as does the gradient from the top of the midpoint between the drains to the drain (Figure 4.3). As this gradient diminishes, the rate of movement of water into the drains diminishes. This is because the component of gravity that is pulling the water sideways is diminished as the height of the free water zone diminishes. When a situation is reached where all of the water has been removed from the free water zone, by the combination of the drains and the base pulling water downwards, all movement of water into the drains ceases in spite of the fact that the capillary fringe is still saturated. All further drainage after this point is determined by the drainage rate of the subgrade. Gradually more and more water is pulled down into the subgrade, and more of the topsoil above approaches field capacity.

When a permeable topsoil is placed over a slower draining base, and sub-soil drains are installed in that base, the rate at which the saturated free water zone of the topsoil will drain can be calculated using Hooghoudt's formula. This formula calculates the rate at which the free

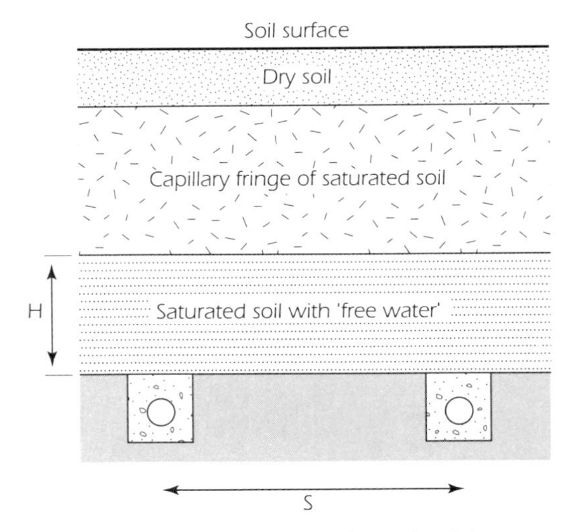

Figure 4.4 The depth (H), of saturated topsoil with free water and the spacings (S) of sub-soil drains

water zone of topsoil drains at the midpoint (M) between the two drains (see Figure 4.4).

Hooghoudt's formula $D = \frac{4KH^2}{S^2}$

where D is the drainage rate in millimetres per hour at the midpoint between the drains of the saturated free water zone; K is the saturated hydraulic conductivity of the topsoil in millimetres per hour; H consists of two components, H_A which is the height of the saturated free water zone in the soil in metres (this is the component which relates to the pressure being exerted by the head of water above it in the soil, and it may be greater than the depth of the saturated free water zone as it also includes the height of the head of water which increases as the slope of the base increases), and H_B which is the height of the saturated free water zone which the water has to pass through to move laterally in metres (this is the height which when multiplied by one metre gives the cross-sectional area through which the water has to flow, and it is the component which takes into account the constraint of flow). S also consists of two components, S_A is the distance between the drains in metres (this is the distance which when divided into H_A gives the gradient on which the pressure of the head is calculated) and S_B is the distance between the drains in metres (this is the distance that when multiplied by one metre gives the area over which the water was previously collected). The formula can now read:

$$D = \frac{4KH_AH_B}{S_A\ S_B}$$

So for most cases where there is very little gradient between the drains, H_A and H_B will be the same, and when multiplied together will be $H_A \times H_B = H^2$ In the same way $S_A \times S_B = S^2$

The distance that the water has to move from the midpoint to the drain is not S, but rather $0.5S$, so the denominator of the formula should now read:

$$\frac{1}{\frac{1}{2} S_A \times \frac{1}{2} S_B} = \frac{1}{\frac{S_A}{2} \times \frac{S_B}{2}} = \frac{4}{S_A \times S_B} = \frac{4}{S^2}$$

When the formula is being used to calculate the drainage rate of a soil between two drainpipes where there is little slope, the first formula can be used. If the area has a slope of 1 in 6 then the components of H can be split to take the slope into account. If there is only one drainpipe on the edge of this area, then the 4 in the formula can be removed as all of the water has to travel all the way across the area to get to the drain exit.

When the formula is used to calculate the rate of drainage for a topsoil depth (H), and a drain spacing (S), this will only give a starting drainage rate. As the water drains down out of the top of the profile and is removed by the drains, the height of the saturated free water zone of soil is decreased. This means that the drainage rate decreases rapidly because the height is decreasing by the square of H (H^2). Because this figure is always less than one metre in sports turf, it means that the H component of the formula decreases rapidly.

For example, if the depth of the saturated free water zone of topsoil is 0.2 metres (200 millimetres) in the initial calculation, then $0.2^2 = 0.04$. If the free water zone has been reduced to 0.15 metres (150 millimetres), then $0.15^2 = 0.0225$. This means that the drainage rate will have been slowed down by about 40 per cent simply by the top 50 millimetres of the free water zone being depleted. The drainage rate of a free water zone of topsoil sitting on a slow draining base decreases rapidly as it drains and as the height of the zone becomes smaller. For example, at 50 millimetres deep the rate at which water moves sideways is extremely slow. For example, let us assume that 0.2 metres (200 millimetres) of the above 300 millimetres profile has a saturated free water zone, the soil has hydraulic conductivity of 20 millimetres per hour, and drains are spaced 10 metres apart. The starting drainage rate at the top of the free water zone is:

$$D = \frac{4 \times 20 \times 0.2 \times 0.2}{10 \times 10} = 0.032 \text{ mm/h}$$

This rate is only the starting rate and will slow as the height of the saturated free water zone diminishes. When this zone has been lowered by 50 millimetres (i.e. the height of the free water zone will be 150 millimetres or 0.15 metres), the drainage rate will be:

$$D = \frac{4 \times 20 \times 0.15 \times 0.15}{10 \times 10} = 0.018 \text{ mm/h}$$

The average drainage rate for the top 50 millimetres will be:

$$\frac{0.032 + 0.018}{2} = 0.025 \text{ mm/h}$$

As can be seen, this is an extremely slow drainage rate. If the drains were spaced only one metre apart, then the rate would be:

$$\frac{4 \times 20 \times 0.2 \times 0.2}{1 \times 1} = 3.2 \text{ and}$$

$$\frac{4 \times 20 \times 0.15 \times 0.15}{1 \times 1} = 1.8$$

$$\frac{3.2 + 1.8}{2} = 2.5 \text{ mm/h}$$

If the porosity of the soil was 20 per cent, then it would require 10 millimetres to be removed from the top 50 millimetres of topsoil for it to reach field capacity (20% of 50 millimetres = 10 millimetres). It would therefore take 4 hours for the top 50 millimetres of this profile to drain to field capacity, i.e. $\frac{10}{2.5} = 4$ h.

This means that the top 50 millimetres of topsoil will only drain if it is above the capillary fringe. In a great number of shallow profiles this does not happen as the saturated capillary fringe will extend to be near or indeed above the surface, in some cases. These same calculations have been carried out for the same soil using three different drain spacings. Twenty millimetres per hour is a very good drainage rate for a soil in the field, but installing drains at a 10 metre spacing is a complete waste of effort. Even with the drains spaced at one metre apart, it would take four hours to reduce the free water zone by 50 millimetres at the midpoint of the drains. If the drains were spaced at 2 metres apart it would take 16 hours for the midpoint to be reduced by 50 millimetres, and this is probably an acceptable drainage rate in some circumstances. The profile would have to be 300 millimetres deep, of a good sandy loam, and be regularly de-compacted. The more sandy loams will have a lower capillary fringe because they have less fine pores than the heavier silt–clay loams. The more sandy, higher draining topsoils which are 250 to 300 millimetres deep will be able to drain the surface 50 millimetres quickly, because the capillary fringe will rarely extend to the surface. However these profiles with sub-soil drains are very costly, and in most circumstances would be uneconomic.

Table 4.3 illustrates the drainage rate of the top 50 millimetres of a soil profile with a free water zone at 100 millimetres and 200 millimetres in depth; a hydraulic conductivity of 20 millimetres per hour; a porosity of 20 per cent; drain spacings at 1.0, 2.0 and 10.0 metres apart; and a non-draining base. The drainage rate for the same facility with a base that is draining at the slow rate of 0.3 millimetres per hour is also shown.

Table 4.3 The effect of the drainage rate of the top 50 mm of a soil profile with 'free water' zones of 100 mm and 200 mm deep; a hydraulic conductivity of 20 mm per hour; a porosity of 20%; and drains spaced at 1, 2 and 10 m on a base that does not drain. This is compared to the rates where all of the above conditions are the same except for a base that drains at 0.3 mm/h as shown by *.

	Depth of saturated free water zone of topsoil (mm)			
	200		*100*	
Distance drains are spaced apart (m)	*Drainage rate of top of free water zone (mm/h)*	*Time taken to drain top 50mm of free water zone (h)*	*Drainage rate of free water zone (mm/h)*	*Time taken to drain top 50mm of free water zone (h)*
10	0.025 (0.325)*	400 (30.800)*	0.005 (0.305)*	2000 (32.800)*
2	0.63 (0.930)*	16 (10.800)*	0.125 (0.525)*	80 (23.500)*
1	2.5 (2.800)*	4 (3.600)*	0.5 (0.800)*	20 (12.500)*

Using the same data as that in Table 4.3, with a 200 millimetres free water zone, the slow draining base was able to drain the top 50 millimetres of the profile equal to the drainage rate of sub-soil drains spaced 2.8 metres apart. It also drains the top 50 millimetres of the 100 millimetre deep free water zone almost as quickly as the sub-soil drains spaced one metre apart. Although 0.3 millimetres per hour is a very slow rate of drainage, it still drains the free water faster than drains spaced at 2.8 metres apart. On more normal silt–clay soils, where the hydraulic conductivity is far more likely to be 5 millimetres per hour, this base will drain faster than sub-soil drains spaced at 1.5 metres apart. Given this information, it becomes very clear that it is far more important to have a good draining subgrade than to instal sub-soil drains. Sub-soil drains are most effective when the topsoil depth is more than 250 millimetres, and as the depth of topsoil increases so too does the effectiveness of the drains. Drains have to be spaced quite close together, as close as a metre apart, to be of any great effect in most soil profiles.

The relationship between the drain spacings, the height of free water zone in the topsoil, and the drainage rate is best explained in Figure 4.5. In gradient A, from the midpoint on the closer drain spacing at 5 metres, the gradient is steeper than for gradient D for the same free water depth with the wider drain spacing of 10 metres. Since this gradient reflects the component of gravity involved, it means that there will be a stronger 'pull' by

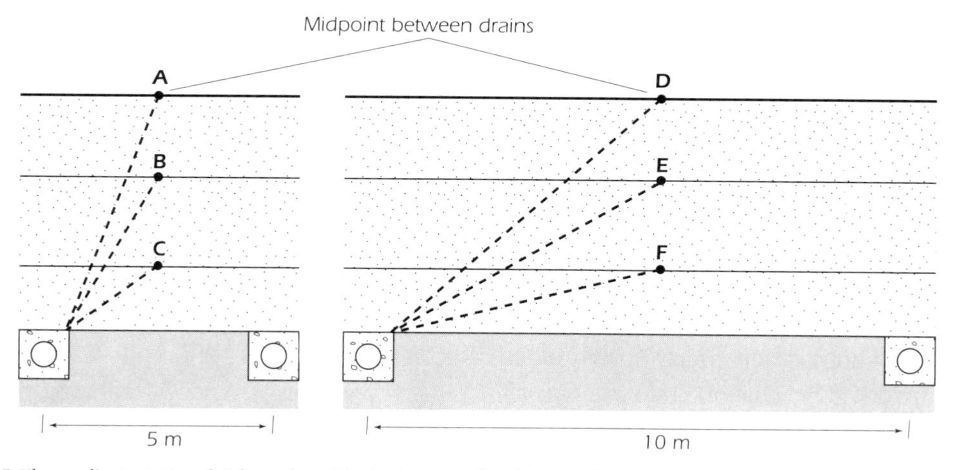

Figure 4.5 The gradients A, B and C from the midpoint between the drain spacings at 5 metres and 10 metres apart for three different heights of saturated free water zones in a topsoil. Gradients D, E and F are for the same heights of free water in the same topsoil but with the drains spaced 10 metres apart.

Figure 4.6 Playing pressure can reduce porosity and restrict water movement on: (a) a bowling green and (b) a sports field

gravity in A than in D. Therefore for the same free water depth, water will move faster along gradient A than D, reinforcing the point that the closer the drains, the faster a profile will drain. It also shows that as the free water depth decreases, so too does the gradient and the rate of drainage.

These principles also apply to the total depth of topsoil. If the topsoil is shallow, then drains will have to be spaced very close together to get any effective drainage rate. If the topsoil depth is much deeper, e.g. 300 millimetres, then the drains can be spaced much further apart to achieve the same drainage rate simply because there is greater storage space in the profile for the free water zone. As both S and H are squared in the formula, as they vary they will change the drainage rate by the square of their distance apart or height, not just by their change in magnitude. Similiarly, as the drain spacings are halved their effectiveness is increased four-fold.

The same applies to the depth of topsoil. As a result, shallow topsoil and widely spaced drains will never work, other than to drain a very small area immediately adjacent to the drains. In most situations the drains only remove a small part of the water. The above calculations are a good estimate of the drainage rate, if it is only the drains that are removing the water. For example, let us take a base that drains at the very slow rate of 0.3 millimetres per hour, and see how much effect this will have on the results shown in Table 4.3.

Drains can be installed in various configurations, but the most important thing is they must be spaced evenly apart to provide uniform drainage of the surface. The best known conventional design pattern for sub-soil drains is the herringbone (Figure 4.7).

Figure 4.7 Herringbone drainage system installed as part of sub-base construction. Uniform spacing is of importance to ensure satisfactory drainage.

References

Adams, W.A. and Gibbs, R.F. 1994. 'Natural turf for sport and amenity: science and practice', *CAB International*, Cambridge.

Blake, C. 1967. 'The mechanical and structural properties of soil', (ed.) C. D. Blake, *Fundamentals of Modern Agriculture*, Sydney Univerity Press.

McIntyre, D.K. and B.F. Jakobsen. 1991a. 'Drainage—Soil structures and their effect upon water movement', *TurfCraft*, May, no. 22, 14 pp.

McIntyre, D.K. and B.F. Jakobsen. 1991b. 'Drainage—Water intake and water retention of water in soils', *TurfCraft*, July, no. 23.

McIntyre, D.K and B.F. Jakobsen 1992. 'Sub-soil drainage—the lateral movement of water in soils', *TurfCraft*, May/June, no. 27.

McIntyre, D.K. and B.F. Jakobsen. 1993. 'The perched water table and its use in sportsturf', *TurfCraft*, March, no. 31:48–9.

McIntyre, D.K. and B.F. Jakobsen. 1998. *Drainage for sports turf and horticulture*, Horticultural Engineering Consultancy, Canberra, 170 pp.

Russell, E.W. 1962. *Soil conditions and plant growth*, Longman, London.

Stewart, V.I. 1980. 'Soil drainage and soil moisture', (ed.) I.H. Rorison and R. Hunt, *Amenity grasslands: an ecological perspective*, John Wiley & Sons, pp. 119–24.

CHAPTER 5

Turfgrass establishment, revegetation and renovation

H. LIEBAO, College of Plant Science and Technology, China Agricultural University, Beijing, China, and **D.E. ALDOUS,** Institute of Land and Food Resources, University of Melbourne, Melbourne, Australia

Introduction

Poor quality turfgrass surfacing, and the revegetation of grasslands generally, may often be traced back to inadequate site preparation. Modern establishment procedures and guideline documentation often have common criteria such as the selection of appropriate species, site evaluation and preparation, seed germination and growth requirements, sowing rates, methods of seed sowing and planting, and adequate post-planting management. The relative ranking of each of these components will vary from site to site and be based on the specific management objectives. Whether the site is a highly intensively managed monoculture, or a native or naturalised grassland, the system requires careful planning, implementation and after-care to achieve a successful outcome.

Turfgrass performance selection characteristics

The primary turfgrass characteristics include its suitability to the local environment, the purpose for which it is being managed, and the expected level of playing quality. In Britain, important qualities in turfgrass selection are persistence under close mowing, durability under wear, disease tolerance, compactness, slow vertical growth rate, good winter colour, and cleanness of cut (Shildrick, 1980). Cannaway *et al.* (1990) lists twenty-six factors that influence playing quality, twenty-four of these being agronomic in nature. Selection characteristics for New Zealand turf include per cent turf cover, thatch depth, root zone conditions, and tolerance to disease and pests. Climatic and cultural conditions aside, the purpose of the surface and the desired level of playing quality dictate even more specific selection characteristics. For example, the best grasses for international sports fields will be those species that are not only suitable climatically and culturally, but also offer good traffic tolerance, shear resistance and recuperative ability. The comparative characteristics of a number of cool and warm-season turfgrasses are summarised in Table 5.1.

Turfgrass quality, or the degree of excellence offered by a natural grass surface, may include components of uniformity, density, texture, growth habit, smoothness and colour (Beard, 1973), as well as elasticity, recuperative capacity, resiliency, rooting capacity, rigidity, verdure and yield (Turgeon, 1980). Turfgrass uniformity is an estimate of the surface's even appearance and is often visually measured. Density, or the number of shoots per unit area, is measured by physical count or with a densitometer. Texture is a measure of leaf blade width. Narrow leafed or fine-textured grasses (< 1 millimetre) include the fescues and bentgrasses, whereas kikuyu grass and tall fescue (> 4 millimetres) are examples of coarse-textured grasses. A textural range between 1.5 and 3.0 millimetres is generally preferred for most turfgrass situations. The turfgrass growth habit is descriptive of the type of shoot growth, which may be exhibited as bunch-type (rye grass), surface stoloniferous stems (Queensland blue couch), or below ground rhizomes (Kentucky blue). Growth habit can be qualitatively valued through visual estimate, as well as through the trueness and distance of ball roll. Smoothness of surface is a feature of turf that has both visual and playing quality value. Poor mowing can leave ragged leaf ends which reduce

Table 5.1 Grass performance characteristics in selecting cool- and warm-season turfgrasses (adapted from Beard, 1973; Turgeon, 1980; Youngner *et al.*, (undated); Drane, 1993)*.

Leaf texture—blade width		**Mowing height**		**Heat tolerance**		**Fertiliser requirement**	
Coarse	Kikuyugrass	Low cut	Creeping bent	High	Kikuyugrass	Low	Kikuyugrass
	Tall fescue		Hybrid bermuda		Zoysiagrass		Centipede
	St. Augustine		Bermudagrass		Bermudagrass		Perennial rye
	Perennial rye		Colonial bent		St. Augustine		Meadow fescue
	Kentucky blue		Zoysiagrass		Centipede		Zoysiagrass
	Colonial bent		Centipede		Meadow fescue		Tall fescue
	Centipede		Kentucky blue		Kentucky blue		St. Augustine
	Zoysia grass		Perennial rye		Colonial bent		Hybrid bermuda
	Bermudagrass		Kikuyugrass		Tall fescue		Bermudagrass
	Creeping bent		St. Augustine		Red fescue		Colonial bent
	Red fescue		Red fescue		Perennial rye		Kentucky blue
Fine	Hybrid bermuda	High cut	Tall fescue	Low	Creeping bent	High	Creeping bent

Winter colour		**Salinity tolerance**		**Drought tolerance**		**Wear tolerance**	
High	Perennial rye	High	Hybrid bermuda	High	Hybrid bermuda	High	Zoysiagrass
	Kentucky blue		Bermuda		Bermudagrass		Hybrid bermuda
	Creeping bent		Creeping bent		Tall fescue		Bermudagrass
	Colonial bent		Zoysiagrass		Zoysiagrass		Tall fescue
	Red fescue		St. Augustine		St. Augustine		Perennial rye
	Tall fescue		Kikuyugrass		Kikuyugrass		Kikuyugrass
	Meadow fescue		Perennial rye		Perennial rye		Kentucky blue
	Kikuyugrass		Meadow fescue		Meadow fescue		Red fescue
	St. Augustine		Kentucky blue		Kentucky blue		St. Augustine
	Hybrid bermuda		Colonial bent		Centipede		Meadow fescue
	Zoysiagrass		Tall fescue		Colonial bent		Creeping bent
Low	Bermudagrass	Low	Centipede	Low	Creeping bent	Low	Colonial bent

Compacted soil tolerance		**Frost tolerance**		**Shade tolerance**		**Establishment rate (against weeds)**	
High	Tall fescue	High	Zoysiagrass	Shade	Red fescue	Fast	Perennial rye
	Hybrid bermuda		Creeping bent		Zoysiagrass		Bermudagrass
	Bermudagrass		Red fescue		Colonial bent		Hybrid bermuda
	Zoysiagrass		Meadow fescue		Kikuyugrass		St. Augustine
	Kentucky blue		Perennial rye		Kentucky blue		Kikuyugrass
	Perennial rye		Kentucky blue		Tall fescue		Tall fescue
	Meadow fescue		Tall fescue		Creeping bent		Meadow fescue
	St. Augustine		Centipede		Kentucky blue		Red fescue
	Red fescue		Kikuyugrass		St.Augustine		Centipede
	Dichondra		St. Augustine		Perennial rye		Creeping bent
	Colonial bent		Hybrid bermuda		Hybrid bermuda		Kentucky blue
Low	Creeping bent	Low	Bermudagrass	Sun	Bermudagrass	Slow	Zoysiagrass

(*) Differences between species selection will be dependent on cultivar, method of culture and environment.

the putting quality and the velocity of ball roll. Colour is a measure of the light reflected by turfgrass and is one of the better indicators of turfgrass health and condition. Different species and cultivars vary in colour from light (annual bluegrass) to very dark (red fescue cv. Dawson). Turfgrass colour can be measured as a rating, by colour charts or chlorophyll index. Elasticity is a measure of leaf recovery once a compressing force has been lifted from the leaf surface, and is a function of moisture content and growing conditions. Elasticity is dramatically reduced under freezing growing conditions. Recuperative capacity is the grass's ability to recover from damage, the rate being a function of the degree of injury, the growing environment and carbohydrate reserves (Younger and Nudge, 1976). Turfgrasses with vigourously growing rhizomes (Kentucky bluegrass) and stolons (bermudagrass) have a superior recuperative ability because they can quickly spread in a horizonal nature (Beard, 1973). Turfgrass resiliency is the capacity of the turf to absorb surface shock without altering the surface contours, and is largely a function of thatch level and soil type. Resiliency is important in reducing player injury.

Healthy root growth will be influenced by mowing height, thickness of the thatch layer, and an imposed stressed environment. The amount of root growth generated at any one time measured during the growing season is a measure of rooting capacity. Leaf rigidity is the resistance of the turfgrass leaves to compression and is related to wear resistance of a turf surface (Turgeon, 1985). The chemical composition of the leaf tissue, water content, growing temperatures, and the particular grass species can influence wear resistance. The zoysiagrasses, St. Augustine grass and bermudagrass form very rigid turf and offer high wear resistance, whereas winter grass and rough bluegrass rank very low in rigidity. Softness is the opposite to the characteristic for rigidity.

Another important measure of quality is the verdure, which has been described as the fresh green colour of turf standing after the mowing process (Adams and Gibbs, 1994). Good correlation exists between the amount of verdure, and increasing resiliency, rigidity and wear resistance (Turgeon, 1980). Verdure can also be an important concept in assessing turf density, as this characteristic provides turf with the degree of ball roll, wearability and repairability. Shoot yield can provide a good indication of the plant's response to culture and growing environment, but is less important as a measure of turfgrass quality.

Many of the previously identified selection characteristics have both an aesthetic and functional orientation. In selecting native and naturalised grasses for revegetation and restoration projects, other attributes to be encouraged should include low biomass production, a high tolerance to environmental stress, improved colour and texture, and a capacity to withstand heavy traffic (Hitchmough, 1994). It is also useful to match these characteristics and those of the dominant grasses with those required with its proposed use, or by the site at which it is to be used. For example, the choice of species may well be decided by the availability of seed, its viability, and the surrounding characteristics of the site (Lodge and Whalley, 1981; Stephens, 1997). In many countries there are native and naturalised grasses to select from, so it is impossible to generalise about such grasses as most species will be distinct in their requirements and characteristics. Table 5.2 lists the more common native grasses suggested for use in revegetation and restoration programs in Australia.

Site planning

Turfgrass establishment and revegetation programs need to be planned within a broader framework to ensure that the various operations progress in a systematic manner. This framework

Table 5.2 Agronomic and cultural criteria for selecting common Australian native grasses (adapted from Craigie, 1994).

Tolerance	*Bothriochloa*	*Danthonia*	*Dichantheum*	*Elymus*	*Microlaena*	*Poa*	*Stipa*	*Themeda*
Drought	Y	Y	Y	medium	medium	Y	Y	Y
Frost	medium	Y	Y	Y	Y	Y	Y	medium
Water	low	low	low	low	low	low	low	low
Waterlogging	seasonal	seasonal	low	seasonal	seasonal	seasonal	varies	low
Low light	sun	sun	sun	sun	semi-shade	varies	varies	sun
Nutrition	low	low	low	medium	medium	low	low	low
Mowing	2–3 p.a.	2–3 p.a.	N	2–3 p.a.	2–3 p.a.	N	annual	N
Trampling	Y	Y	Y	N	Y	N	Y	N
Deep rooted	Y	Y	Y	Y	Y	Y	Y	Y

Y = high tolerance; N= low tolerance

should include information on the nature of the site, objectives of the project, and the allocation of materials, equipment and labour. Good planning also involves the establishment of a set of working drawings which show the design, dimensions, contours, elevations and grade levels, as well as location points of features such as drainage, electricity, irrigation systems and plantings. Complete drawings are particularly useful for sport facility complexes such as athletic fields, golf courses and soccer fields where contours and grades must meet specific requirements. For other facilities, cross-sectional drawings may also be required to show all features and materials associated with the profile, including the location of drainage and irrigation lines, the depth of the subgrade and root zone mix. Associated with these drawings should be a complete set of specifications which provide a schedule on how the work has to be done, the methods to be followed, and the quality and quantities of all materials to be ordered. The landscape or golf course architect usually has the responsibility for plans and specifications, with much of the construction work sub-contracted out.

Site preparation

The chronological inputs associated with establishment include clearing and rubbish removal, elimination of persistent perennial weeds, preparation of the subgrade, installation of drainage works, irrigation and power systems, application of additional fertiliser and conditioners, and concluding with a final grading, contouring and levelling of the site. Clearing involves the removal of building rubble, stones, tree stumps, and any other material that might interfere with construction, or cultivation. Buried organic materials eventually decay and may leave undesirable depressions that destroy grade uniformity, allow for ponding during rains or irrigation, or in other cases contribute to the development of disease. If impervious rock outcrops are involved, cut to at least 30 to 40 centimetres (12 to 15 inches) below the fixed grade levels and introduce sufficient topsoil for satisfactory drainage and grass establishment.

Persistent perennial weeds may be effectively controlled by a nonselective herbicide or soil fumigation. In some countries methyl bromide, as a fumigant, is being phased out as it is suspected of being detrimental to the environment. Difficult-to-control weeds may require treatment by a licensed pesticide contractor. Consult your department of agriculture or district extension officer regarding appropriate pre-emergent or post-emergent herbicides for weed control. Pre-emergent herbicides are incorporated into the top 2.5 to 7.5 centimetres (1 to 3 inches) of soil by rain or irrigation, where they are taken up by the roots and shoot growth of the emerging weeds. Post-emergent herbicides can either be translocated systemically or applied as a contact. Both types of post-emergent herbicides must not be washed from the leaves for at least 48 hours after application. Some applicators make provision for a surfactant, which is added to foliar sprays to help penetrate leaves, as well as conserve water on application (Kostka *et al.*, 1997). Weed control in native grass establishment is usually achieved by the use of non-selective herbicides, such as glyphosate, applied as a spot spray (Hitchmough, 1994). Non-chemical weed control options include manual removal, mechanical control (cultivation, removal of the surface layer with a sod cutter, the use of physical barriers, thermic or heat radiation, and the use of naturally occurring chemicals (Morgan, 1988). The most effective non-chemical control for native grass establishment involves grading the topsoil to a depth of between 25 to 50 millimetres (1 to 2 inches) to remove any viable weed seed, and then cultivating the seedbed in the scalped area. On shallow, compacted sites, limited cultivation can increase soil oxygenation and drainage, and relieve compaction.

If grading is necessary, remove 10 to 15 centimetres (4 to 6 inches) of topsoil and stockpile it for future use. The underlying subgrade can be shaped to the desired contour to assist with drainage, and the topsoil redistributed uniformly on the subgrade. If the subsoil has been compacted as a result of heavy traffic or machinery, it may require correction with a deep chisel plough or mole drain to a minimum depth of 200 to 300 millimetres (0.8 to 1.2 inches) applied in the same direction as the slope to provide for movement of water. Provide for satisfactory surface drainage with a two to three per cent slope. The installation of adequate surface and subsurface drainage should follow. The final surface drainage slope should not be less than one per cent. Subsurface drainage pipes are installed by trench digging. The topsoil should be returned and worked with a disc cultivator or rotary hoe to a depth of at least 15 to 20 centimetres (6 to 8 inches), provided normal slope and subgrade conditions apply. The final steps of seedbed preparation may be a set of harrows, steel chain mats, or wooden drag boards, which are often attached to a tractor to smooth out and fill in surface irregularities.

Soil samples should be taken before seeding or planting from several locations on the site to determine soil pH and nutrient requirements. Samples should be taken to a depth of 8 to 13 centimetres (3 to 5 inches) from a number of locations and mixed to produce a composite sample. More than one sample may be necessary if there is soil diversity on the site. Soil test kits are available, or samples could be forwarded to various government agencies or laboratories for testing and interpretation. Stephens (1997) related soil type with the establishment of significant Australian native grasses. For example wallaby grass (*Danthonia caespitosa*), kangaroo grass (*Themeda triandra*), and blue tussock grass (*Poa labillardieri*) prefer a range of soil types, whereas mat grass (*Hemarthria uncinata*) prefers damp clays and swamps, and redleg grass (*Bothriochloa macra*) and Queensland blue grass (*Dicanthium sericeum*) clay, sandy, shale and rocky soils.

Soil pH readings determine the availability of soil nutrients to the turf plant. Whereas a soil pH reading of 6.5 to 7.5 is desirable, many acidic soils can do with about 150 to 200 grams per square metre of lime incorporated into the upper ten centimetres (4.0 inches) prior to cultivation. Any soil having a soil pH higher than 7.5 is considered alkaline, and light applications of sulfur compounds or frequent applications of sulfate of ammonia will assist in lowering the soil pH. Major nutrients required by juvenile grasses at establishment are nitrogen (N), phosphorus (P) and potassium (K). The soil test can be used as a guide in determining the amount of these nutrients. The rate of NPK application can vary from

0.45 to 0.9 kilogram (1 to 2 pounds) per 93 square metres (1000 square feet) (Emmons, 1984). If nitrogen in the starter fertiliser is in a soluble form, the quantity applied should be adjusted to supply a maximum of 0.25 to 0.50 kilograms (0.5 to 1.1 pounds) of actual nitrogen per 93 square metres (1000 square feet). This would require 5 to 10 kilograms (11 to 22 pounds) of a 10.5.5 complete fertiliser or 2.5 kilograms (5 pounds) of sulfate of ammonia.

Physical soil amendments can increase the cation exchange and water holding capacity on sandy soils, and improve soil porosity and infiltration rate on heavily textured soils (Nus, 1994). Australian research has shown that the addition of amendments, particularly peat moss, pine bark, and porous ceramic beads increased moisture retention, while zeolite improved potassium retention (Neylan and Robinson, 1993, 1997). Domestic peat and colloidal phosphate have also been shown to increase moisture holding capacity and fertiliser retention in soils of a sand texture (Smalley, 1961). Domestic peats are usually applied at ten per cent by volume per cubic metre. Natural zeolites have a high cation exchange capacity (100 to 200 milliequivalents per cent) and can be used to increase nutrient storage in turf soils, as well as absorb water (Ferguson *et al.*, 1986; Kaapro, 1993). Natural composts, such as municipal solid waste, leaves and grass clippings, sewage sludge (biosolids), animal manures, and various paper mill and food waste products, can be used as soil amendments at establishment, topdressing, and as low-analysis fertilisers. Recently composts, such as sugar cane bagasse, municipal solid waste, and spent mushroom medium, proved acceptable in the production of St. Augustine grass (*Stenotaphrum secundatum*) sod (Dudeck, 1997). Final grading serves to work the lime, fertiliser, and soil amendments into the surface soil to the required depth and to smooth and condition the surface for seeding. Hand raking is the usual practice in the final grading of small areas of 600 square metres or less, whereas tractor mounted grading and fitting tools, such as soil blades, wire mats and plank drags, can be used on larger areas. Complete seedbed preparation so that it coincides with the optimum time to sow or plant the selected grass.

Establishment procedures

Seeding operations

Many cool-season turfgrass species are planted by seed, either as monocultures or mixed with other genera or species for improved performance. These grass species may be sown out as a mixture, which involves bringing two or more genera together, or as a blend, when grasses of the same genus (cultivars of the same species of grass) are brought together. Common mixtures include the turf type ryegrasses and Kentucky bluegrass, tall fescue and bahiagrass cv. Pensacola, tall fescue and zoysiagrass, and Kentucky bluegrass and common bermudagrass. In the transitional zones of the United States and Australia, cool-season/warm-season polystands are becoming increasingly popular as they have been found to persist longer than their individual monostand equivalents (Dipaola, 1993). To obtain a high density stand of tall fescue, the sowing rate needs to be four kilograms per 100 square metres, and must constitute at least 80 per cent by weight of the final mixture (Eade, 1990). Oversowing rates of tall fescue at renovation will depend on mowing height, variety and condition of the sports turf, but is generally in the order of two kilograms per l00 square metres. Similiarly, to gain maximum benefit from Kentucky bluegrass, a 100 per cent bluegrass turf is desirable, with the percentage by weight of the mix as high as 60 to 70 per cent. To sustain Kentucky bluegrass in this state the turf should be mown around 25 millimetres. To establish a fine leaf ryegrass turf surface, a proportion of 70 per cent plus by weight is desirable in mixtures with an initial sowing rate of 250 to 300 kilogram per hectare (two to three kilograms per 100 square metres). Fine fescues, such as chewings and creeping red fescue, are useful filler grasses and are included in mixes to add density to establishing turf. Due to the fact that fine fescues are not very wear tolerant, a five to ten per cent inclusion in mixes is usually desirable. A higher percentage can be used if a finer more aesthetic turf is required. English mixtures have traditionally comprised mixtures such as seven parts chewings fescue to three parts New Zealand browntop, or four parts chewings fescue, three parts creeping red fescue, one part fine-leaved fescue and two parts New Zealand browntop (Cobb, 1936).

In Australia, strawberry clover (*Trifolium fragiferum*) is often included to seven per cent by weight in cool-season mixtures. White clover (*Trifolium repens*) is generally considered a weed in quality turfgrass areas. Turfgrass blends are valuable and effective particularly where there is wider variation in the environment, or where a number of different disease and/or insect problems exist (Turgeon, 1980). Other turfgrass varieties, such as creeping bentgrass, are often packaged alone as they perform well enough in these situations.

Good quality seed is essential in establishment. The three basic components of quality are purity, viability (germination) and trueness to type. Purity refers to the actual percentage by weight of pure seed, other crop seed, weed seed, and inert matter within the species, variety or specific mixture (Musser and Perkins, 1969; Beard, 1973). The sum of these percentages should equal 100. Viable seeds are those that are capable of germinating and producing a normal seedling under standard laboratory tests (Table 5.3).

Viability is calculated from the number of seeds germinating in that test sample. For example, if a sample of seed contains 95 per cent perennial ryegrass and this seed has a germination rate of 98 per cent, then 93 per cent (0.95×0.98) of the contents by weight is pure live seed. The last component, varietal purity or trueness to type, becomes important when specifying seed of a given variety. Quality seed should also conform to the seed labelling laws in each country.

Cool-season turfgrasses are usually sown in the cooler times of the year, such as late summer

Table 5.3 Days to germination and establishment rate of major cool- and warm-season turfgrasses (after Stephens (1998)).

Turfgrass species	*Scientific name*	*Days to germination*	*Establishment rate*
Italian ryegrass	*Lolium multiflorum*	6–10	R
Perennial ryegrass	*Lolium perenne*	6–10	R
Tall fescue	*Festuca arundinaceae*	7–14	M
Chewings fescue	*Festuca rubra var. commutata*	10–21	M
Hard fescue	*Festuca longifolia*	10–21	S/M
Red fescue	*Festuca rubra*	10–21	M
Kentucky bluegrass	*Poa pratensis*	7–21	S
Rough bluegrass	*Poa trivialis*	14–28	M
Creeping bentgrass	*Agrostis stolonifera*	6–14	M
Colonial bentgrass	*Agrostis capillaris*	4–14	M
Zoysiagrass	*Zoysia spp.*	21–28	M
Queensland blue couch	*Digitaria didactyla*	7–21	S/M
Kikuyu grass	*Pennisetum clandestinum*	10–21	M
Bermudagrass	*Cynodon dactylon*	14–18	M

R = rapid, M = medium, S = slow.

and early autumn, while for warm-season turf grasses, late spring to early summer is preferred for best results. With cool-season grasses the greatest stress is in midsummer, whereas for warm-season grasses this period is experienced in midwinter. Therefore such seeding needs to provide for the longest establishment period before the temperature changes adversely influence growth and development. Optimum temperature range for seed germination is 21 to 35°C (70 to 95°F) for warm-season turfgrasses, and 16 to 30°C (60 to 85°F) for cool-season turfgrasses (Emmons, 1984). Warm-season grasses can be sown as long as soil temperatures exceed 12.7°C (55°F). Dormant seeding is sometimes practised with cool-season grasses in temperate climates. Such seed is planted before the soil freezes and does not germinate until the soil warms up the following spring. Overseeding during *in situ* dormant growth with cool-season grasses is also sometimes practised in temperate climates. Overseeding is carried out to maintain an active cool-season turfgrass surface and to protect the underlying warm-season dormant grass. The process enables the maintenance of a continuously actively growing surface and helps to reduce the time and expense of re-turfing following the winter season. Germination of seeds can be advanced by pregerminating the seeds (Lush and Birkenhead, 1987) or osmoconditioning the seed in a osmoticum (Copeland *et al.*, 1992). Seed may also be purchased as primed seed, which means the seed is brought through the germination process but stopped just before the root and shoot emerge. Such seed can be planted dry. Other seed, such as bermudagrass, can be purchased as hulled seed. This is where the outer seed coat has been removed to hasten germination.

Similarly most native grasses have distinct growth periods. Exceptions are species such as Australia's wallaby grasses (*Danthonia spp*) which although predominantly cool-season growing in autumn and spring, also respond well to summer rainfall. For this reason, as well as native grass sensitivity to weed competition, it is important to sow each grass species at the appropriate time of the year. For example, summer growing species such as kangaroo grass (*Themeda triandra*) and silky bluegrass (*Dichanthium sericeum*) should be sown in late winter to early spring so that they have the subsequent summer period in which to grow.

Turfgrass seeding rates will be influenced by grass species and cultivar, seed quality, germination percentage and conditions, seed costs, seedling vigour, level of post-planting care (Beard, 1973) and planting depth (Newman and Moser, 1988). A living stand of between 1000 to 2000 seedlings per square foot (10 760 to 21 520 seedlings per square metre) has been shown to compete successfully with weed ingress (Musser and Perkins, 1969). For example, perennial ryegrass with 250 000 seeds per pound (551 150 seeds per kilogram) and 93 per cent live seed (95 per cent purity, 98 per cent germination) should be sown out at 2 to 4 kilograms per 93 square metres (4.3 to 8.6 pounds per 1000 square feet). Additional seed may have to be included to account for seed loss due to seed quality and unfavourable environmental conditions following seeding. Similarly, seeding rates of mixtures are based upon the percentages of the various seeds they contain. In any seed mixture, grasses mixed by weight will not result in the same number of plants of both species. For example, a 1:1 mixture by weight of Kentucky bluegrass and red fescue may contain as much as four times Kentucky bluegrass seeds per kilogram than red fescue (5 500 000: 1 320 000). Seeding rates of mixtures should always account for the primary species that will predominate in the sward. Some mixtures are made up for specific purposes, such as low light, high wear or low maintenance.

In most countries the seed laws require that the percentages of ingredients in a mixture be stated on the seed label. Certified products indicate that the seed or plants have met certain standards to assure high quality and low levels of contaminants. Australian native grasses are best sown out at approximately 1000 viable clean seed per square metre (Powell *et al.*, 1974). The sowing rate may vary from approximately four grams per square metre for kangaroo grass to 0.7 grams per square metre for wallaby grass, to account for differences in seed weight between species. Native grass seed is available for use as individual dispersal units or florets which contain the pure seed or caryopsis within them. Although some of the caryopses within these dispersal units are soft, most are of a nut-like consistency and a fill rate test simply involves squashing them with a thumbnail to feel whether they contain seeds or not. This fill rate is normally expressed as a percentage and can vary with species, source, and even within harvests, so each seed batch should be tested. These tests normally involve sowing a 20 to 50 seed sample (often with several replicates) in controlled environment chambers at established temperatures, with germinants being recorded weekly for up to six weeks.

Seed size and weight are important characteristics associated with seedling vigour (McKell, 1972; Patchett, 1982). Generally larger seeds have greater vigour both within and between species and cultivars. For example, ryegrass has greater vigour than bentgrass, and the tetraploids tend to be of greater vigour than the diploids. Relative vigour of such seedlots can be obtained from the interim germination (IG) on the seed test certificate. Lots with a high IG, that is approaching the same value for final germination (FG), have shown the greatest vigour. Seeds per unit weight differ widely for different species. With large seeds there are less seeds per gram, and therefore more weight of seed is required to achieve the same number of plants. Seed rates in Table 5.5 are used only when the single species is involved. The method of propagation, establishment, and seeds per kilogram for 42 cool- and warm-season turfgrasses are summarised in Table 5.4.

Calculating an area and amount of seed for seeding turfgrass

The following methods can be used: measure the length (*l*) and breadth (*b*) of the proposed turf area, for example 120 × 100 metres = 12 000 square metres. Use the scale on the landscape plan and add up the number of squares required under turfgrass, or determine the total area of land in square metres and subtract from it the total area of non-turf in square metres, for example, the total area under turfgrass will equal the total area of land minus the total area of non-turf area (clubhouse, machinery shed, fairways, waterways). If the total area, for instance, equals 225 000 square metres, and the total area of non-turf equals 9000 square metres, the total area to be seeded equals 225 000 minus 9000, which equals 216 000 square metres. If seed is sown at 3.0 kilograms per 100 square metres, approximately 6480 kilograms (2160 × 3.0) will need to be purchased. If the turf area is of an irregular shape, divide the area up into a number of geometric shapes, such as rectangles and circles, calculate their respective areas and add them together to give the total area.

Calculating an area and amount of seed for seeding native grasses

Native grass seed consists of florets which contain the pure seed or caryopsis within them. The area is calculated in a similar way to turfgrass. To calculate the sowing rate, the weights of the filled and empty florets are required. These weights are then used in combination with the fill rate to calculate the number of filled seeds per kilogram, then combined with the germination rate to give a number of germinable seeds per kilogram. The final calculation is based on seedling survival rates that 1000 germinable seeds are required per square metre to result in 100 established plants per square metre. One hundred plants per square metre is considered a desirable turfgrass rate.

Sowing depths for most turfgrass seeds range from 1.25 to 6.5 millimetres (0.2 to 0.4 inch) below the soil surface (Beard, 1973). Sowing depths are critical for many species, such as Kentucky and Canada bluegrass (Nelson, 1927), and annual bluegrass (Engel, 1967) will only germinate in the light. Four-month-old seed of corkscrew grass (*Stipa scabra ssp. falcata*) has 40 per cent germination in the light, but zero germination in the dark (Paget, 1990). Still other species have different requirements, for example weeping grass (*Microlaena stipoides*) seed germinates almost 100 per cent in either light or dark (Paget, 1990). This means that certain seeds sown too deeply may not germinate, and indicates that surface-sowing is required.

Three general classes of mechanical seeding equipment may be found: the hopper type distributor, the seed drill, and the broadcast seeder or spreader. The hopper types are calibrated to deliver seeds at a pre-determined rate, then cover and firm the seed and soil in one operation. The grass seed drill is useful for pasture and native grass establishment, but least desirable for sowing turfgrasses as they sow the seed in rows. The direct drop-type spreaders are preferred over all others as distribution is often more accurate. To ensure uniformity of coverage divide the total quantity of seed to be used on an area into two equal parts, and distribute it in two directions at right angles to each other. The cultipacker seeder, a machine drawn by a tractor, distributes the seed as well as firms the soil around the seed, and is best suited to large, reasonably level areas. Where the slopes are too steep for conventional equipment, seed may be established by hydroseeding, in which the seed and fertiliser are delivered in a suspension of water and mulch under pressure. Small seeds, such as bermudagrass, bentgrass and Kentucky bluegrass, may also be sown with water via a diffuser. Native grass seed can either be sown directly into a non-cultivated seedbed, broadcast onto a cultivated seedbed, or drilled directly into a non-cultivated seedbed (Hitchmough, 1994). In Australia, direct seeding has been found to be the most efficient way to re-establish native and naturalised vegetation as the seed is dispersed into the cultivated area and firmed-in by a trailing press-wheel (Robinson, 1994).

Areas sown down to smaller areas may be consolidated with the back of the rake, or given a controlled rolling with a light corrugated roller to firm the soil about the seed. Larger areas can benefit by being consolidated with a roller which provides shallow furrows in the surface of the seedbed to trap surface moisture. Newly seeded areas may also be mulched to aid establishment and reduce runoff and soil erosion, especially on sloping areas. Mulching materials may vary from organic materials, such as the loose wood materials, bark chips, wood shavings, bagasse, grassy hay and straw, to inorganic forms, such as clear polyethylene covers, seed cloth and hessian (burlap) strips. The inorganic sources need to be removed before the seedlings become too large. Mulching rates are one to one-and-a-half bales of clean (weed free) straw per 300 square metres (1000 square feet), applied so that no more than 50 per cent of the soil is covered. Too much mulch can inhibit seed germination. Adams (1966) found that a five centimetre (two inch) thick application of straw eliminated runoff, decreased evaporation, and reduced raindrop action. Results appear to be better if straw mulching is applied after seeding rather than together (W. Leech, B. Stephens, personal communication, 1998). In the case of native grass establishment Stafford (1991) and McDougall (1989) significantly improved the germination of

Table 5.4 Primary methods of propagation and seeding rates for major cool- and warm-season turfgrass species (adapted from Musser and Perkins, 1969; Beard, 1982; Drane, 1993; Stephens, 1998).

Scientific name	*Common name*	*Establishment method*	*Weight (kg) of seed 100 sq.m.*	*Approx. seeds per kg. (×1000)*
Cool-season grasses				
Agrostis alba	Redtop	seed, sod	1.0–1.5	11–19000
Agrostis capillaris	Browntop bent grass	seed, sod	1.0–2.0	13–15500
Agrostis canina	Velvet bent grass	seed, sod	1.0–1.5	15–17000
Agrostis stolonifera	Creeping bent grass	seed, sod	0.5–1.0	13–15000
Agrostis tenius	Colonial bent grass	seed, sod	1.0	19 000 000
Cynosurus cristatus	Crested dogstail	seed	1.0–1.5	13–15000
Dactyloctenium australe	Sweet smother grass	seed, stolonise, sod		
Danthonia setacea	Wallaby grass	seed, plug		
Eremochloa ophiuroides	Centipede grass	sod	0.25–2.0 (*)	1760
Festuca arundinaceae	Tall fescue	seed, sod mixes	4.0–5.0	456
Festuca pratensis	Meadow fescue grass	seed, sod mixes	3.0–4.0	750–1123
Festuca longifolia	Hard fescue grass	seed, sod mixes	2.0–3.0	750–1123
Festuca ovina	Sheep's fescue	seed, sod mixes	2.0–3.0	750–1000
Festuca rubra var. commutata	Chewings fescue grass	seed, sod mixes	2.0–3.0	900
Festuca rubra subsp. rubra	Red fescue grass	seed, sod mixes	2.0–3.0	770–1320
Lolium perenne	Perennial rye grass	seed, sod mixes	3.5–5.0 (**)	440–661
Lolium multiflorum	Italian ryegrass	seed, sod mixes	4.0–8.0	460
Microlaena stipoides	Weeping grass	seed, plugs	3.0–5.0	308
Phleum pratense	Timothy grass	seed, sod	0.5–1.0	
Poa pratensis	Kentucky bluegrass	seed, sod mixes	2.0–3.0	5500
Poa annua	Annual bluegrass	seed, sod mixes		
Poa trivialis	Rough bluegrass	seed, sod mixes	2.0–5.0 (***)	5065
Zoysia japonica	Japanese lawngrass	seed, sod	2.0–3.0	22–2860
Zoysia matrella	Manila grass	seed, sod	2.0–3.0	
Zoysia tenufolia	Mascarenegrass	seed, sod	2.0–3.0	
Warm-season grasses				
Axonopus affinis	Narrow-leafed carpet grass, native cow	seed, sod	1.5–5.0	2480
Axonopus compresses	Broad-leaved carpetgrass	seed, sod	1.5–5.0	2640
Bothriochloa macra	Red-leg or red grass	seed, plug		
Buchloe dactyloides	Buffalo grass	burs	1.0–1.5	220
Chloris truncata	Windmill grass	seed, plugs	1.0–1.5	4400
Cynodon dactylon	Couchgrass, bermudagrass	seed, stolonise, sod	1.0–2.0 (h) 1.5–2.0 (u)	4500 3900
Cynodon transvaalensis	South African couchgrass	seed, sod	1.0–2.0	
C.dactylon X C. tranvaal	Hybrid bermudagrass	sod, stolonise		
Dactyloctenium australe	Sweet smother grass	seed, stolonise		
Digitaria didactyla	Queensland blue couch	seed, stolonise, sod	1.0–2.0	4200
Panicum maxum		seed, sod		
Paspalum distichum	Mercer grass, water couch	sod, stolonise		
Paspalum notatum	Bahia grass	seed, sod	3.0–8.0	374–594
Paspalum dilatatum	Paspalum, dallis grass	seed, sod	2.0–4.0	352
Paspalum vaginatum	Saltwater couch	sod, stolonise	2.0–4.0	
Pennesetum clandestinum	Kikuyu grass	seed, stolonise, sod	0.1–0.5	340
Stenotaphrum secundatum	St. Augustine grass	sod, stolonise		

(*) higher rates are best but lower rates commonly used as seed is expensive; (**) seldom sown alone, 5 to 10% in mixtures seeded at 2.2 to 8.8 kg (1.0 to 4.0 lbs); (***) seldom seeded alone, 10 to 25% in mixtures seeded at 4.4 to 8.8 kg (2.0 to 4.0 lbs); (h) hulled, (u) unhulled.

kangaroo grass swards by spreading the culms at commencement of seed shedding and immediately broadcasting them over the seeding site.

Vegetative planting operations

The primary planting methods for propagating turfgrasses include sprigging, stolonising or chaffing, plugging, the use of turves, and sod. Vegetative planting is often confined to those grasses that produce numerous creeping stolons and/or rhizomes, such as zoysiagrass, St. Augustine grass and kikuyugrass, or in other cases, grasses such as Queensland blue couch and buffalograss, where seed may not be readily available. Late winter to early spring or late spring to early summer are considered the better times for planting vegetative material, although in areas with moderate winters, autumn may also be considered an alternative (Chamblee *et al.*, 1989).

Sprigging involves the planting of stolons and rhizomes into furrows or small holes. Generally one square metre of dense turf sod will produce sufficient material to cover ten square metres where the furrows are placed 15 centimetres apart with at least 25 per cent of the leaf protruding from the soil. Larger areas are broadcast with sprigs at a rate of 0.2 to 0.4 metre per 100 square metres with mechanical equipment, and then lightly cultivated with a rotary hoe. Carroll *et al.* (1996) found that monthly applications of nitrogen had no influence on sprig establishment in the first year, but did slightly increase zoysiagrass cover in the second year. Chaffing consists of spreading the shredded planting stock in a thin layer over a slightly moistened seedbed, then rolling the stoloniferous material into the soil.

Chaffing can be carried out by hand on small areas, or with spreading equipment such as manure spreaders on larger areas. Rates of planting vary from one to two square metres of dense turf producing enough vegetative material to chaff approximately eight square metres. If chaffing is done towards the end of the growing season, planting rates should be increased to account for losses. Another method incorporates the chaffed material with topdressing, which is broadcast over the measured area and followed by a light rolling. Some chaffed materials may also be distributed by hydromulching, in which material is sprayed on in combination with a mulch and a binder.

Plugging or spot sodding is carried out with a specially designed tool called a plug cutter, which provides round or square plugs of grass attached with soil from the turf area. These small pieces of grass may be five to ten centimetres (two to four inches) in diameter and two to ten centimetres (three-quarters to four inches) deep and are planted into previously prepared plug holes of the same dimension in the new seed bed. Planting dimensions range from 30 to 40 centimetres (12 to 16 inches) centres in each direction and are best suited to grasses that spread by means of stolons or rhizomes. The plugs are firmed by pressing them into the seedbed by the foot or light roller. In addition to introducing a new species into prepared soil, plugging can also be used to introduce a new species into an existing turf or a non-cultivated seedbed. Native and naturalised grasses now use cell plugs as a viable alternative to the direct sowing of native grass seed (Aldous, 1995). Grass plugs make greater use of the current small lots of native grass seed commercially available, are easier to transplant, and provide for improved survivability and uniformity. Sowing rates for plugs range from 25 plants per square metre for small tussocks such as wallaby grass, through to nine to sixteen plants per square metre for medium- sized tussocks such as kangaroo grass, to two to four plants per square metre for large tussocks such as blue tussock (Hitchmough, 1994).

In New Zealand, bowling greens have been successfully established using vegetative buds or bulbils of *Cotula dioica* and *C. maniototo*. The bulbils are worked over the green during final levelling in winter when they are available from the donor greens, and sown at a rate of 15 grams per square metre (Haycock, 1982). If stolons are preferred they are distributed at a rate of 350 to 450 grams per square metre, spread evenly over the soil surface and followed by a light rolling. A topdressing of a well structured, screened and sterilised topsoil is then applied through a soil distributor to cover the stolons. A second rolling is commonly used to firm the soil onto the Cotula.

Larger turf pieces called turves are 25 centimetre square slabs of turf grass of 2.5 centimetres in thickness. These have been removed with a cutting knife or turing iron from a nursery or other grassed area and are placed into similarly marked out areas in the new planting area. Turves need to be watered and rolled in to ensure consolidation.

Sodding is the term used to describe the planting or covering of an entire area with pieces or rolls of grass. The process is the most expensive method of turfgrass propagation, but it can be planted out at virtually any time of the year resulting in an 'instant turf'. If the site is sloped, sodding offers immediate protection from erosion and runoff and may be a better option than waiting for seed to germinate. Sodding is best practised during the autumn and early spring, the most favourable time for root initiation and growth. Commercial sod is from a number of sod nurseries as either rolled or folded pieces of 0.8 to 1.25 square metres (1.0 to 1.5 square yards). The sod is harvested at a soil depth of less than 1.3 centimetres (0.5 inch), although good quality bentgrass, bermudagrass and zoysiagrass sod can be cut even thinner. Still larger pieces are now being used to lay sod on racetracks and golf greens. Sodding is particularly useful for the repair of turfgrass on sportsfields where injury has been severe and replacement is necessary within the shortest possible time. Still larger turf sods that have been reinforced with mesh elements (Beard and Sifers, 1989) are now being used in sports field restoration work.

Seedbed conditions for the laying of sod are similar as for seeding and vegetative planting. Ensure that the sod is delivered fresh and kept in the shade if a delay is expected. Prior to installing, firm the planting site by light rolling, fill any depressions, and irrigate the soil to cool the surface and provide early moisture to the roots. When laying sod, the first strip should be laid along a straight edge. For better knitting place individual pieces together in a brickwork pattern, so that as few as possible of the joints are in line. Instant lawn can be fitted by using a sharp knife, rather than tearing it apart. To assist in consolidating the sod and to distribute body weight, lay out a plant or board and work from that in laying out the material. Following installation fill in gaps with clean soil or sand to reduce weed encroachment. Smooth the surface and encourage rooting by rolling. When applying sod to sloping areas commence from the bottom, gradually working across the slope. Sod should show new root growth in one to two weeks and can be checked by lifting and inspecting a corner of the grass section. Warm-season grass sod will not produce roots unless soil temperatures exceed 12.7°C (55°F) for several weeks. When sod is laid on terraces or steep slopes where slippage is likely to occur, each piece may be held in place with a peg pin driven through the sod deep enough to be free from mower damage. The principal means of propagation and establishment are illustrated in Figure 5.1.

To calculate an area for sodding

Calculate the turf area which requires sodding, then the area of the sod roll, and then divide the area of sod into the total area. For example, in an area of 350 square metres, divide by the area of the sod roll (0.75 square metre) to give 467 rolls of sod. If a 5% allowance is made for wastage, fitting and edging, the quantity required would be 490 rolls of sod.

Post-establishment management

Post-establishment care practices need to be in place for at least four to eight weeks following sowing, planting or sodding. The immediate priority is irrigation, particularly in the case of vegetative placement and native grass plantings, to avoid loss by desiccation. The initial irrigation should be long enough to fully wet the root zone and frequent enough to keep the area moistened. When the plant is at the three to four leaf stage, or at least 30 millimetres high, reduce hand watering and transfer to an oscillating or rotary sprinkler that will provide for a better distribution of water. Chamblee *et al.* (1989) found that too much moisture in late winter or early spring, or drought in late spring and summer, can create problems in the vegetative establishment of warm-season perennial grasses. Initiate mowing when the young shoots are firmly rooted, and when new leaf growth has reached 40 to 50 millimetres (1.6 to 2.0 inches). This may be in the order of 100 to 150 millimetres (4 to 6

Figure 5.1 Forms of establishment by (a) seed, or vegetatively by means of (b) sod, (c) bulbils and (d) plugs

inches) for native and other amenity grasses. Over a period of one to two months gradually work back to the recommended mowing height, or in the case of native grasses, the sustainable grazing height. No more than one-third of the leaf stem should be removed at any one cut during the growing season. Maintaining the correct mowing height and frequency at this time will develop strong turf by promoting a high root:shoot ratio.

For cool-season turfgrasses, fertiliser application should be made at a rate of 0.1 to 0.3 kilogram (0.2 to 0.6 pounds) of N per 93 square metres (1000 square feet) on a 10 to 20 day interval, the more frequent rate being available on coarser textured soils. Inorganic rates of nitrogen can vary from 10 to 15 grams per square metre. Fertiliser programs should continue until adequate turf establishment is achieved. Subsequent fertilisations will depend on soil texture and turfgrass growth. Irrigate immediately after each fertiliser application. Although not all new turfs require topdressing, it does favour the development of shoots and roots from nodal tissue of stoloniferous grasses, as well as providing a means for levelling the surface. Initial topdressing rates may be as high as 0.3 cubic metre per 100 square metres, gradually being reduced to 0.1 cubic metre per 100 square metres as the surface smoothness improves; the frequency depending on the existing degree of surface smoothness. The topdressing material must be the same as the soil used in supporting the turfgrass growth. Topdressing is typically applied with a powered mechanical topdresser and worked in with rotating brooms or a flexible steel drag mat. A postplanting cultural practice whose use varies greatly is that of rolling. Rolling firms the soil, as well as bringing the crowns and vegetative material into contact with the soil. Rolling may need to be carried out from one to four times, depending on the smoothness and firmness of the surface.

Herbicide usage on juvenile turf and young broad leaf weeds should be delayed for as long as possible, usually for a period of eight to twelve weeks, if root inhibition is to be avoided. McCarty and Weinbrecht (1997) found that pre-emergence applications of oxadiazon WP, metolachlor, pendimethalin, prodiamine, and dithiopyr delayed establishment of *Cynodon dactylon x C. transvaalensis* cv. Tifway sprigs, but these herbicides had varying control on a range of weed species such as goosegrass (*Eleusine indica*), and nut grass (*Cyperus rotundus*). Combination pesticide–fertiliser products are now emerging in the marketplace. Cooper (1998) has found that Pendimethalin Ferticide 22-0-5 can be used for pre-emergent control of annual bluegrass in autumn as well as summer grass in spring. Hitchmough *et al.* (1994) studied the efficacy of the grass specific herbicides, ethofumesate, sethoxydim and fluazifop, in removing exotic grass seedlings among native grasses, again with variable control results. To date the best chemical control of grass weeds in semi-natural vegetation has involved the use of non-selective herbicides such as glyphosate, applied selectively to weed foliage by spot spraying, hand or boom mounted wick wipers (Hitchmough, 1994).

Renovation

Renovation is a process of restoring the turf, beyond that achievable through routine cultural practices, but without complete tillage of the soil (Turgeon, 1980). Deterioration of the surface may have resulted from one or more of the following factors: excessive thatch accumulation; a predominance of unadapted or undesirable grass and/or weed species; excessive shade; or injury due to disease, insects, nematodes, toxic chemicals or nutrients. Under temperate climates, cool-season grass surfaces are often renovated under cool, moist growing conditions that occur in the spring and autumn. In subtropical climates, warm-season turfgrasses are overseeded with cool-season grasses, during autumn, to ensure playability while the warm-season turfgrasses are dormant (Ward *et al.* ,1974). In the southern United States and the transitional zones in Australia, bermudagrass greens and athletic fields are commonly overseeded with creeping bentgrass or turf-type fine leaf perennial ryegrasses. The timing of the overseeding operation must be conducted when the warm-season species are entering dormancy, and therefore not competitive, as well as providing a favourable germinating environment for the cool-season grass. The two critical transition phases occur in autumn and spring, as they cover the two periods during which the growing turf surface is altered in composition from warm-season to cool-season species and vice versa. Under North American conditions, one recommendation is to overseed 20 to 30 days before the first expected killing frost, or when the soil temperature at 10 centimetres (4 inches) depth are in the high 20°C range, or when the midday air temperatures decline to the low 20°C (Emmons, 1984). In a warm-season/cool-season polystand the transition period needs to occur gradually, especially in spring. Chemical control of cool-season species should not be considered in managing this transition period as it often proves unslightly, does not allow for a gradual effective change in composition from cool to warm season species, and will result in excessive wear to the unprepared warm-season turf surface.

Renovation, as a process, can either be partial or complete in nature. Complete renovation involves resowing or planting the entire site, whereas partial renovation requires only spot replacement without disturbing the turfgrass surface. The sequence of operations required in the renovation process includes site preparation (elimination of weed population, thatch control, mowing, aeration and nutrition), turfgrass selection, planting procedures, and postplanting care. If substantial weed populations are present (the surface should contain least a 60 per cent cover of desirable perennial grasses), the entire area may have to be treated with a non-selective herbicide. When seeding into a *Poa annua* dominated area, nitrogen sources should be withheld for two to three weeks prior to and following seeding in order to reduce competition. Sulfate of iron may also be applied as a foliar spray at 30 to 60 grams per 100 square metres for colouration (Baker, 1987). Following a satisfactory kill of the weed population and dissipation of the residue, the site is ready for cultivation. Thatch control is important because the layer can affect the growth and survival of the cool-season grass. Thatch layer should be reduced with a vertical mower, or in excessive situations a sod cutter, and removed from the site. Turgeon (1980) found that frequent topdressings and light vertical mowings, and a final coring 50 to 60 days prior to overseeding, assists in the control of thatch in warm-season grass swards. Generally the area is vertical mowed (dethatched or verticut) in several directions before overseeding. Warm-season grass swards are often closely mown to reduce their competitive advantage and give additional time for the cool-season grass to germinate and establish. Low mowing height and vertical mowing were found to control thatch and sponginess in zoysiagrass (*Zoysia spp.*) cv. De Anza and Victoria (Cockerham *et al.*, 1997). Applications of growth retardents when applied to warm-season grasses prior to seeding can also be beneficial in establishment. Compacted areas will need to be relieved with a hollow-tine corer. Complete fertiliser requirements, and lime to correct pH and supply calcium, as well as organic matter and soil conditioners to correct soil surface and physical condition problems should be applied if required. Most renovations involve overseeding with seed, rather than vegetative material, using a broadcast or disc seeder. The rate at which a grass surface is overseeded is of great importance.

Table 5.5 Suggested sowing rates for cool- and warm-season turfgrasses for use in renovation (adapted from Stephens, 1994).

Turfgrass	*Overseeding ability*	*New sowings kg/100 sq.m.*	*Oversowing into existing * cool season turf kg/100 sq.m.*	*Transition oversowing kg/100sq.m.*
Ryegrass	1	3–4	2–4	5–10
Tall fescue	5	4–5	4–6 (nr)	nr
Chewings fescue	2	2–3.5	2.5–3.5	2–3
Red creeping fescue	2	2–3.5	2.5–4	2–3
Bentgrass	3	0.5–1	1–1.5	1–1.5
Kentucky bluegrass	4	1.2	2–2.5	nr
Rough bluegrass	1	nr	1–3	4–5
Hard fescue	3	2–3	2–3	nr
Sheep fescue	3	2	2–3	nr
Bermudagrass	3–4	1	1	—

nr= not recommended
* Current turf density of the surface to be oversown will determine the appropriate sowing rate.
1= excellent, 2 = good, 3 = average, 4 = poor, 5 = very poor.

Differences in oversowing ability, new, over and transition sowing rates are summarised in Table 5.5.

Seed is sown with a direct drop spreader, and vegetative material into the thin turf using a sprig planter. Partial renovation also involves the seeding, sprigging or plugging, either carried out by hand or machine. After the seed is spread it is worked into the turf with a rake or drag mat and consolidated by matting and rolling. Seeding should be followed by topdressing, and adequate moisture provided for germination and growth. It may also be necessary to apply a fungicide to avoid the seedling disease damping off (*Pythium spp.*) which can be most severe during the first four weeks of establishment.

References

Adams, W.A. and R.F. Gibbs. 1994. 'Natural turf for sport and amenity: science and practice', *CAB International*, Cambridge.

Adams, J.E. 1966. 'Influence of mulches on runoff, erosion, and soil moisture depletion', *Soil Sci. Soc. Am. Proc.*, 30:110–14.

Aldous, D.E. 1995. 'Grass plug technology: establishment criteria for urban horticulture settings', (ed.) D.E. Aldous. *Proceedings of the Changing Face of Environmental Horticulture, University of Melbourne-Burnley College, Melbourne, August*, 123 pp.

Baker, M.W. 1987. 'Autumn turf management-renovation', *New Zealand Turf Management Journal*, January, pp. 13–6.

Beard, J.B. 1973. *Turfgrass: science and culture*, Prentice-Hall, 658 pp.

Beard, J.B. 1982. *Turf management for golf courses*, US Golf Association, Burgess, Minneapolis, Minn., 642 pp.

Beard, J.B. and S.I. Sifers. 1989. 'Stabilization and enhancement of sand-modified root zones for high traffic sports turfs with mesh elements', *The Texas Agricultural Experiment Station. B-1710*, February, 40 pp.

Canaway, P.M., M.J. Bell, G. Holmes and S.W. Baker. 1990. 'Standards for the playing quality of natural turf for Association Football', in *Natural and Artifical Playing Fields: Characteristics and Safety Features. ASTM STP 1073* (eds.) R.C. Schmidt, E.F. Hoerner, E.M. Milner and C.A. Morehouse, American Society for Testing and Materials, Philadelphia, USA, pp. 29–47.

Carroll, M.J., P.H. Dermoeden and J.M. Krouse. 1996. 'Zoysiagrass establishment from sprigs following application of herbicides, nitrogen and a biostimulator', *HortScience*, 31(6):972–5.

Chamblee D.S., J.P. Mueller and D.H. Timothy. 1989. 'Vegetative establishment of three warm-season perennial grasses in late fall and late winter', *Agron. J.*, 81:687–91.

Cobb, A.J. 1936. *Modern garden craft, Volume 1 Lawns and sports grounds*, The Gresham Publishing Company, pp. 82–114.

Cockerham, S.T., V.A. Gibeault, S.B. Ries and R.A. Khan. 1997. 'Verticutting frequency and mowing height for management of De Anza and Victoria zoysia' in *International Turfgrass Society Research Journal*, vol. 8:419–26, University of Sydney, NSW.

Cooper, R. 1998. Personal communication.

Copeland, D., T.K. Danneberger, M.B. McDonald, Jr., C.A. Geron and P. Kumart. 1992. 'Rate of germination and seedling growth of perennial ryegrass seed following osmoconditioning', *HortScience*, 27(1):28–30.

Craigie, V. 1994. 'Grass and forb selection characteristics for open space' in D.E. Aldous and T. Arthur, *Trends in Sports Turf and Amenity Grassland Management, Proceedings of Royal Australian Institute of Parks and Recreation, Melbourne*, August, pp. 5–12.

Dipaola, J.M. 1993. 'Cool- and warm-season turfgrass mixtures', *Grounds Maintenance*, May, pp.26, 28, 87.

Drane, D.S. (ed.) 1993. *On your home turf*, Australian Turfgrass Research Institute, Concord West, NSW, 50 pp.

Dudeck, A.E. 1997. 'Influence of compost root zone media on growth of *Stenotaphrum secundatum*', (ed.) J. Hall, *International Turfgrass Society Research Journal*, vol. 8:87-99, University of Sydney, NSW.

Eade, M. 1990. 'Mixtures and blends for the establishment of turf from seed', (eds.) T. Arthur and D. Aldous, in *Management of Amenity and Sports Turf, Proceedings of Conference of the Royal Australian Institute of Parks and Recreation*, March, Melbourne, pp. 46–9.

Emmons, R.D. 1984. *Turfgrass science and management*, Delmar Publishers Inc.

Engel, R.E. 1967. 'Temperatures required for the germination of annual bluegrass and colonial bentgrass', *Golf Superintendent*, 35:20–3.

Ferguson, G.A., I.L. Peppers and W.R. Kneebone. 1986. 'Growth of creeping bentgrass green on a new medium for turfgrass growth: clinoptilolite zeolite-amendment sand', *Agron.J.*, 78:1095–8.

Haycock, B. 1982. 'Cotula: Methods of establishment and management', *Second New Zealand Sports Turf Convention, Massey University, Palmerston North, May 24–7*, pp 25–7.

Hitchmough, J.D. 1994. 'The management of semi-natural and natural vegetation', in *Urban Landscape Management* (ed.) J.D. Hitchmough, Chapter 14: 192–422, Inkata Press, 594 pp.

Hitchmough, J.D., R.A. Kilgour, J.W. Morgan and I.G. Shears. 1994. 'Efficacy of some grass specific herbicides in controlling exotic grass seedling in native grassy vegetation', *Plant Protection Quarterly*, vol. 9 (1): 28–34.

Kaapro, J. 1993. 'Beware–the zeolites are coming', *A.T.R.I. Notes*, vol.12, no. 2, Winter, pp.1–2.

Kostka, S.T., J.L. Cisar, J.R. Short and S. Mane. 1997. 'Evaluation of soil surfactants for the management of water repellency in turfgrass', in *International Turfgrass Society Research Journal*, vol. 8:485–94, University of Sydney, NSW, Australia.

Leech, W. (1998). Personal communication.

Lodge, G.M. and R.D.B. Whalley. 1981. 'Establishment of warm- and cool-season native perennial grasses on the north east slopes of New South Wales. 1. Dormancy and germination', *Australian Journal of Botany*, 29,111–9.

Lush, W.M. and J.A. Birkenhead. 1987. 'Establishment of turf using advanced ('pregerminated') seeds', *Aust. J. Exp. Agric.*, 27, 323–7.

McCarty, L.B. and J.S. Weinbrecht. 1997. '*Cynodon dactylon x C. transvaalensis* cv. Tifway sprigging establishment and weed control following pre-emergence herbicide use', in *Proceedings of the International Turfgrass Society Research Journal*, University of Sydney, NSW, Vol. 8, pp. 507–15.

McDougall, K.L. 1989. The re-establishment of (Kangaroo grass): implications for the restoration of grasslands. Arthur Rylah Institute for Environmental Research, *Technical Report Series No. 89*. 53pp.

McKell, C.M. 1972. 'Seedling vigour and seedling establishment', (eds.) V.B. Youngner and C.M. McKell, *The Biology and Utilization of Grasses*, Academic Press.

Morgan, W.C. 1988. 'Alternatives to herbicides,' *Plant Protection Quarterly*, vol 4(1):33–6.

Musser, H.B. and A.T. Perkins. 1969. 'Guide to planting', (ed.) A.A. Hanson and F.V. Juska, *Turfgrass Science*, American Society of Agronomy, Madison, Wisconsin, Chap. 18:474–90.

Nelson, A. 1927. 'The germination of *Poa* spp.', *Annals of Applied Biology*, 14(2):157–74.

Neylan, J. and M. Robinson 1993. 'Sand amendments for turf construction—laboratory trials', *Horticultural Research and Development Report TU203*, 6650.

Neylan, J. and M. Robinson. 1997. 'Sand amendments for turf construction', in *International Turfgrass Society Research Journal*, vol. 8:133–47, University of Sydney, NSW.

Newman, P.R. and L.E. Moser. 1988. 'Grass seedling emergence, morphology, and establishment as affected by planting depth', *Agron. J.*, Vol 80:383–7.

Nus, J. 1994. 'Soil amendments', *Golf Course Management*, August, pp. 54–8.

Paget, D. 1990. 'Establishing native grasses', (eds.) T. Arthur and D. Aldous, in *Management of Amenity and Sports Turf, Proceedings of Conference of the Royal Australian Institute of Parks and Recreation, March, Melbourne*, pp. 147–56.

Patchett, B.J. 1982. 'Sowing rates for species and cultivars', *Second New Zealand Sports Turf Convention, Massey University, Palmerston North*, pp. 7–9.

Powell, R.H. 1974. *Australian native grasses for amenity purposes*, Australian Parks and Recreation, August, pp. 25–7.

Robinson J. 1993. 'Direct seeding of native vegetation', *Land for Wildlife News*, vol. 1:9:1–6.

Shildrick, J.P. 1980. 'Species and cultivar selection', (eds.) I.H. Rorison and R. Hunt, *Management of Amenity and Sports Turf, Amenity Grasslands: An Ecological Perspective*, John Wiley & Sons Ltd.

Stafford, J.L. 1991. 'Techniques for the establishment of kangaroo grass in South Australia conservation reserves', *Plant Protection Quarterly*, vol. 6 (3):120–2.

Stephens, B. 1994. 'Cool-season turfgrasses: new sowings and oversowings for sports turf', (ed.) D.E. Aldous and T. Arthur, *Trends in Sports Turf and Amenity Grassland Management, Proceedings of Royal Australian Institute of Parks and Recreation, Melbourne, August*, pp.16–17.

Stephens, B. 1997. 'Vegetative establishment of native grasses (Part II)', *Victorian Golf Course Superintendents Association Inc. Newsletter*, June/July, pp.8–9.

Stephens, B. (1998). Personal communication.

Smalley, R.R. 1961. 'Effects of amendments on soil properties and on growth of bermudagrass on putting greens', Ph.D. diss., Univ. of Florida, Gainsville.

Turgeon, A.J. 1980. *Turfgrass management*, 3rd edn, Regents Prentice Hall.

Turgeon, A.J. 1985. *Turgrass Management*, Rev. Ed., Raston Publishing, Raston, Virginia, 416 pp.

Ward, C.W., E.L. McWhirter and W.R. Thompson, Jr. 1974. 'Evaluation of cool-season turf species and planting techniques for overseeding bermudagrass golf greens', in *Proceedings of the Second International Turfgrass Research Conference*, E.C. Roberts (ed.), Madison, Wisc., American Society of Agronomy, pp. 480–95.

Youngner, V.B. and F.J. Nudge. 1976. 'Soil temperature, air temperature, and defoliation effects on growth and nonstructural carbohydrates of Kentucky bluegrass', *Agron. J.*, 68: 257–63.

Youngner, V.B, J.H. Madison, and W.B. Davis (undated). *Which is the best turfgrass?* University of California, Agricultural Extension Service, AXT-227, 2 pp.

CHAPTER 6

Turfgrass construction materials and methods

D.K. McINTYRE, Horticultural Engineering Consultancy, Canberra, and **B.F. JAKOBSEN,** Rootzone Laboratories International, Canberra, ACT, Australia.

Introduction

Two types of construction are generally used, the traditional and the sand based. The traditional profile, which accounts for the majority of projects, involves 100 to 300 millimetres of a uniform loam soil overlaying a slow draining base. In sand-based construction, high draining, uniform, low-compacting sands are laid down over a gravel drainage layer. This method relies on almost all of the water that reaches the surface travelling down through the profile and being removed through a sub-soil drainage system. In this method a perched water table is created, which allows grass to survive and grow in otherwise very droughty sands. The principles of sand-based construction have been highly developed by the United States Golf Association (USGA) in their USGA specification profile (Anon. 1960, 1973; Hummel, 1993). Other specifications include the University of California (Madison, 1982), the PAT (Prescription Athletic Turf) and a later version, the PURR-WICK (Plastic Barrier Under Reservoir Rootzone with Wick Action) (Daniel and Freeborg, 1980). The history of green and sports turf modification has been documented by Ferguson (1968) and Beard and Sifers (1993) among others.

Traditional construction based on surface drainage

Surface drainage is extremely important in the design, construction and maintenance of the sporting surface. Fifty millimetres of water or rain falling on one hectare generates 500 000 litres of water. If water reaching the playing surface is not rapidly removed by the surface slope, the topsoil can become saturated and vulnerable to damage. When the top of the soil profile is saturated, the "fines" become mobile and, with play, are redistributed within the soil voids. The process which causes this damage is called puddling. Water drains down through the profile more slowly, hence the surface 50 millimetres remains saturated for longer periods compared to adjacent uncompacted areas.

Ponded or stagnant surface water only remains on the surface when the topsoil is saturated right down to the subgrade. It therefore only drains downward through the profile at the drainage rate of the subgrade. This rate may be only one millimetre per hour or less (25 millimetres of surface water will take 25 hours to drain away) under a heavy clay subsoil base. Shallow rooted grass occurs on these areas, and often its growth is poor because of reduced availability of air. Air is essential for the survival and development of roots. This process most commonly develops under high use and water logged conditions.

Surface slopes of 1:100 for sportsfields are insufficient. For sportsfields a surface slope of 1:70 has been found acceptable for play and sufficient to remove excess water quickly, provided the length of any slope does not exceed 70 metres in any one direction. Slopes of greater than 1:70 tend to begin to influence play. Players begin to feel they are running up or down hill, and the ball tends to roll too much down hill. Golf fairways do not have such a slope restriction. The minimum of 1:70 should prevail, but much steeper slopes can occur. These steeper slopes will help in removing surface water more quickly off the playing surface. Insufficient slope will mean a large percentage of this water has to pass down through the profile so it can often take weeks for soils to dry out, particularly in the winter. It is also very important on golf courses to make sure that water does not flow all the way down long fairways. Long fairways should never slope in the direction of the fairway for more than 60 to 70 metres, because wet areas develop after this distance. They should be tilted or domed to deflect water off the playing surfaces. Surface problems can often be solved by changing surface slope, not by introducing underground drains.

Surface water should be collected by surface drains and removed as quickly as possible. One strategy is to collect this water in dish drains with sumps at regular intervals, and then pipe it to a stormwater system, either above or below the ground. This ensures a minimal depth of ponded water, a drained surface soon after rain ceases, and a rate of flow which never reaches a velocity that will cause erosion. Water also stays on the surface in depressions and these areas quickly develop into mud. If these depressions are located in high use areas they may stay in a wet and muddy condition for the whole winter.

Sporting surfaces never have perfectly uniform slopes when they are constructed, as there are always small undulations, depressions and 'walls'. 'Walls' are created by uneven topdressing, wheel ruts made by service vehicles, footprints made by players, indentations made by golf buggies, and sometimes by poor construction methods. Consider the effect of a 5 millimetre high wall on areas with different slopes. On a 1:70 slope such a wall would retain water for a distance of 0.35 metre behind it, while for a 1:150 slope water would be retained out to a distance of

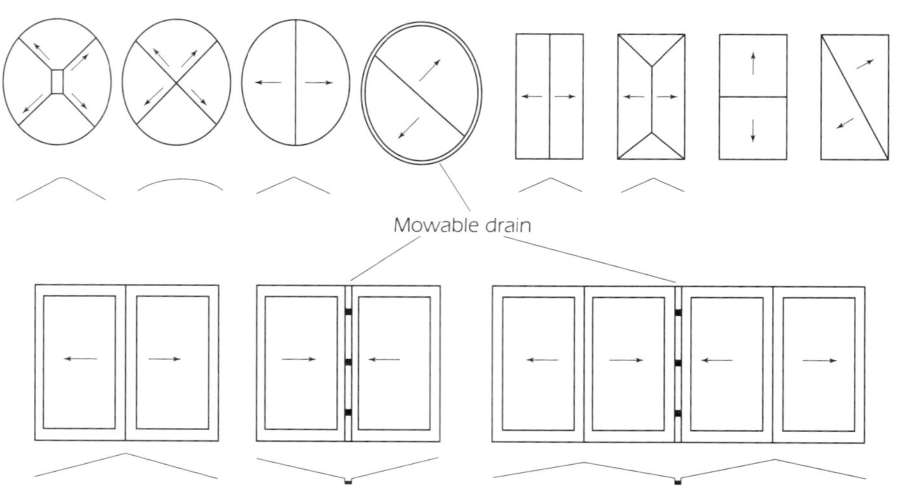

Figure 6.1 Possible shapes and design options for base and location of mowable drains, using surface slopes of 1:70 and no slope exceeding 70 m in length

0.75 metre. With the 1:150 slope, the surface area of the stagnant water is more than twice as large as with a 1:70 slope. The volume of water left on the surface to slowly drain down through the profile is also doubled (Figure 6.2). When water lies on the surface for long periods it creates a situation where the risk of further walls being formed is greatly increased. A gradient of 1:70 is considered necessary to minimise the amount of stagnant water that remains behind walls after the surface flow has ceased.

Slope length is also very important because long slopes in excess of 70 metres retain water at the bottom for longer periods than further up the slope.

After surface run-off has ceased, deep drainage through the profile is necessary to remove any ponded surface water as well as excess water in the topsoil. Water may also flow onto well-designed facilities from banks, car parks or surrounding areas. Such sources can be easily diverted from the playing surface by installing cut-off drains (surface drains) to collect the water before it reaches the vulnerable area. Small diversion banks can also be used to deflect water. These are cheap and effective and can be made part of the landscape. Mowable drains are cheap, easily maintained, and effective. They are essentially surface drains which collect water and transfer it to underground pipes, usually through grated sumps. The grass cover in the drain is the same as the surrounding playing surface and is mown with the same mowing equipment. The bottom of mowable drains must be easily mown by the mower that mows the sportsfield. Typical locations for mowable drains occur around the perimeter of ovals and in between fields, and for golf courses, between fairways. The bottom of the drain must be very uniform in slope but not be flatter than 1:50. This is essential to allow the very rapid removal of surface water. The length of the slopes running into the grated sumps should be not more than 20 metres, otherwise the tops of the sumps will become too far below the surrounding area. This will mean the sumps can be spaced about 40 metres apart if the slope goes towards them in both directions. A small concrete V-shaped strip, about 300 to 400 millimetres wide, in the middle of a mowable drain works extremely well and ensures an even slope on the bottom of the drain and the rapid delivery of water to external sumps.

Preparing the subgrade

The shaping and preparation of the subgrade is usually carried out by engineers who have experience in building roads and the preparation of other bases such as for car parks. In these situations the engineer is trained to consolidate the base to a high degree so that it will be a platform for heavy weights, and more importantly, it must be as impermeable as possible to water. In the preparation of the subgrade, it should be as

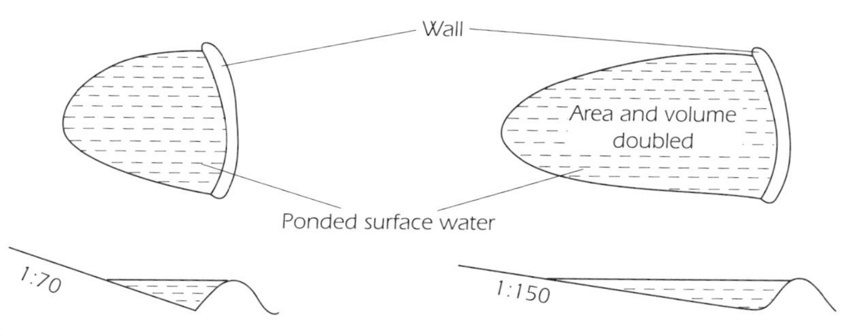

Figure 6.2 The difference in size and volume of ponded water behind a small 5 millimetre wall for slopes of 1:70 and 1:150 (after McIntyre and Jakobsen, 1991)

permeable as practicable, and drain at a rate of at least one millimetre per hour or higher. It is of paramount importance that the subgrade drains as quickly as possible because once the topsoil becomes saturated, the only way water can be removed from the profile is for it to drain down through the subgrade. If it is slow, then the profile will remain saturated, the surface soil will be unstable and grass roots will die.

Ripping

Once the subgrade has been shaped to reflect the surface slopes and lightly compacted (this compaction should not exceed 90 per cent), then it should be ripped to a depth of 300 to 400 millimetres with ripping tines. The surface should be restored to a reasonable uniformity by dragging harrows over it. This breaks up any soil clods and restores the levels. Under no circumstances should it be rolled at this stage. No work should be carried out on the subgrade when it is wet otherwise it can become glazed, and this can seal it to a point where water penetration can be reduced. If the clay material can be easily moulded in the hands or rolled out into long strings, then it is too wet to work.

Addition of gypsum

Under Australian conditions the ripped subgrade will benefit greatly from an application of gypsum, which will ensure that the drainage rate of the ripped sub-soil will stay high for a long time. The recommended rate is 500 grams per square metre, applied to the surface immediately prior to the spreading of the topsoil. Other soil should undergo a soil test to determine its nutrient status.

Selection of rootzone (topsoil) mixture

Rootzone mixtures must not over-compact under heavy use in the wet; and must not be too droughty in the summer time. Under hot dry climates like Australia, it must continue to drain at a rate in excess of 5 millimetres per hour even under heavy use. Successful specifications for topsoils were developed at the Technical Services Unit of the Australian Capital Territories (ACT) Parks Department in Canberra, Australia (Department of Urban Services, 1993), and include measurements of the water holding capacity at field capacity, hydraulic conductivity and particle size distribution (Table 6.1). The water holding capacity and the hydraulic conductivity under compaction are the most important parameters, and if these are within specification then the particle size distribution can be outside the specification.

Table 6.1 Rootzone specifications used for the construction of football fields and other sporting facilities (after McIntyre and Jakobsen, 1998).

Soil specification	*Soil for football fields, golf fairways and other sporting facilities*		*Soil for racecourses*	
Water holding capacity at 1 m suction	>12% by weight		10–20% by weight	
Hydraulic conductivity at 50 kPa (16 drops); at 130 kPa(32 drops)	>5 mm/h –		>50 mm/h >15 mm/h	
Wet sieve mechanical analysis	USDA Retained sieves by weight		USDA Retained sieves by weight	
	(mm)	(%)	(mm)	(%)
	>2.0	0	>2.0	0
	1.0–2.0	0–10	1.0–2.0	0–10
	0.10–1.0	55–70	0.106–1.0	60–80
	<0.10	30–45	<0.106	15–30
	<0.002	2–15	<0.002	2–10

The deeper the rootzone mixture, the better the surface will drain, and the better the playing performance. A 300 millimetre profile at 20 per cent porosity, which is dry at the time of a rainfall event, will accept 60 millimetres of rain before it becomes saturated at the surface. If the rootzone mixture is only 100 millimetres deep, it will only take 20 millimetres of rain to saturate the soil to the surface. Once any soil becomes saturated to the surface, then it will only drain downwards at the rate of the hydraulic conductivity of the subgrade. For example, if a 300 millimetre topsoil with a hydraulic conductivity of 10 millimetres per hour receives 20 millimetres of rain, it would wet the soil to 100 millimetres. Once the rain has ceased, it would take 60 minutes for the top 50 millimetres to drain back to almost field capacity. If, however, the depth of the topsoil was only 100 millimetres, and the drainage rate of the base was one millimetre per hour, then the same 20 millimetres of rain would saturate the whole profile and take at least 10 hours for the top 50 millimetres of the soil to drain back to field capacity. In this example the shallower profile is very vulnerable to damage for at least 10 hours after only 20 millimetres of rain. If there is regular light rain, a shallow topsoil can be continuously vulnerable to damage, because the top 50 millimetres will be near saturation for very long periods. If the field is used during this time, then considerable damage can occur to the surface and grass. Therefore football fields should have a minimum depth of topsoil of 200 millimetres and wherever possible it should be increased to 250 to 300 millimetres. Although 100 millimetres depth of topsoil is too shallow for most sportsfields, it has proved to be adequate for golf fairways. This is usually for two reasons. First the surface slopes on most golf fairways is at least 1:70, and secondly, the intensity of use is nowhere near as damaging as sports such as rugby union.

Construction based on the perched water table

Since the USGA first published specifications for its USGA profile, they have been revised three times, the latest in 1993. USGA specifications specify a 300 millimetre layer of medium sand over a 100 millimetre layer of specially selected compatible gravel or crushed stone drainage bed which overlays a drain line network. The UC profile consists of a 450 millimetre layer of medium sand, and the modified profile consists of 300 millimetres of medium sand overlying 150 millimetres coarse sand. In Great Britain, the preferred method of greens construction includes a 300 millimetre rootzone of a sand blend or a suitable soil and sand, overlying a coarse sand or grit (0.5 to 3.0 millimetre diameter) layer, over a 100 to 150 millimetre drainage layer of 5 to 10 millimetre diameter gravel (Baker, 1997). Another method of greens construction uses randomly oriented interlocking Reflex® mesh elements (formerly Netlon) incorporated into a specialised sand rootzone mix (Beard and Sifers, 1993; Robinson and Neylan, 1994). Research has also been conducted into the use of fibres to reinforce and stabilise sand rootzones (Baker, 1991).

The concept of a perched water table (PWT) is one of the most difficult to understand in sportsturf drainage, yet it is one of the most important because it is the basis for the construction of most modern golf and bowling greens and sand-based football fields, and more recently some racecourses. A good PWT construction provides very fast drainage during and immediately after rain, and drainage will stop quickly when the excess water is removed from the profile but still retain a reservoir of water for grass growth. The profile also provides ideal aeration and moisture conditions in the grass rootzone.

The theory behind the PWT has been explained in great detail by McIntyre and Jakobsen (1998), but the relevant aspects are briefly described here.

Consider a tube filled with a sand (or soil) with a coarse gauze covering placed over the bottom of the tube to prevent the sand falling out, as shown in Figure 6.3. The tube is then filled with water, placed on a sink and allowed to drain

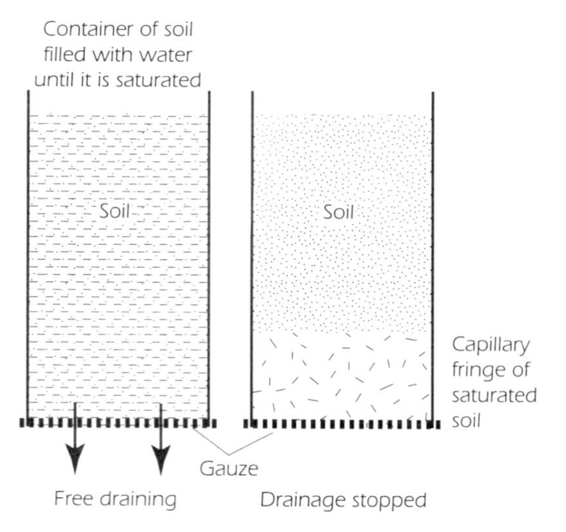

Figure 6.3 A column of soil which has had water added, allowed to drain, and eventually formed the saturated zone or perched water table at the bottom of the tube

freely until no more water comes out the bottom. One would expect gravity to pull water out of the sand column until all the sand in the tube reaches field capacity, but it does not happen. The sand at the top does reach field capacity provided the column is high enough, but water does drain out of a zone at the bottom of the column, and this zone will always remain saturated (or for some purists quasi-saturated) and is called a perched water table. This saturated zone from the bottom of the tube up to the air entry point is also called the capillary fringe. In the tube the suction being applied at the base is zero, and there is no 'free water' present at the bottom of the tube.

The top of the PWT is the air entry point, where there is sufficient pull by gravity to break the menisci of some of the large pores, so they begin to drain. Above the PWT the water content decreases with height until field capacity is reached, at which point all water movement stops and there is no change in the water content of the sand. The height of the PWT is determined by the pore size distribution of the sand, which is related to the size of the particles and how densely they are packed together (McIntyre and Jakobsen, 1993). A fine sand will have a deep PWT and a coarse gravel will have a shallow one. For example, in a USGA sand with an average particle size of 0.4 millimetre, the PWT may occur at about 180 millimetres (can vary from 120 to 250 millimetres). A coarse washed river sand with an average particle size of 1.5 millimetres may have a PWT of about 50 millimetres, and gravel with an average particle size of 4 millimetres may have a PWT of about 18 millimetres.

When the moisture status of the sand column in the tube is expressed as a percentage at various heights in the column, a curve is generated which is known as a moisture release curve. The height of the PWT for a particular sand can be determined from its moisture release curve. Different sands will have different shaped moisture release curves, and the shape and characteristics of each curve is determined by the particle size distribution of that sand and its degree of compaction (McIntyre and Jakobsen, 1993). Aspects of the behaviour of a sand in relation to future use can be predicted from these curves. Table 6.2 shows four different sands and their particle size distribution expressed as a percentage by weight. Three of these have been used to build high profile stadiums in Australia. The fourth sand (D) should not be used for any PWT construction because it does not meet the USGA specification as it has too many fines in the 0.10 to 0.25 millimetre range. Figure 6.5 shows two moisture release curves for each of these four sands, one prepared on a lightly compacted sample, and the second on a heavily compacted sample of the same sand. If the two curves are very divergent as in D, it means that the sand has a different water content at different levels of compaction. In practical terms this can mean that lower use areas of a green can be much drier at the surface than the high use areas, and can make water and grass management difficult. Sands with these characteristics should not be recommended for use in the construction of PWT facilities.

How does the perched water table work in a sand gravel profile?

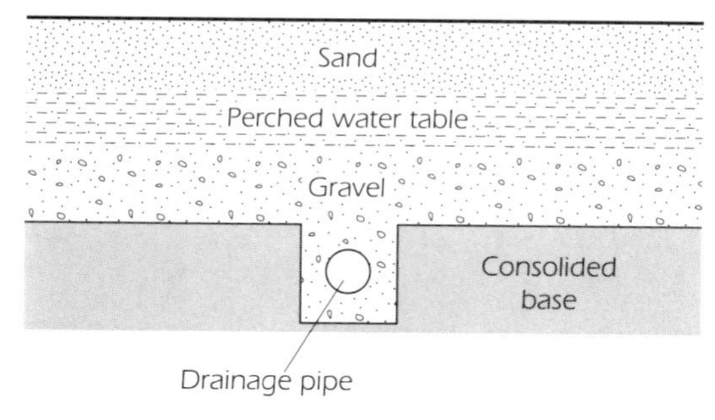

Figure 6.4 A typical perched water table profile when a USGA sand is placed over a gravel layer. Sub-soil drains are cut into the base (after McIntyre and Jakobsen, 1993).

Let us now consider the typical sand gravel profile shown in Figure 6.4. During rain or irrigation water is entering the top of the profile very quickly and moving down through it at the rate of the saturated hydraulic conductivity of the sand, which is usually several hundred millimetres per hour. When the rain stops, water ceases to enter from the top and the water is drained down out of the large pores in the upper part of the profile because gravity is able to break the surface tension of the menisci at the top of these pores. The upper part of the profile drains rapidly and quickly approaches field capacity.

In the zone of sand just above the air entry point, water is held by the menisci in pores of increasing diameter, as gravity can only pull the water out of the largest of the pores. At the air entry point gravity is no longer able to break the menisci at the top of the largest pores, and for the remainder of the bottom of the column the sand is saturated, as all pores are full of water. From that point onwards, only small amounts of water drain down out of the sand into the gravel.

At the interface between the sand and the gravel there will be a large number of pores in the sand, and only a few in the gravel. In other words there will be a large number of pores and particles in the sand which will not be touching any particles in the gravel below. These pores will be in contact with air, which means there is only a fraction of the sand particles actually in contact with gravel particles. When there is a lot of water reaching the interface from above, water mainly enters the gravel from the large pores in the sand as the menisci cannot hold the water back and it flows or drips down from the spaces between the gravel particles. Most of the water enters the gravel by this means when the upper part of the PWT has been raised by the extra water from above. The rate at which most of this water drains out into the gravel is at the saturated hydraulic conductivity of the sand. This will always be slower than the gravel's ability to accept it and remove it downwards, provided there is somewhere for this water to be stored or otherwise removed by drains. When the input of water from above ceases, the PWT begins to drain back to its equilibrium state, and the rate of downward movement of water slows quickly. This is because most of the large pores in the top of the profile have drained. It is then increasingly difficult for gravity to drain water down from the narrower pores against the combined forces of surface tension of the menisci in the smaller pores and the adhesive forces holding the water to the particles. As flow slows to a point where the menisci at the top of the perched water table are beginning to bend, then all further flow from the bottom of the sand layer into the gravel will occur at the contact points between the sand and gravel particles.

The water continues to flow down the gravel particles because it is being pulled down by gravity and the adhesive forces on the gravel particles. These two forces pull water from the pore spaces adjacent to the sand particles in touch with the gravel particles. This 'flow' along the surface of the touching particles will continue until the combined forces of surface tension in the menisci of the pores in the sand, and the adhesive forces of the sand particles in contact with the gravel, equal that of those forces trying to pull the water down. At this point an equilibrium will have been reached, there will be a constant PWT above the interface, and no further water will move downward out of the PWT.

The height of the PWT is reduced, as some of the water from the PWT is drawn down by the sands contact with the gravel. The extent to which the height of the PWT is reduced is really determined by the number of gravel particles that are in contact with the sand above. If the gravel is fine, or has many fine particles, there will be much more contact with the sand above and more water will flow down across these contacts. Gravel particle shape is also important, because if the gravel is rounded there will be only a small point of contact at the top of each particle with the sand. If the gravel particles are flat and narrow, then when they pack together some of them will lie horizontally, thus presenting a much bigger surface area in contact with the sand. The extent to which a gravel can reduce the height of the PWT is known as the suction of the gravel, and must be considered when determining the depth of sand to be used.

Any two gravels may have a very similar sieve analysis, yet they can have a very different influence in lowering the height of a PWT. Taking an actual example, a particular rounded 3 to 5 millimetre gravel had a suction of 35 millimetres, whereas an angular, sharp 3 to 5 millimetre gravel had a suction of 70 millimetres. When a good USGA sand was placed over these two materials, the depth of the perched water table was 170 millimetres in the profile with the rounded material, and 135 millimetres with the sharp gravel. If the two profiles had been constructed with the same depth of sand, the one using the sharp gravel would have been droughty at the top of the profile. This principle is essential in the consideration of design and construction of golf and bowling greens and sand-based sportsfields. There will always be a PWT in the sand above the gravel, and this will occur right up until the two materials are so similar that they contain the same amount of water at field capacity.

Table 6.2 Four different sands and their particle size distribution as a percentage of weight (after McIntyre and Jakobsen, 1993).

Particle size	*Sand A % by wt*	*Sand B % by wt*	*Sand C % by wt*	*Sand D % by wt*
> 2.0 mm	0	0	0	0
1.0–2.0 mm	0	1.4	1.6	0.4
0.5–1.0 mm	3.0	23.5	24.0	15.9
0.25–0.5 mm	93.0	60.4	49.4	40.4
0.10–0.25 mm	4.0	12.4	15.5	41.4
< 0.10 mm	0	2.3	9.5	1.9

From the particle size analyses of the four sands and from their moisture release curves (Figure 6.5), it can be seen that each sand profile behaves differently. Sand A which is extremely uniform is not affected by compaction as the two curves are almost identical. The PWT is 250 millimetres deep; its field capacity about four per cent; and if this sand was laid too deep the surface would be extremely droughty because the top of the profile will only hold four per cent water. Sand B is an excellent sand and was used in the reconstruction of Queen Elizabeth II (ANZ) stadium in Brisbane. It compacts a little, but it has a field capacity of about eight per cent, and the PWT is 200 millimetres deep. Sand C was used in the Bruce Stadium in Canberra (since reconstructed), and one can see that the shape of the curve is steeper than B because it has more fines. The depth at which this sand is laid is less critical than for A or B because it takes more depth for field capacity to be reached. Sand D has a much higher fraction in the 0.10 to 0.25 millimetre range (41.1 per cent) and as a result, the PWT is at about 300 millimetres. This sand is too fine for a PWT construction because there are often problems of poor root growth due to too many small pores. Each of the four sands have different shaped moisture release curves which are mainly influenced by the amount of fines in each. The more fines, the deeper the PWT. The narrower the range of particles, the flatter the section of the curve between the air entry point and field capacity. In sand A, this is flat because it is an extremely uniform sand, while in sand C this part of the curve is steep because there is a wide range of particle sizes.

In Australia there has been a highly successful modification to the USGA specifications used for the construction of greens, and in particular for the construction of stadiums, bowling greens and one racecourse. The main difference being that the depth of the sand layer is determined for each particular sand based on its moisture release curves, and also the influence of the suction of the chosen gravel which lowers the height of the PWT. The sands used fit a much tighter specification than the USGA, while still being inside it. In particular, there has to be 55 per cent or more of particles in the range between 0.25 and 0.5 millimetre.

Selection of materials

Selecting the correct sand is probably the most important aspect of constructing a PWT facility. If it is done correctly, the facility has a good chance of being successful. However if the wrong sand is chosen, then it can be heading for disaster as there is virtually no room for compromise in this area. The materials used should meet the specifications in Table 6.3.

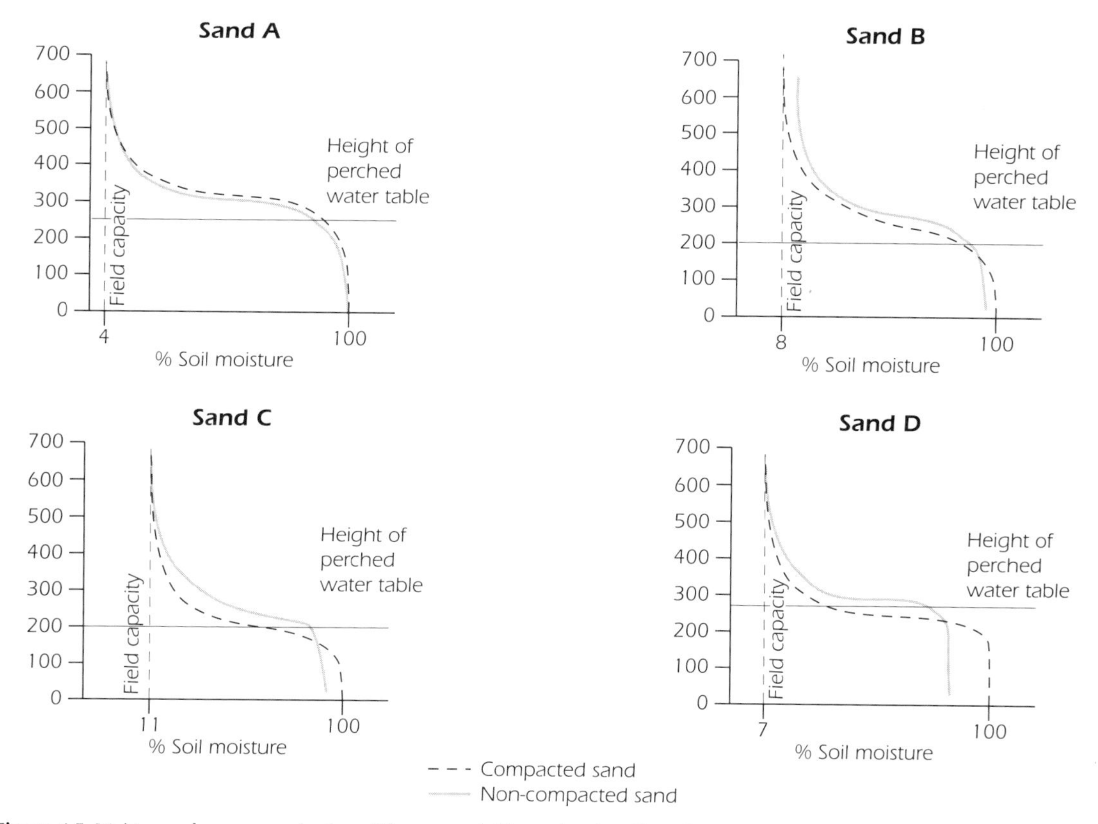

Figure 6.5 Moisture release curves for four different sands illustrating the effect of compaction on the shape of the curve (after McIntyre and Jakobsen, 1993)

Table 6.3 The specifications for the topsoil and underlying gravel for sand-based constructions; analysis is by wet sieve.

Sand specification			*Drainage gravel specification*	
USDA sieves	*% Retained by weight*		*(used as a guide for gravel selection)*	
> 2.00 mm	0		> 5 mm	<10%
1.0–2.0 mm	0–10		2–5mm	>80%
0.5–1.0 mm	0– 20		1–2mm	<10%
0.25–0.5 mm	55–90 *		< 1 mm	0
0.1–0.25 mm	< 20	Maximum combined proportion of these fractions shall not exceed 25%		
< 0.1 mm	0–10			
< 0.002 mm (clay)	0–4			

*If a sand has more than 90% in this range it must have proven stability.

Sands used should meet the specification in Table 6.3 for the mechanical analyses, but they must also meet the following specifications which are more important than the gradings; compacted hydraulic conductivity at 16 drops must exceed 600 millimetres per hour (method described in McIntyre and Jakobsen, 1998); bulk density at 16 drops must not exceed 1.6. There must also be less than five per cent of any particles soluble in hydrochloric acid.

Those familiar with the USGA specifications will see that the sand specification, while it fits within their specifications, is much tighter, particularly in not allowing as many coarse particles. The authors believe that the USGA is too broad, and as a result allows sands which are too coarse and with too wide a particle range. If profiles are built with these coarser sands, they have a tendency to be droughty at the surface if the profile depth is not carefully calculated. Sands coarser than those specified in Table 6.3, should never be laid at a depth of 300 millimetres as this is too deep. The selected sand must meet the specification outlined in Table 6.3. Once a sand has been selected, its hydraulic conductivity and bulk density should be determined and two moisture release curves made, one on a lightly compacted sample and the second on a heavily compacted sample. These measurements provide the necessary information on which to make several decisions.

Selection of the gravel layer

To ensure that the sand does not migrate down into the drainage layer, the gravel used should have a mean diameter no more than ten times the mean diameter of the sand, and also the finer fifteen per cent of the gravel must have a diameter no more than five times the 85 per cent diameter of the sand. To achieve the fastest drainage and the maximum amount of perched water, the gravel should be close to the above limits. A gravel with these specification (see Table 6.3) will do an excellent job under the specified sand. A sharp gravel, for example crushed rock, is preferable because of its stability and ease of handling during construction, allowing a flat base to be achieved easily and quickly. This makes it easier to accurately place the sand layer onto the gravel. The relationship between the sand and the gravel should be strictly adhered to as outlined in the USGA Record (Hummel, 1993). If the compatibility index is maintained close to 4 the PWT will not be drawn down excessively, and there will be little chance that the index will go over 5 with small variations in both materials.

The value of suction in the gravel at its field capacity, which is measured at one metre suction (i.e. the point where drainage stops), must be determined in the laboratory. This property of the gravel has a direct effect on the height of the PWT in the sand layer above it. If the gravel is coarse as described above, the value for its suction may only be about 50 millimetres. This means it may lower the PWT, as shown on the moisture release curve, by 50 millimetres. If the gravel is finer its suction will be higher, and the PWT will be lowered even more. If, for example, fifteen per cent of the gravel has a diameter of less than 2 millimetres, then the suction could be 100 millimetres or more. This would lower the perched water table by 100 millimetres and significantly reduce the amount of water held in the sand. The depth of sand must then be reduced to avoid the surface becoming too dry, otherwise shallow rooted grass may die and the soil surface would become very loose. This is often the cause of dry patch, particularly when the sand depth is too great and the depth of the sand layer has not been reduced to take into account the effect of the suction of the gravel.

Rootzone mixture

The depth of the layer of sand above the gravel has to be determined for each profile based on the characteristics of both the sand and the gravel to be used. From the moisture release curve for sand B in Figure 6.5, it can be seen that the depth of the PWT is 200 millimetres. Let us assume that a gravel is being used that has a suction of 50 millimetres, then the depth of the PWT will now be only 150 millimetres in this profile because the gravel will have lowered it by 50 millimetres. The optimum depth of the profile can be determined by using the following criteria. From the moisture release curve for the sand, find the suction where it starts to drain, i.e. the top of the PWT. At this point the soil should contain at least ten per cent air, which will be sufficient for grass roots at all times. This would probably also represent the bottom of the rootzone during prolonged wet periods. Next check at a point 100 millimetres higher on the curve and determine whether there is still sufficient water in the sand to support the grass and to give the sand some cohesion. This value should exceed ten per cent moisture. If these criteria are met then this should be the top of the profile. In sand B the depth of the profile would be 250 millimetres. If there is less than ten per cent moisture at the top of the profile, then organic amendment is needed.

The moisture content of sand B at a point 100 millimetres above the top of the PWT is about fifteen per cent, indicating that when the profile reaches equilibrium the surface will hold sufficient water to support good grass growth without the addition of an organic amendment. Sand A, however, would have a water content of below ten per cent at equilibrium, and would need the addition of an organic amendment such as peatmoss to increase the water content in the top of the profile to prevent it becoming constantly droughty.

This is where the authors believe that the USGA specification can cause problems by combining a range of differing sands over differing gravels, but still specifying the sand depth to be a constant 300 millimetres. The three different sands A, B and C in Figure 6.5 would all have different profile depths for optimum performance. The suction of the gravel can vary greatly, and gravels with similar gradings can be significantly different because of their shape. Two gravels were used successfully on two major projects: the first was a rounded smooth gravel with a suction of 35 millimetres, and the second was an angular crushed rock with more flat surfaces which had a suction of 65 millimetres. In this case the profile depth had to be varied by 30 millimetres using the same sand to achieve optimum profile depth. By using a constant depth of 300 millimetres, situations can arise where grass roots have to grow up to 170 millimetres just to reach the top of the PWT, and this is very difficult to achieve on closely mown golf and bowling greens. The extra unnecessary depth of sand can add as much as $A40 000 to the cost of an eighteen hole golf course or football field for no gain in efficiency.

Depth of the gravel layer

The thickness of the gravel drainage layer should always exceed the value of suction at field capacity. This will ensure a uniform value of suction at the interface of the sand in all places.

Even where the drainage layer is much deeper, the suction at the top of that layer will not be greater than its field capacity value. However for all practical purposes the gravel layer must be at least 100 millimetres, even though the suction may be lower than that. This is to enable the construction to be carried out in a practical way.

Blinding layer

The authors consider that a sand 'blinding' layer between the gravel drainage layer and sand is completely unnecessary if the correct sized gravel and sand is used in the first place. It is extremely difficult to lay a uniform 50 millimetre layer of coarse sand over a coarse gravel as recommended by the USGA, and for a large facility it is impossible to lay uniformly with machinery because 50 millimetres is not deep enough to support machines without causing rutting into the underlying gravel. This results in a blinding layer of differing depths and affects the behaviour of the profile, resulting in patchy turf as a result of differing moisture levels at the surface.

It also interferes with the PWT in a complex way, usually reducing the water holding capacity of the profile. If the sand is laid to a depth of 300 millimetres over the blinding layer, the effective depth of these two layers becomes 350 millimetres, as there is rarely a PWT caused by a blinding layer with the sand above it. This will cause the top of the profile to be very droughty for most sands that meet the USGA specification and could mean that grass roots have to travel 200 millimetres to reach the top of the PWT. The green has then to be managed as if there was no PWT, and could often be subject to dry patch.

Organic amendment

If the sand has a low field capacity and contains little clay and silt, then it is essential that the material used for the top 100 to 150 millimetres of the profile be amended with organic matter. This will ensure there is sufficient water in the top of the profile at field capacity to support grass growth. The organic matter should not be chicken manure or other composts which breakdown quickly. Peatmoss, composted pine bark fines, and coconut fibre have all proved to be successful materials. The amount added is usually in the order of ten to fifteen per cent by volume, resulting in one to two per cent organic matter by weight in the mixture. This amount is not likely to affect the depth of the PWT, but it will increase the amount of water held in the top of the profile as well as enhancing root growth. This mixture should have a slightly lower bulk density and higher porosity than the sand alone, but not enough organic matter to make it spongy. Amendment levels of four per cent are considered by the authors to be too high as the surface can become spongy and the hydraulic conductivity is often greatly reduced.

The authors believe that an organic amendment should only be added to the top 100 to 150 millimetres. There appears to be little benefit in having organic materials in the saturated perched water table zone, indeed there may be some very strong negatives in that it may well add to black layer (although we have no direct evidence of this). Even if there is no proof, why risk it if there is no advantage? To add an organic amendment to a 300 millimetre profile, when only the top 100 millimetres is necessary, is very expensive and, based on our experience, unnecessary. It may add tens of thousands of dollars to the cost of a golf course or football field.

Sand-based construction methods

There are several essential steps in the planning of a sand-based construction. These are as follows.

Preparing the subgrade

The subgrade should be shaped to give maximum efficiency to the drainage system, and consolidated to prevent slumping later on. It is not essential to maintain the natural drainage of the subgrade because a gravel drainage layer with pipes will be installed. However if wet soft spots occur in the subgrade, these areas must be covered by a geo-textile before the drainage gravel is applied. This will prevent subsoil being worked up into the gravel and, in time, interfering with water movement through it. Geo-textile materials should only be used between the subsoil and drainage gravel, and this includes lining the bottom and sides of trenches. It should not be used to wrap pipes or placed in a layer between the sand and the gravel or over the top of the pipes in trenches. Drainage pipes must be laid in trenches which are cut into the subgrade and surrounded by drainage gravel (Figure 6.4). Irrigation pipes can be installed into the subgrade before the gravel layer is laid.

Subsurface drainage and drain spacings

Drainage pipes should be dimensioned to cope with a specified rainfall event likely to occur in that location. For example, in Sydney, pipes should be dimensioned and spaced to cope with a rainfall event of at least 60 millimetres per hour. In Melbourne, which does not receive rainfall events of the same intensity, it need only be about 30 millimetres per hour. These figures take into account the fact that the gravel layer can store about 30 millimetres of water if it is well drained when the rain starts. The drain spacings are calculated using Hooghoudt's formula which is as follows:

$$S = \sqrt{\frac{4\,KH^2}{D}}$$

where S is the drain spacings in metres; D is the drainage rate in millimetres per hour (equivalent to the rainfall event chosen); K is the saturated hydraulic conductivity of the gravel; and H is the depth of the gravel in metres.

Once the drain spacing has been decided any number of patterns can be used including herringbone and gridiron, but care should be taken to ensure that drains are evenly spaced over the whole area. Often the capacity of the pipes to discharge water, rather than the hydraulic conductivity of the gravel, becomes the limiting factor. This is especially so where the lateral lines are long. A closer spacing of the pipes must then be used to ensure that the water is removed quickly and the gravel layer never becomes saturated. This may mean bringing lateral line pipe spacings in to five to six metres, thus allowing the use of standard slotted or perforated drainpipes. For example, if a 100 millimetre corrugated agricultural pipe on a 1:100 slope is considered, it can discharge 3.1 litres per second or 11 200 litres per hour. If a drainage rate of 50 millimetres per hour is desired, this means that 50 litres per square metre per hour must be removed from an area equal to the pipe length times the spacing. If the laterals are 70 metres long, then the spacing must not exceed:

$$\frac{11200}{(50 \times 70)} = 3.2 \text{ m}$$

A smooth slotted pipe with a 100 millimetre inner diameter would discharge up to 6.7 litres per second, and it could be used at more than twice the above spacing at:

$$\frac{6.7 \times 60^2}{(50 \times 70)} = 6.9 \text{ m}$$

Drains should always be cut into the base at as shallow a depth as possible and completely surrounded by the drainage gravel. USGA specifications (Hummel, 1993) suggest dimensions of 150 millimetres wide and a minimum of

200 millimetres in depth, however it is preferable to have about 50 millimetres of gravel on either side of the pipe as well as below it. There is definitely no need to use socks around the drainage pipes, and definitely no need to have large sized gravel around the pipes as this allows surrounding soil to migrate into the gravel (McIntyre and Jakobsen, 1998). The same sized gravel as specified in Table 6.3 and used in the gravel drainage layer should be used to surround the drain pipes. If the subsoil is unstable, such as may occur with expanding clays and sand, geo-textile fabrics may be used as a barrier between the subsoil and the gravel layer.

Gravel

The drainage gravel must be applied to at least the minimum thickness of 100 millimetres in all places, and then levelled and lightly consolidated. This may be carried out with light machinery, but care must be taken at all times to prevent the gravel being pushed into the subgrade. The surface of the gravel must be firm and smooth, and must be parallel to the final surface of the sand. This point cannot be over-emphasised, because if the sand is not a uniform depth over the whole facility, then the water content at the surface will always be different and this will adversely affect grass growth and often stability.

Rootzone (sand) mixture

The particle size distribution of the prepared rootzone mixture should conform to the specifications in Table 6.3 or, at the very least, requirements provided by the USGA Green Section (Hummel, 1993). If the sand selected requires an organic amendment as discussed earlier, then the amount of peatmoss or other material should be mixed off site before application. It should never be spread over the top of the green and rotary hoed in, as this practice results in a great deal of variation of thickness of the amended layer which will affect grass and water management. The bottom of the sand layer, the PWT layer, should consist of pure sand. It should be laid from one side of the green or field by piling sand up and then pushing it over with a suitable small machine, making sure that all traffic is limited to areas covered by at least the final depth of sand. A small tracked machine with a bucket is ideal. This layer is then graded and rolled to an even, stable density and to the correct thickness over the whole area. It is important during this stage of the project to avoid any sand entering the drainage gravel. The sand must be moist but not saturated when applied, and work should not be carried out in the rain. When the sand is dry and when it is saturated, the sand grains move freely, while in the moist stage they cling together due to the surface tension of water and will not enter the gravel below. Later on, when the sand layer is fully installed and compacted, the sand grains are bridging and unable to move. If during the placement of the first sand layer the front is allowed to dry out, or if heavy rain falls onto such a front, then significant amounts of sand may move into the gravel layer. This could create an area where there is a difference in the drainage rate, and a possible loss of the effects of the PWT. Watering must not be carried out until the sand has been lightly compacted as this locks all the particles together and prevents the fines migrating. If water is applied before this happens then fines can migrate, and often they will lodge in a band lower in the profile, significantly reducing hydraulic conductivity.

The top layer of the amended sand is applied in a similar manner. In this process the risk of particle migration into the layer below does not occur. However the sand should still be moist, but not wet, to make it easier to work and to prevent a dry mixture from separating during handling. This layer should be graded and rolled to an even, stable density of between 1.45 and

1.50 grams per square centimetre and to the correct thickness in all places. A thorough watering will help with this final consolidation. If appropriate, the final surface is best achieved by laser levelling, but even on the biggest of projects hand levelling has proven to be efficient once the final levels are being approached.

Grass establishment

The best way to establish grass on soil-based constructions is by using washed turf. Conventional turf with soil attached must never be used, unless it has been grown in a sand that is identical to that being used. Even small amounts of fines from soil attached to this turf can negate all the advantages of the high draining sand. The infiltration rate of the soil at the top of the profile will determine the rate of water entry, not the soil below it. The infiltration rate could change from 150 millimetres per hour to 10 millimetres per hour by using turf with soil attached. Once soil has been introduced to the top of the profile, there is virtually nothing that can be done in a remedial way to fix it other than to remove the top 100 millimetres of sand and start again. Coring and rubbing good sand into the core holes will make only minor differences.

The irrigation system must be fully functional before grassing is commenced. Make sure that the field is ready for immediate laying of the turf as soon as it arrives at the site. The surface should be flat and firm so there is good contact between turf and soil for quick and even establishment. The best results are achieved by lightly rolling the turf immediately after it is laid. Newly laid turf must be irrigated every few hours, or hourly in hot weather, until it develops an adequate root system. This may take a couple of days in warm weather and a little longer in cooler weather. Washed turf on pallets can become very warm if stored for too long, and the grass will die from heat stress. The signs of heat stress may not be obvious immediately, but the grass will die from it a few days after it has been laid. If there is a need to store turf for more than one day, it must be spread out and irrigated frequently. However, turf treated in this way may take longer to produce roots and be generally slower to establish when finally laid in its permanent place.

If seeding is used to establish grass, remember that the surface of the soil can dry out rapidly in hot weather and very frequent light watering is required. Shade cloth and other germination enhancing cloths work well. Establishment of grass from seed on these profiles is slow, and a long period is required before sufficient grass density is achieved so that they can be used for play. Frequent and low applications of fertiliser, particularly phosphorus, are needed to retain good root growth. Overwatering should be avoided as it leaches out plant nutrients. It is not essential to fill up the profile when watering. Even from a light watering some of the water will reach the perched water table and a deep root system can be maintained.

Maintenance requirements differ in some ways from that of conventional turf. For example, fertilising should be little and often (once every 10 to 14 days). Excessive watering should be avoided even though it will not cause any drainage problems. The thatch layer must be controlled so it never exceeds 10 to 15 millimetres in depth. Topdressing must be kept to a minimum, as the original depth of sand must be maintained and not increased over time. Frequent shallow coring has proven to be beneficial in keeping up a high infiltration rate and preventing slime and algal growth in the winter in some areas.

Engineering tolerances and practices during construction

When building these facilities great care must be taken to ensure that each layer is installed to very high tolerances. There should not be any more than a few millimetre difference in the specified levels and the finished product, and this should be expressed as ± 5 millimetres. It will not perform properly if strict adherence to these levels is not maintained with all strata in the profile. All too often we see the correct sand laid over the correct gravel and a poor product resulting from poor workmanship in laying the materials. Common faults are not laying the gravel parallel to the finished surface; driving over the gravel layer and leaving ruts in it; turning machinery on shallow sand which forces sand down into the gravel layer; proper care not taken to get the top sand surface exactly at the designed level; and uneven compaction of sand before it is finally levelled. The construction supervisor must be aware of all of these pitfalls to ensure they do not occur. Rushing the job often results in many of the above problems and causes the future manager problems for the life of the facility.

Inclusion of mesh element reinforcement into sand profiles

The inclusion of Reflex® mesh elements (formerly known as Netlon) into sand profiles in Australia has proven to be very successful. This material is incorporated into the top 100 millimetres of the profile for football fields and the top 150 millimetres for racecourses at a rate of six kilograms per cubic metre. All of these small plastic pieces of mesh lock together in a random fashion and give the top of the profile great strength. It reduces divoting and reduces wear from heavy use. Grass recovery is quicker and it also allows such activities as pop concerts to be held on these venues with minimal damage to the surface. It has a resilience which absorbs energy, hence reducing strain on the athlete or the legs of horses. In Australia five stadiums, including the Olympic stadium which is under construction, and one racecourse have this material in their profiles.

The wearability of these surfaces is quite remarkable. Parramatta stadium, in Sydney, Australia which has a bermudagrass (*Cynodon dactylon* cv. Santa Anna) surface, has been in use for more than 150 games of rugby league and soccer in one football season, as well as club training for four nights a week, and the grass surface was intact after eight months. The Melbourne Cricket Ground also has heavy use in a much harsher climate. There have been more than 110 matches played in a season. Seventeen of these were sets of three matches played on Friday night, Saturday and Sunday afternoons. There were also several concerts during the non-cricket season.

References

Anon. 1960. 'Specifications for a method of putting green construction', *USGA Green Section Record*, 13(5):24–8.

Anon. 1973. 'Refining the green section specifications for putting green construction', *USGA Green Section Record*, 11(3):1–8.

Baker, S. 1991. 'Sports turf reinforcement', *Grounds Maintenance*, Feb., p. 74.

Baker, S. 1997. 'The construction and maintenance of golf greens in Great Britain', *Golf & Sports Turf Australia*, Aug./Sept., pp 25–7.

Beard, J.B. and S.I. Sifers. 1993. *Stabilisation and enhancement of sand-modified root zones for high traffic sports turfs with mesh elements. A Randomly Oriented Interlocking Mesh Inclusion System*, Texas Ag. Exp. Sta. College Station, Texas, B-1710, 40 pp.

Daniel, W.H. and R.P. Freeberg. 1980. *Turf managers handbook*, Harvest Publishing Co., N.Y.

Department of Urban Services. 1993. *Canberra Landscape Guidelines*, ACT Government, Canberra, 67 pp.

Ferguson, M.H. (ed.). 1968. *Building golf holes for good turf management, United States Golf Association*, New York, 55 pp.

Hummel, N. 1993. 'Rationale for the revisions of the *USGA greens construction specifications*', USGA Green Section Record, Mar./Apr. 7–21.

Madison, J. 1982. *Principles of turfgrass culture*, Robert E. Krieger Publishing Company, Florida, 431 pp.

McIntyre, D.K. and B.F. Jakobsen 1991. 'Drainage—soil structures and their effect upon water movement', *TurfCraft*, no. 22, May.

McIntyre, D.K. and B.F. Jakobsen 1992. 'Sub-soil drainage—the lateral movement of water in soil', *TurfCraft*, May/June, no. 27.

McIntyre, D.K. and B.F. Jakobsen 1993. 'The perched water table and its use in sportsturf', *TurfCraft*, March, no. 31: 48–9.

McIntyre, D.K. and B.F. Jakobsen. 1997. 'The science of surface drainage and its importance in construction and maintenance', *Proceedings of the 3rd Annual Turfgrass Seminar, Australian Golf Course Superintendents Association, Royal Pines Resort, Gold Coast, Queensland*, July, 6 pp.

McIntyre, D.K. and B.F. Jakobsen. 1998. *Drainage for sportsturf and horticulture*, Horticultural Engineering Consultancy, Canberra, 170 pp.

Robinson, M. and J. Neylan 1994. 'Options for bowling green drainage and construction', *TurfCraft*, July, pp. 41–5.

Sifers, S.I. and J.B. Beard. 1992. 'Monitoring surface hardness', *Grounds Maintenance*, 27:60,62,90.

CHAPTER 7

Turfgrass irrigation

G.J. CONNELLAN, Institute of Land and Food Resources, University of Melbourne, Burnley College, Victoria, Australia

Introduction

The growth and condition of all plants, including turf, is dependent on a ready supply of water and nutrients. It is water that has the most immediate effect on plants. Short-term deficiencies under demanding conditions can have disastrous consequences. The management of the crop, soil and water supply is therefore of paramount importance in achieving turf surfaces that satisfy the desired aesthetic and functional performance requirements.

The movement of water through a plant system can be simplified to consider the water entering the roots and moving up through the trunk or stem to the leaves as a continuum of a solution as a liquid. Within the leaf there is a change from a liquid to a water vapour, and water is released to the atmosphere as a vapour. This is the process of evaporation, and when it specifically refers to the evaporation of water from within the foliage to the atmosphere it is called transpiration. In order to maintain the plant in a healthy condition, all water requirements of the plant need to be satisfied. This includes water that evaporates from the soil between the plants and also any water that may be on plant surfaces. The total water use by a plant, including both transpiration and evaporation from other surfaces is referred to as evapotranspiration (ET). The ET rate is generally measured in millimetres depth of water per day (mm/day).

Factors influencing water use

The rate of movement of water from the leaves is influenced by several factors. These include the weather conditions, the water use characteristics of the turf species, and the availability of water. Water use rates are understandably very variable as the atmospheric conditions driving the process are also very variable.

The main weather factors that influence plant water use are solar radiation, air temperature, relative humidity and wind. Some conditions, such as sunny, hot and dry air, encourage high rates of evaporation from plants, soil and water surfaces. The atmosphere surrounding the plant under these conditions is said to have a high evaporative demand. How the plant responds to these conditions depends on the particular characteristics, stage of growth and development of the plant. The size and leaf area of the plant is particularly important in influencing the ET rate. The transpiration rate is strongly related to the total leaf surface area, and hence potential transpiration area, of the plant. The Leaf Area Index (LAI), which is the ratio of the total leaf area to the unit area of ground beneath the leaves, is a strong indicator of plant transpiration rates (Meyer, 1985).

The desired condition of plants is another factor influencing water use rates. For example, to maintain turf in a lush condition requires a higher water use rate than to maintain it in a moderate or acceptable condition. The nutrient program will also influence water use rates. If the turf is accessing optimum levels of nutrient, then growth and hence water use will be high compared to growth under less favourable conditions.

There is a variation in water use rates between similar types of plants for the same level of evaporative demand. Turf species can exhibit a wide range in water use rates for similar weather conditions. Knowledge of this variation is important in designing and managing irrigation systems.

For turf to grow strongly it must have access to an ample supply of quality water. Should water not be available under demanding conditions the turf will experience stress, possibly damage and even death. The amount and availability of water is dependent on the soil properties, the moisture status of the soil, and the root zone depth of the turf. The availability of water stored in the root zone is strongly influenced by the soil properties. The ease with which the turf roots can access the water is dependent on the tension of the water in the soil. Heavy soils, for example clays, which comprise many small particles, hold the water with considerable force and the plant must overcome these forces in order to extract the moisture. Water availability needs to be considered both in terms of amount and the tension with which it is held.

Turf species water use rates

While turf species exhibit varying water use rates for similar growing conditions, it is common to categorise the major turf species according to two broad climate based categories. They are warm-season grasses and cool-season grasses. The cool-season grasses are generally considered higher water users (Table 7.1).

Table 7.1 Typical water use rates for selection of turfgrasses growing in temperate climate in Australia.

Category	*Species*	*Typical daily ET (mm/day)*
Warm-season grasses	couch	5–6
	kikuyu	5–6
	zoysia	5–6
	buffalo	5–7
Cool-season grasses	bluegrass	6–8
	ryegrass	6–8
	tall fescue	7–9
	bentgrass	9–10

The water use rates indicated in Table 7.1 are a guide only and considerable variation from these values can be expected at a particular location and in particular growing conditions.

Estimating turf water use

Value of water use estimation

Knowing the amount of water used by the turfgrass is fundamental to the responsible management of turf areas and the design of efficient irrigation systems. Without this knowledge irrigation management is a matter of guesswork rather than informed decision making.

There are numerous reasons for knowing the amount of water required by turf. They include:

- The design of an irrigation system requires the designer to match the application capacity of the system to the peak water demands of the plants.
- The scheduling of the irrigation system, so that the correct amount of water is applied at the right time, is dependent on knowing the water demand.

- The evaluation of the overall effectiveness and efficiency of an irrigation system can only be carried out by comparing the actual water supplied to the site to the amount required by the plants.
- The selection of storage capacities requires the assessment of weather data, including rainfall, and the estimation of water used over the irrigation season.
- The economic analysis of turf irrigation systems is dependent on knowing the amount of irrigation water that needs to be supplied and the cost of this water.

Knowledge of water consumption is not an option, it is a necessity for good irrigation management.

Water use estimation techniques

It is generally not practical to directly measure turf water use with instruments as they tend to be expensive and complex. Rather than directly measure, the approach commonly used is to determine turf water use by one of a number of estimation techniques available.

Turf researchers do however employ a direct measurement type device to study turf water consumption characteristics and other aspects of turf performance. The device is called a lysimeter and it allows the water used by a sample area of turf to be measured. The three types of lysimeter (Whithers and Vipond, 1974) used are:

1. non-weighing lysimeter (inflow and outflow volumes are measured)
2. weighing lysimeter (total crop water use measured) and
3. water table lysimeter (water required to maintain constant water table measured)

Lysimeters are generally not used as irrigation management tools for turf areas. They are more suited to turf water use studies.

The underlying principle used to estimate turf water use rates is to initially determine the water use rate or evaporation of a reference surface, either vegetation or water, and then adjust this value for the particular crop under consideration.

The techniques used to determine the value of the reference water use rate for turf can be broadly grouped as follows:

- Weather based water use prediction models
- Evaporation data (evaporimeters)

Weather data based models

If a reference crop is used then the reference water loss rate by evapotranspiration is designated by ET_o. The following definition of ET_o is presented in Doorenbos and Pruitt (1977): 'the rate of evapotranspiration from an extensive surface of eight to fifteen centimetres tall, green grass cover of uniform height, actively growing, completely shading the ground and not short of water'.

The ideal prediction model, to calculate ET_o, would take into account all climate and plant factors that would influence water use rates. In this way an accurate estimate could be undertaken. Unfortunately, at times, not all of the necessary information is available. There have been a range of models developed which take into account various climatic and plant factors. The water use prediction models consider the climate and plant factors only. The role of soil is taken into account when considering the optimum irrigation schedule.

The models in use range from simple to very complex ones. A very popular model for irrigation management purposes is the Penman method. This model takes into account the energy available for evaporation, that is the solar energy, and the ability of the atmosphere to evaporate water and remove the water vapour from the leaf. It is a comprehensive model and provides good accuracy. It does however require several weather parameter values to be available for the calculations. It therefore requires a number of weather instrument readings to be recorded in order to estimate water use.

Some models, for example Blainey-Criddle, only measure air temperature and so do not have the accuracy of Penman models. Other models, such as the Radiation Model, measure solar radiation and air temperature but also do not provide the accuracy of Penman and similar models. A guide to the accuracy of the various models is presented in Table 7.2.

Table 7.2 Accuracy of various water use predictive and evaporation models (adapted from Doorenbos and Pruit, 1977).

Model	*Measured climate data*	*Estimated climate data*	*Accuracy %*
Blainey-Criddle	Air temperature	Relative humidity Wind Sunshine hours	25
Radiation	Solar radiation Air temperature	Relative humidity Wind	20
Penman	Solar radiation Air temperature Relative humidity Wind		10
Evaporation	E_{pan} reading		10–15

The estimation of the water use for a particular crop (ET_c) is found by first determining the evapotranspiration rate for the reference crop (ET_o) at that site, and then adjusting this value using an adjustment coefficient. This coefficient is called the crop coefficient and is designated K_c. It generally has values in the range of 0.4 to 0.9 for turfgrass species.

The estimated value of crop water use (ET_c) is calculated as follows in millimetres per day:

$$ET_c = K_c \times ET_o$$

Evaporation pan

The evaporation from a free water surface approach uses evaporation values available from the Bureau of Meteorology network or privately owned evaporimeters as the reference data. The standard evaporation instrument used in Australia is the Class 'A' Evaporation pan.

The availability of evaporation data from Bureau of Meteorology stations provides access to low cost data that allows plant water use to be estimated quickly and simply. The evaporation pan is somewhat limited in its accuracy of measurement (refer to Table 7.2). Reasons for this include effects due to localised wind disturbances, interference from animals and birds, microclimate variations and human error in reading.

Evaporation pans are a very useful tool for irrigation management. In localities where predictive models are not available they provide excellent information on potential crop water use. They do however have some limitations which should be recognised.

The water used by a particular crop (ET_c), measured in millimetres depth of water, can be determined using the following expression:

$$ET_c = F \times E_{A\ pan}$$

where F is the crop factor or proportion of water used by the crop or plant compared to the water used by the Class 'A' pan. $E_{A\ pan}$ is the depth of water evaporated from the Class 'A' pan (millimetres per day).

The amount of evaporation that occurs at a particular locality varies on a daily and seasonal basis. Summer values are higher than winter values and some sites are exposed to climatic conditions that produce more evaporation than others. Across Australia the annual evaporation values can vary from around 950 millimetres to over 4000 millimetres. Mean daily evaporation rates for Australian capital cities are presented in Table 7.3.

The actual evaporation for a particular location and day can be obtained from the Regional

Table 7.3 Mean daily evaporation values (millimetres) for Australian capital cities.

Location	*Summer January*	*Autumn April*	*Winter July*	*Spring October*	*Annual*
Adelaide	8.7	4.3	1.9	5.5	1898
Brisbane	7.4	4.4	3.0	6.5	1960
Canberra	7.0	2.7	1.2	4.0	1375
Darwin	6.7	7.1	7.3	8.6	2696
Hobart	4.9	2.1	0.9	3.1	994
Melbourne	6.1	2.8	1.3	3.8	1268
Perth	8.1	4.0	2.0	5.0	1764
Sydney	7.0	4.1	2.7	5.7	1788

Source: National Climate Laboratory, Bureau of Meteorology, Melbourne.

Bureaus of Meteorology. It should be noted that the use of average evaporation data provides only a guide to the actual water use rates of plants.

Crop factor

The value of *F*, the crop factor, varies principally with plant type and the stage of growth. A value of 1.0 for *F* would represent a water use rate equal to the rate water is evaporated from the Class 'A' pan. This is a very high rate and is not usually encountered with turf. Crop factor values for turf are typically in the range of 0.3 to 0.8.

A guide to typical crop factor values is presented in Table 7.4.

Table 7.4 provides a guide to the appropriate crop factor (*F*) values for both warm-season and cool-season grasses. The effect of turf condition is evident in the range of values indicated for moderate growth condition to vigorous growth condition.

There is some confusion in the literature about the various coefficients and crop factors. The reason for the confusion is that there are a number of different water use references. These include the evapotranspiration from a reference crop (referred to as ET_o) and the evaporation from a free water surface (referred to as E_{pan}). If different references are used, then the value of the adjustment factor must change in order to achieve the same value of estimated water use for the particular plant under consideration. The term crop coefficient (K_c) is commonly used to determine ET_c when the water use reference is another crop (ET_o), and the term crop factor (*F*) when the reference is the evaporation pan ($E_{A\ pan}$).

Table 7.4 Guide to crop factor (F) values for various turf conditions (adapted from Handreck and Black, 1991).

Turf category	*Crop factor (F)*
Warm-season	
(e.g. buffalo (St. Augustine), couch, (bermudagrass) kikuyu, zoysia)	
Vigorous growth	0.55–0.70
Strong growth	0.45–0.55
Moderate growth, just acceptable	0.25–0.40
Cool-season	
(e.g. bentgrass, bluegrass, tall fescue, ryegrass)	
Vigorous growth	0.80–0.85
Strong growth	0.70–0.75
Moderate growth, just acceptable	0.65–0.70

Water use estimation example

The application of water use estimation is demonstrated through the situation where the daily water requirements of a turf area are to be determined. The details are:

Site:	Sports oval
Area:	1.2 hectares (12 000 m^2)
Turfgrass:	Cool-season mixture
Condition:	Good, strong growth
Crop factor (*F*):	0.7
$E_{A\ pan}$:	10 millimetres per day

Calculation example:

$ET_c = F \times E_{A\ pan}$
$= 0.7 \times 10 = 7$ millimetres per day

volume = area × depth (ET)
= 12 000 × 7 = 84 000 litres per day

Note: 1 m^2 × 1 millimetre (depth) = 1 litre

Irrigation requirements

Available soil water

The soil acts as the water storage reservoir for plant growth. If rainfall does not maintain this storage at adequate levels then it is necessary to supplement the storage through irrigation. Both the quantity to apply and the timing of the application are important irrigation management decisions.

The amount of water that is stored in the soil is dependent on the water holding properties of the soil and the root zone depth of the crop, in this case turf. A particular problem for turf is that the root zone is relatively shallow, often in the range of 150 to 200 millimetres, and the storage is small.

The amount of water that is present in the soil and available to the plant depends on the soil structure and particle size distribution. Not all water in the soil is available to the plants. Some of it is bound or held so tightly that it cannot be removed by the plant. The soil water condition reached when the plant cannot remove additional water is called wilting point. There is also an upper level of soil moisture that can be stored in the soil and available to the plant. If soil is flooded and so becomes saturated some of the water will drain from within the spaces between the soil particles. Most of the water will remain. The amount of water in the soil profile following drainage is called field capacity.

Available water (W) of a soil is the maximum water that can be stored and is available for plant growth, it is the soil water between field capacity and wilting point. It is expressed in millimetres depth of water that is available in each metre depth of soil (mm/m). Indicative vales of W for a range of soil types is shown in Table 7.5.

Table 7.5 Available water (W) values for selected soil types.

Soil type	*Available water (W) (mm/m)*
Sand	60
Fine sand	90
Sandy loam	110
Loam	170
Silt loam	170
Clay loam	165
Clay	140

As previously mentioned, the water available to the plant is dependent on the available water (W) property of the soil and the root zone depth (R). The amount of water available at a particular site is referred to as plant available water (PAW). This amount can be calculated in the following way:

plant available water (PAW) = available water (W) × root zone depth (R)

Example

Using the previous example of the sports oval and assuming the following details, the PAW can be determined.

Site	Sports oval
Turf:	Cool-season mixture
Soil:	Sandy loam
Root zone:	200 millimetres

$$\text{PAW} = W \times R$$
$$= 110 \times \frac{200}{1000}$$
$$= 22 \text{ mm}$$

It is most important to know the soil properties and the root zone depth when planning and managing irrigation systems. The root zone is often very shallow and so the size of the water storage reservoir is also small.

Irrigation depth

The amount of water applied at each irrigation is largely dependent on the depth of water required to fill the soil reservoir to field capacity. If the soil reservoir had depleted to wilting point, and this is not recommended, then it would be possible to completely refill the root zone. In the example above, this would represent a refill depth of 22 millimetres. Good irrigation practice is to manage soil water so that the turf is not stressed. It is generally recommended that only 50 per cent of the soil reservoir should be depleted prior to irrigation. In the example above, where the stored water was determined to be 22 millimetres, irrrigation would be required once 11 millimetres (50 per cent of 22 millimetres) had been removed by evapotranspiration.

In some situations, where it is desirable to maintain high levels of readily available water, smaller depletion amounts (20 to 30 per cent) can be used. Soil water should not be depleted by more than 75 per cent of the stored capacity.

The actual depth to be applied by the irrigation system needs to be greater than the depth required to refill the root zone. An allowance needs to be made for the inefficiencies in the application of water by the sprinklers. There are losses associated with evaporation of droplets, wind drift and uneveness (or non-uniformity) of application depths. There is also the possibility of some water moving through the soil profile and beyond the active root zone of the turf. These are referred to as percolation losses. A detailed analysis of these potential losses should be undertaken by the irrigation system designer.

If an overall allowance of 75 per cent is made for the irrigation system in the above example, the actual depth to be applied by the irrigation system would be:

$$\text{Irrigation system application} = \frac{11}{75\%} = 14.7 \text{ mm}$$

This means that nearly 15 millimetres (14.7 millimetres) of water would need to be applied to ensure that 11 millimetres was deposited in the root zone of the turf.

The amount of water that would be required to be supplied by the sports oval irrigation system in the above example at each irrigation can be calculated as follows:

Irrigation volume
= irrigated area × irrigation system application
= 12 000 × 14.7 = 176 400 litres

Irrigation interval

The irrigation frequency or interval, in days, between consecutive irrigations is dependent on the rate at which water is removed from the soil reservoir, by ET_c, and the depth of water that is allowed to be removed (refill depth). If the daily ET_c rate was determined to be 7 millimetres and the refill depth was 11 millimetres in the example above, then the irrigation interval would be:

$$\text{Irrigation interval} = \frac{\text{refill depth}}{\text{daily } ET_c}$$

$$= \frac{11}{7} = 1.6 \text{ days}$$

In this example, the soil water storage would not last two full days if these conditions prevailed. In most situations there is daily variation in ET_c, so it would be likely that the turf would in fact be able to be irrigated at two day intervals at times. The limitations of small storage reservoirs with turf should be acknowledged.

Irrigation systems

Irrigation system requirements

The control of the level of soil moisture through irrigation is a vital element in the overall management of quality turf areas. It is important to recognise that the effective irrigation of turf is a particularly demanding task due to the special needs and characteristics of turf sites. These needs impact on both the design and the management of the irrigation system.

Special needs of turf are:

1. Uniformity—Individual turf swards draw water from only very small lateral distances, so the irrigation system must therefore apply water with a high degree of evenness or uniformity to ensure all plants are watered.
2. Limited root depth—The active root systems of turf occupy only a comparatively shallow depth of soil, often 100 to 200 millimetres, so in order to avoid wastage through over watering the application depth must be precise.
3. Responsiveness—Turf responds very rapidly to changes in the soil moisture status, so that if inadequate levels of moisture are present in the root zone, during periods of high water demand the turf will rapidly become stressed.

Turf irrigation must therefore be capable of applying predetermined amounts of water to the turf root zone, with a high degree of uniformity, and with minimum wastage.

There are several factors that contribute to a successful irrigation system.

- Accurate determination of plant water requirements and site details.
- Selection of quality water distribution equipment.
- Sound hydraulic design.
- Selection of appropriate system components.
- Installation of the system to a high standard.
- Competent system management.

Inadequacies in any one of these areas will result in a major deficiency in the effectiveness of the whole system. For example, expert hydraulic design without due consideration to sprinkler head selection and soil properties can render an irrigation system totally ineffective or useless.

Irrigation system efficiency

The term efficiency can be used in many different ways when describing the performance of an irrigation system. The efficiency, for example, can be assessed in terms of the amount of water delivered, the energy consumed and the labour required. There are numerous efficiency indicators based on water supplied, applied and consumed. A useful efficiency indicator is the field application efficiency, which is the ratio of the crop water use to the water delivered to field.

The efficiency of the irrigation system should be assessed not only over the short term but also over the long term, one or more seasons. Issues, such as the reliability of equipment, should be incorporated into the measure of efficiency of the system.

All water efficient turf irrigation systems should satisfy the following two criteria:

- Water should be applied with a high degree of uniformity.
- The system should be managed so that water wastage is kept to a minimum.

It is not possible to have an efficient turf irrigation if the application is not uniform. Uniformity in isolation does not guarantee efficiency. Water use efficiency is dependent on a system that is able to apply water uniformly and is also managed so that the correct depth is applied.

Irrigation managers and others responsible for maintaining irrigated turf areas would benefit from the greater adoption of an indicator that would describe how well the system has performed over a complete irrigation season or other designated periods of time.

It is the responsibility of the irrigation system designer to identify the required performance specifications of the system so that the specific needs of the plants, soil and site are satisfied. The designer also needs to either identify the degree of uniformity required or ensure that the system does meet some minimum stated standard.

The irrigation industry has expended much effort in achieving improved uniformity in application from turf sprinklers. High uniformities are achievable for all situations. There is a temptation by some site owners to place a greater emphasis on the initial cost of the system, rather than on systems that incorporate equipment and layouts that produce higher uniformities. Turf managers should insist on the highest possible uniformity standards for all sprinkler systems.

Irrigation methods

The irrigation of turf, using techniques other than sprinklers and sprays, have only become commonly available in the last few years. Applying water in droplet form through the air has been the traditional irrigation method for turf. It has many characteristics in its favour, including simplicity in equipment, reliability and reasonable effectiveness. The limitations of this method, including lack of eveness of application, losses due to evaporation and wind and high energy requirements, have led to the development of alternative methods that deliver the water directly to the root zone rather than trajecting it through the air.

Subsurface drip technology, adapted from agricultural cropping, has been applied to the irrigation of turf by irrigation companies and research organisations. The suitability of various drip products for subsurface irrigation of turf was investigated by Zoldoske *et al.*, (1995) at the Center for Irrigation Technology, Fresno, California. The permanent nature of turf, as distinct from annual or short life agricultural crops, is a particular challenge for subsurface drip. The system obviously needs to keep functioning effectively and reliably for a reasonable life, presumably not less than ten years. A potential problem is the intrusion of roots into the drip system outlets. The root systems of turf are generally very active and dense under irrigation. The irrigation companies have improved the design of drip outlets to minimise the risk of root intrusion. Preventative strategies, using chemicals either incorporated into the dripper ('Geoflow') or injected into the irrigation system (for example 'Treflan'), have been successful. It is important that any chemicals applied through the irrigation system be approved and comply with local regulatory requirements.

The design and installation of subsurface drip systems is critical to their success. A thorough understanding of the soil wetting characteristics, hydraulics and product performance is required to ensure that adequate and uniform wetting of the turf root zone is achieved.

A relatively new method of turf irrigation has been described by Hung and Byle (1996). Water is delivered from an underground pipe system, through small diameter tubes, so that the total turf area is saturated. It is basically a microflooding technique. The method is referred to as wick irrigation and is still being developed. The wetting properties of the soil, the soil slope and the flow rate from the emitter all have a strong influence on the appropriate spacing of emitters, which has to be considered along with operating times to ensure uniformity and correct depth of application.

The irrigation of turf should no longer be assumed to be best achieved using sprinklers. The technology is now available to give turf managers options in the method of application.

Characteristics of sprinkler systems

The turf manager must decide on the most appropriate type of irrigation system for each situation. As a general rule, the level or degree of control and uniformity of application will increase as the cost of the system increases. The initial cost of the system is therefore an important consideration.

The labour component of an irrigation system is also important. In many instances, as the level of human participation in operating irrigation systems increases, there is a corresponding decrease in the effectiveness of application of water. As well as being involved in the actual manual operation and shifting of the system components, the field person is also the decision maker who turns on and off the supply valves. The operating duration can be variable and influenced by factors other than the soil moisture needs of the turf. Automation is far more reliable for the precise control of irrigation equipment.

Some comments on the various types of sprinkler systems are:

A. Travelling sprinkler supplied by hose

1. Relatively inexpensive and portable (a target for theft!)
2. Larger pressure and discharge rates are required and the precipitation rate is generally high and sensitive to wind.
3. Labour required for setting up, storage, handling etc.

B. Quick coupling valves

1. Labour required for setting up.
2. Inefficient watering often occurs.
3. Less expensive than in-ground pop-up sprinklers.
4. Sprinklers and fittings prone to damage and wear due to use and handling.

C. Automatic pop-up sprinklers

1. High capital cost.
2. Potentially high uniformity of application and efficiency.
3. Maintenance of equipment is essential to maintain system effectiveness.
4. Very low labour requirement.

Sprinkler performance

Water distribution from sprinklers is often compared to rainfall. There are similarities, however the sprinkler technique has some important differences. One consequence of delivering water in a radial manner is that the precipitation rate decreases as the distance from the sprinkler head increases. There is a much greater area to be covered by the water falling from the stream at the extremities of the wetted circle than there is close to the sprinkler head.

Achieving uniformity in precipitation over lateral distances is a major issue with sprinkler irrigation systems.The circular pattern of coverage itself makes it difficult to achieve uniform application as considerable overlapping is required to achieve even watering of the total irrigated area. Other important differences between sprinklers and rainfall is the greater range in droplet size from sprinklers, and also droplet trajectory can be at oblique angles to the ground from sprinklers compared to rainfall. While the average precipitation rate from sprinklers can be relatively low, for example four to eight millimetres per hour, the instantaneous rate can be high as the water stream passes over a particular point on the ground during a rotation.

The precipitation distribution profile of sprinkler heads is influenced by many factors including nozzle orifice size, nozzle shape, nozzle number, operating pressure and discharge rate. The prevailing environmental conditions, in particular wind and evaporative demand, also influence the sprinkler distribution profile. Each sprinkler nozzle has a limited range of pressures over which it will produce its optimum distribution profiles. Low pressures result in doughnut shaped profiles and excessively high pressures result in higher precipitation rates close to the sprinkler head (Watkins, 1979).

Turf sprinklers have to operate in particularly demanding conditions. In addition to the need for high uniformity, the physical operating environment can be very harsh. The sprinkler body must be housed in the ground with the sprinkler head designed to raise to an elevated position to effectively distribute the water, it must withstand top loads (feet and machine wheels), and sand and silt may wash into the seals and working mechanism.

There are a wide range of sprinkler options available for turf. The main areas of differentiation are in drive type, nozzle arrangement and materials used in construction. The majority of turf sprinklers use either impact or gear drive mechanisms. There have been alternative drive systems developed in recent years. The key requirement of the drive is to provide reliable even rotation at specified speeds, under varying environmental conditions (wind).

The single biggest factor that influences the uniformity of a sprinkler system is the sprinkler layout pattern and spacing. Common layout patterns used are triangular, square and rectangular. Large open areas of turf are usually covered using a triangular pattern as it allows greatest spacing between sprinklers while still achieving good uniformities. As a general rule, the cost of sprinkler systems decreases on a per unit area basis as the spacing of sprinklers is increased, as fewer sprinklers and less pipework is required. Triangular patterns should therefore be lower cost systems than square or rectangular, for a given level of uniformity.

The uniformity of systems is very sensitive to the spacing of sprinklers. Uniformity data presented by Pair (1969) shows that for sprinklers with triangular shape distribution profiles, the uniformity decreases markedly when the spacing exceeds the range of 65 to 70 per cent of the sprinkler wetted diameter. The spacing of sprinklers so that the distance between the heads is 50 per cent of the wetted diameter is commonly recommended for situations that require high uniformity, such as golf greens. This spacing is referred to as 'head to head'.

Turf sprinklers operate in a harsh environment under demanding conditions, and are therefore vulnerable. Due to the key role of the sprinkler in determining the overall effectiveness of the system, continual monitoring of the performance of the sprinkler head should be given high priority.

Turf sprinklers checklist

The following features and characteristics of sprinklers should be considered prior to the selection of turf sprinkler heads.

1. Performance: required inlet pressure, effective wetted diameter, discharge rate, spacing and pattern for uniformity of application.
2. Drive mechanism: suitability for particular application, rotation uniformity, rotation speed, wear characteristics, expected life under site operating conditions.
3. Materials: body, nozzle assembly, cover, bearings and seals.
4. Nozzle assembly: design, number, smallest opening size (blockage risk), stream trajectory angle.
5. Part-circle mechanism: effectiveness, reliability, simplicity of adjustment.
6. Quality of manufacture: degree of precision, finish (potentially dangerous sharp protrusions).
7. Retraction method: retraction force and sealing effectiveness.
8. Operating height: clearance of water stream above surrounding turf.
9. Cover: protection of sprinkler, protective safety cover and its attachment, resistance to vandalism.
10. Filter: inlet screen sufficiently fine to prevent blockage of smallest opening.
11. Anti-drain valve: operating conditions, i.e. pressure at which it functions.
12. Serviceability: ability to service and repair from the top.
13. Spare parts: availability and cost.

Pumps, pipes and hydraulics

The maintenance of the correct pressure and flow rate at each outlet is dependent on the selection of the correct water conveyance and transfer components (pipes, valves and fittings). A thorough understanding of hydraulics is essential in making these important decisions as small variations in size can have large influences on the hydraulic performance of the system. A problem in some turf irrigation systems is the use of pipes that are too small, with the result that excessive pressure variations, due to high friction losses, occur. High water velocities are another problem. In addition to high friction losses, high velocities produce water hammer or pressure surges when automatic valves are opened and closed. Many irrigation designers limit water velocities to a maximum of 1.5 metres per second to minimise this potential problem.

Pumping plants for irrigation systems have undergone considerable change in recent years. The pump is required to supply the required flow rate at a pressure that will provide optimum operating conditions at each outlet or sprinkler head. The basic centrifugal pump, driven at fixed speed, has been extensively used to supply water to many thousands of turf irrigation systems in the past. This approach has often resulted in inefficiencies however, because the required flow and pressure conditions of the system vary and the pump plant is designed for one set of operating conditions. Variable speed drives, which allow the pump to be driven at a speed optimum for each set of operating conditions, are becoming more popular. The majority of pumping plants installed in new golf courses now employ variable speed drives.

The selection of the correct equipment, including pumps, power units, drive arrangements and controls, warrants the engagement of specialist expertise to ensure that the full range of operating conditions will be satisfied at the highest energy efficiency and reliability.

Irrigation system control

There has been an explosion in features available on irrigation controllers. Today's microprocessor provides tremendous opportunities in features and functions that controllers may perform. Selection of the most appropriate controller for each situation is a challenge for turf managers. While the irrigation controller may appear to be very sophisticated and complex, the basic task of the device is to operate the irrigation system so that the correct depth is applied at the right time. It is absolutely essential that, no matter the size or operational requirements of the irrigation system, the controller be reliable and readily operated by the irrigation manager.

The success of the irrigation system is dependent on all the key components, including the controller, functioning together in a coordinated way to achieve the desired system objective.

The controllers that are available to turf managers can be grouped or categorised as follows:

1. Stand alone controllers. These are totally self contained units which operate multiple stations, typically six to 50 valves. They may have provision for inputs from sensors, e.g. soil moisture or rainfall. Often one controller is dedicated to a single irrigated site.
2. Master satellite controllers. The control functions are split between a central location (computer based unit) and remote field locations. There may be capacity to control many stations (100s) and they are typically used on golf courses. Communication between field and central control position is via land lines or wireless, e.g. radio or microwave.
3. Central controllers. A large controlling capacity is centrally located in a region, district or very large irrigated site. Distances between central and furthest control site may be ten to 20 kilometres. A control unit at each site receives program instructions via telemetry. The system is capable of incorporating many input and output functions. It is suited to local government which has many remote sites (parks and ovals) to be controlled.

The irrigation controller has traditionally been an open loop type controller. It has worked on the principle that an output signal is sent to operate a valve for a specified period of time and, after the programmed time has elapsed, the valve is turned off. There is no feedback of information from the irrigated area in this type of control approach.

An important area of development in controllers has been the increased capacity to incorporate feedback from the irrigated site (soil moisture, nutrient levels, temperature) and from the system (flow rates, pressure). Climate data from a weather station can also be fed into the control processing unit. The achievement of efficient irrigation is dependent on the availability of quality information on the water status of the soil. Feedback of some sort from the irrigated site is therefore essential.

Another major area of development in controllers is in the area of communication, both between individual controllers and with base or central control stations. The trend towards central control is understandable in an era of organisational restructuring and amalgamations. The replacement of labour with technology is economically attractive. In the future, many irrigated sites will not have resident staff to contribute to the irrigation control process. Relevant information on the site and the system will be collected and transferred by land lines or telemetry to central control stations using one of the various forms of communication now available.

The modern irrigation controller is potentially a powerful tool for turf managers. The functions of the controller extend well beyond an electrical switching device. The measurement of the irrigated site, the prediction of turf water use, and the ability to control the operation of the system so that the hydraulic performance is optimised, are examples of functions that demonstrate the expanding potential value of the controller.

The monitoring and recording capability of controllers should be utilised by irrigation managers. The accurate measurement and recording of soil conditions, the system operation and the climate provide the irrigation manager with a powerful tool to assist in decision making on current turf conditions, and also in evaluating the past performance of the system.

The desirable characteristics (top ten) of an irrigation controller have been identified by Peasley (1992), in order of priority, as:

1. Reliability
2. Durability
3. Ease of programming
4. Sensor inputs
5. Flexibility
6. Program performance
7. Monitoring
8. Recording capability
9. Alarm facility
10. Remote communication

Establishing new irrigation systems

Water sources and supply

Irrigation systems are potentially large users of water. An area of ten hectares of turf requires one million litres to apply a depth of ten millimetres of water. In planning an irrigation project, the total volume of water that will be required and also the security of the supply need to be carefully assessed. The total cost of the water, which is directly dependent on the total volume, is becoming more important for turf managers.

Water sources for turf include rivers and streams, wells and bores, dams and reticulated town water supplies. Each source has its own characteristics and should be evaluated for each irrigation project. Turf irrigation systems demand attention, not only to the total available volume, but also the quality of the water. As a general rule turf irrigation systems have a requirement for high flow rates and reasonable quality water.

The system flow rate is determined from the following: the area to be irrigated, depth to be applied and the time available to deliver the water to the area (pumping time). In the example outlined previously, in which one million litres was required, if that water needed to be delivered in a ten hour period, then the required flow rate would be 100 000 litres per hour (27.8 litres per second). If this flow rate was not achievable, then either the area watered or depth applied would need to be reduced or the delivery time extended. Pump sizes, power unit capacity and pipe sizes are all directly dependent on the system flow rate.

The quality of water is important from the aspect of the turf and soil and also the irrigation equipment. Water quality can be assessed in terms of physical, chemical and biological properties. Should the water source contain sand and debris (physical), the system designer will need to consider the potential wear on components and risk of blockages. Filtration requirements will need to be assessed. The chemical properties of the water, in particular the Total Dissolved Solids (TDS), should be known and any likely adverse impact on the turf considered. The effect on the nutrient or chemical balance in the soil needs to be evaluated and amendment strategies adopted. At times water supplies contain microorganisms (biological) which need to be identified and treated. In some cases it is a matter of screening to remove material such as algae, but in other cases specialist water treatment may be required to remove the risk of disease.

When the potential implications of water quality are appreciated, it is obvious that a detailed water analysis is warranted. There are many laboratories that can competently carry out water analysis. The water analysis could include testing for alkalinity (pH), salinity (TDS), suspended solids, contaminants and specialist elements such as nitrogen, sodium, chlorine, phosphorous and iron.

The laboratory analysis of water is only the first step in the water assessment process. Equally important is the interpretation of the results in terms of the suitability of the water for the designated turf, soil and site. Interpretation services are provided by some laboratories, consultants and government agencies. Generally speaking, analysis services are more available than interpretation services. It is important that interpretation be provided by appropriately skilled experts.

System design

Designing an irrigation system requires many different and sometimes conflicting factors to be taken into account. A well designed irrigation system allows the appropriate amount of water to be applied effectively and efficiently at the correct time. It is much more than simply the selection of hardware such as sprinklers and pipes.

The design process requires the designer to have a full appreciation of issues such as:

- Crop water use characteristics
- Soil properties
- Terrain and area requirements
- Climate of site
- Crop management practices

At times one factor may have a strong influence over the design process. In the case of turf, vigorous plants with shallow roots, growing in relatively light soils, there will be issues relating to the small amount of water stored in the root zone, the rapid drainage of water through the profile and the limited lateral movement of water in lighter soils. It is the designer's responsibility, in consultation with the turf manager, to identify factors that may have a significant influence on the system and to recommend a design solution that optimises all of the competing needs of the irrigation system.

The design of an irrigation system requires the designer to be competent in their knowledge of crops, soils, irrigation hydraulics and irrigation equipment performance. Unfortunately there are examples of existing turf irrigation systems that do not allow the turf to be irrigated to the optimum level.

Turf managers should request that those undertaking the design of their system should be appropriately qualified in irrigation, hydraulics and engineering. The Certified Irrigation Design (CID) program, which is available in the USA and Australia, is recommended as a minimum standard for turf irrigation designers.

The design process

These are the basic steps a designer may follow when developing a design for a new irrigation system.

1. Collection and analysis of all relevant data
2. Determination of total irrigation requirement
3. Selection and positioning of outlets
4. Pipe layout of system and control zones
5. Hydraulic design of pipes and valves
6. Determination of pumping equipment
7. Control system selection and development of control program
8. Preparation of plans and specifications

Installation and commissioning

The quality of equipment, the installation techniques and the competency of the irrigation company in servicing the system, all need to be considered when selecting the irrigation company to supply and install the irrigation system. Producers should insist that all products, including pipes, fittings and valves, comply with the relevant national and international standards.

A well prepared and thorough specifications document is an excellent aid in achieving a quality irrigation system. It also allows alternative systems to be compared and evaluated on an equal basis, and minimum standards in equipment to be established. Independent irrigation design consultants generally prepare specifications and associated documentation as part of their professional services.

An important stage in a new irrigation project is the formal acknowledgment that the equipment has been supplied and installed according to appropriate standards and is capable of performing to the required standards. This is a vital step in the process of acquiring a new irrigation system.

In addition to visual inspections of equipment and installation techniques, specific tests should be carried out to test both the integrity of the complete system and the hydraulic performance.

Suggested tests include:

- Checking that the pumping plant delivers the design duty (flow and pressure) at the designated efficiency
- Pressurising all mainlines and submains to ensure that they are capable of holding designated test pressure, under static conditions, for specified periods of time
- Checking pressure variations along laterals and between laterals throughout the system
- Checking flow variation between outlets along particular laterals to ensure system delivery rates are within acceptable limits
- In the case of sprinkler systems, checking the average precipitation rate and the uniformity of application is essential. Application uniformity coefficients, for example (Distribution Uniformity (DU)) should be determined.

Irrigation system management

Irrigation scheduling

The achievement of the designated quality of turf is dependent on the maintenance of optimum soil moisture conditions. The control and operation of the irrigation system, so that the appropriate depth or amount of water is applied at the correct time, is referred to as irrigation scheduling. Basically it is the determination of how much to apply and when to apply it.

It is becoming more important to know how much water is actually available to the turf and how the turf will respond to the available water.

The techniques available include those that involve direct measurement of soil moisture, using soil moisture sensors, and those that estimate available water using a crop water use estimation approach. In the latter method, the expected water demand is predicted using an evaporation value, either measured directly from an evaporation pan or determined using data from a weather station.

Irrigation scheduling techniques can be categorised and summarised as follows:

1. Turf appearance—irrigate when crop shows visible signs of stress (very basic technique)
2. Monitor turf condition using instrumentation, e.g. leaf temperature measurement
3. Direct measurement of soil moisture status using soil moisture sensors
4. Estimation of turf water use and available soil water using climate information from weather station

The production of turf that meets increasing performance standards is becoming more dependent on improved water management. The adoption of irrigation scheduling techniques, which incorporate both an understanding of turf response to water availability and access to accurate available soil water data, should be a priority for turf managers.

Weather stations

Weather stations are a very valuable tool in the management of turf irrigation systems. Units are available which allow the measurement and recording of all relevant factors including temperature, solar radiation, relative humidity, wind speed and direction, and rainfall. The data can then be fed back to a central computer to allow the information to be collated, processed and acted upon. Some units will calculate the crop evapotranspiration rate (ETc) using mathematical expressions such as the Modified Penman formula, and also transmit alarms to relevant personnel when readings reach particular set or limit points. These can also be used to initiate or cancel the operation of automatic control systems.

Soil moisture sensors

The amount of water that is held in the soil and available to the turf is dependent on the soil type and crop root characteristics. There are large differences in the water properties of sands, loams and clays as a result of differences in particle sizes and particle distribution. In soils with small particles, for example clays and loams, water is stored in many small cavities or voids which are formed by the surfaces of many small angular particles. The forces holding the water (water tension) in contact with the surface of the particles are potentially large in soils with small particle sizes. Continual removal of water from these soils becomes increasingly difficult because soil water tension increases.

Both soil water tension and soil water content are relevant in the measurement of water in the soil. The relationship between soil water content and associated water tension, for a particular soil, can be described through a characteristic curve for that soil. This relationship needs to be appreciated when managing the water in the soil for turf, as both the accessibility of the water and the total amount of water available are important in making irrigation management decisions.

Some requirements for soil moisture sensors used for turf are:

1. Accurate and rapid measurement in reasonably open soils, e.g. sands and sandy loams
2. Work effectively in relatively shallow depths, e.g. from 100 to 300 millimetres
3. Not significantly affected by soil nutrient, soil salinity and temperature
4. Repeatable behaviour following wetting and drying cycles
5. Able to be readily calibrated for particular soil

6. Output signal to be readily understood and compatible with monitoring equipment
7. Robust and reliable

Types of soil moisture sensors

Soil moisture sensors can be grouped according to two broad categories:

- those that measure water tension
- those that measure a soil property which changes as a result of changes in the water content of the soil.

Soil moisture sensors provide valuable data for improved irrigation management, however the limitation that readings reflect only one point in the soil profile needs to be recognised. It is a sampling device. The selection of the soil moisture sensor position, in relation to the soil profile, root system and system distribution, is critical to ensure that readings are representative. The installation of the sensor is also vital to the success of this technology.

Evaluating system performance

A quantitative measure of performance of an irrigation system is a valuable tool for those responsible for irrigated areas. An appropriate seasonal water consumption indicator is the Irrigation Index (Ii) (Kah and Willig, 1992) which compares the depth of water actually applied to the estimated depth of water required over the complete irrigation season. This simple measure provides the manager with a visible, readily understood measure of how well or how efficiently the system is performing and how the performance compares with other sites. An irrigated area that is being well managed would have an Irrigation Index value of 1.0 or less. In a study carried out by Keig (1994) on ten irrigated sports grounds in Melbourne, Victoria, it was found that Ii values ranged from 0.63 to 2.75. These performance values suggest there was an opportunity for improvement in some of these systems through the adoption of technology and smarter management.

Irrigation managers and others responsible for maintaining irrigated turf areas would benefit from the greater adoption of an indicator that would describe how well the system has performed over a season or designated period of time.

The evenness of application of a sprinkler system can be checked through the sampling of the system during operation. A number of catch cans are placed within the irrigated area and readings taken of the depth of water deposited at each position. Both the uniformity coefficients Christiansen (Cu) and Distribution Uniformity (DU) are used to provide a statistical measure of the eveness of application across the pattern (Zodolske and Solomon, 1986). The preferred measure of uniformity for turf is the DU (Connellan, 1997) which compares the average of the lowest 25 per cent of can readings to the average of all readings. A DU of 100 per cent would indicate that the application was perfectly even. In practice, this does not happen. It is generally accepted that sprinkler systems for turf should have a minimum DU of 75 per cent.

The regular assessment of the performance of sprinkler systems is recommended. The Certified Landscape Irrigation Auditor (CLIA) program, which trains and accredits industry personnel, in USA and Australia, is an important resource for turf managers. Auditors not only assess the system but also provide recommendations on strategies and techniques to improve the performance of the system.

Water conservation strategies

The turf manager is often confronted with the conflicting requirements for higher quality surfaces, which generally means more water, and containment of site costs, which generally means using less water. It is environmentally and economically sound practice to ensure that all water applied by the irrigation system is necessary and is used efficiently. Water conservation should be at the forefront of all irrigated turf management plans.

The overall approach to water conservation should include attention to the following aspects of irrigated turf:

1. Selection of water efficient turf species
2. Management of turf to maximise turf water use efficiency
3. Selection of well designed and well engineered irrigation systems
4. Management of irrigation system so that irrigation efficiency is maximised

The starting point in water conservation is to assess the suitability of the grass species to the site. It may be that low water use alternative turf grasses, for example warm-season grasses or native grasses, can fulfill all of the vegetation requirements for the area. The ultimate water conservation technique for turf is to select a species that does not require irrigating at all.

There are several turf management techniques that can be adopted to improve irrigation efficiency. It is recommended that grass be mowed at the highest allowable height for the turfgrass species (Gibeault, 1992). The additional height does provide a greater leaf area, which will produce an increased transpiration area; however the greater the height the deeper and more extensive will be the root system. This produces a stronger plant with greater access to soil water. Frequent mowing should be avoided if possible as this encourages increased water use.

The management of the soil water level to encourage greater root depth is another technique in water conservation. By allowing the soil moisture level to be depleted to low levels and applying deep waterings, there is a tendency for turf root systems to extend lower in the soil profile. In addition to providing increased storage the greater root depth provides the turf with greater reserves of soil nutrients. The effectiveness of rainfall is also increased as the soil water reservoir will on average have a greater capacity to benefit from rainfall (McIntyre, 1992). If the soil water is managed so that it is nearly always close to the full point, rainfall may be wasted through runoff.

Maintaining the soil in good condition is an essential part of good irrigation management. Compacted layers can be broken by splicing, mole ploughing and hollow tyning according to Neylan (1992). The mangement of the nutrient program so that excessive lush growth, which has a higher water requirement, should be carried out with due consideration to water consumption implications. Removal of thatch, which restricts water entry into the soil, is also recommended.

Developments in irrigation technology which assist water conservation are numerous. They include:

- Improved flow and pressure control devices
- Better sprinkler distribution profiles for improved uniformity
- Lower precipitation rate sprinklers for low infiltration rate sites
- Sensors for system, soil and environment
- Improved control capability
- Communication systems for remote monitoring and control

Turf managers should ensure that irrigation system performance is to the highest irrigation efficiency. Emphasis should be placed on uniformity of application and the precise operation of the system so that water is not wasted. Continual monitoring of the performance of the irrigation system is a key part of a total water conservation program.

System maintenance and failures

The components of irrigation systems are designed and constructed to control water flow under very demanding conditions. The environment, both internal and external, is harsh and equipment reliability is essential. Single failures, for example blockages, failed electrical connections and breaks in seals, can be very costly in terms of turf areas ineffectively watered. Maintenance is therefore a key requirement in the overall management of the irrigation system. Electrical, mechanical and hydraulic equipment needs to be maintained in first grade working order. It is to be expected that parts of the system will degrade over time and require adjustment or replacement. In addition to the regular execution of the maintenance program, it is important to monitor the performance of the irrigation system. Keeping records of the completion of regular maintenance activities and repair work carried out, and replaced equipment, is valuable.

Some examples of failures of turf irrigation systems that may be due to poor maintenance include:

- Sprinkler not functioning at all
- Flooding around sprinkler head
- Inadequate distance coverage by sprinkler
- Poor uniformity of system
- Misting of sprinkler stream
- Overspray on to non-irrigated areas
- Excessive water consumption by system

Turf managers should ensure that staff undertaking maintenance and repairs should be appropriately trained. It is often more cost effective to hire a trained repairer than it is to attempt a trial and error approach.

Summary

Automatic inground sprinkler systems can provide the means of establishing and maintaining high quality turf areas. However they do require good design, careful equipment selection, sound installation and ongoing maintenance.

Turf managers now have available the technology to gain much more information about the irrigated site and the system. In addition to the benefits of this increased knowledge, which will significantly aid overall site management, the turf industry will also benefit from the greater utilisation of irrigation performance indicators. Both uniformity (DU) and water use efficiency indicators (Ii) should be used.

Developments in turf irrigation are mirroring other areas of turf management in that the challenges in getting it right are increasing, and so are the rewards.

References

Connellan, G.J. (1997) 'Technological challenges in improving water use efficiency in urban areas', *Proceedings of Irrigation Association Technical Conference, Nashville, Tennessee, 2–4 Nov.*, Irrigation Association, USA.

Doorenbos, J. and Pruitt, W.O. (1977) 'Guidelines for predicting crop water requirements', *FAO Irrigation and Drainage Report 24*, Rome.

Gibeault, V.A. (1992) 'Water conservation in turfgrass irrigation', *Proceedings of Irrigation Australia National Expo and Conference, Melbourne, 12–14 May*.

Handreck, K.A. and Black, N.D. (1991) *Growing media for Ornamental Plants and Turf*, UNSW Press, Kensington, NSW.

Hung, J.Y.T. and Byle, J. (1996) 'Wick irrigation for lawn', *Proceedings of the Irrigation Association International Exposition and Technical Conference, San Antonio, Texas, 3 Nov.*

Kah, G. and Willig, C. (1992) 'Irrigation Efficiency—How to make it work for you', *Landscape and Irrigation*, pp. 44, 45.

Keig, S. (1994) *Water Use Efficiency—Evaluation of Selected Sports Grounds*, Horticultural Project Report, Burnley College (VCAH), University of Melbourne, p. 20.

McIntyre, D.K. (1992) The need for water conservation, Technical Paper, Horticultural Services Unit, ACT Parks and Conservation Service, Canberra, ACT.

Meyer, W.S. (1985) 'Irrigation scheduling—principles', *Irrigation Association of Australia J.*, August, pp 7,8,10.

Neylan, J. (1992) 'Irrigated turf—Perceptions and

Realities', *Proceedings of Irrigation Australia National Expo and Conference, Melbourne, 12–14 May.*

Pair, C.H.(ed) (1969) *Sprinkler Irrigation, 3rd edn,* Irrigation Association, Washington, USA.

Peasley, B. (1992) 'What do you need in an irrigation controller?', *Proceedings of National Irrigation Convention, Melbourne, 12–14 May,* Irrigation Association of Australia, 35 pp.

Watkins, J.A. (1979) *Turf Irrigation Manual,* Telsco Publications, Dallas, Texas, USA.

Withers, B. and Vipond, S. (1974) *Irrigation–design and practice,* B.T. Batsford, London.

Zoldoske, D.F. and Solomon, K.H. (1986) *Coefficient of Uniformity—What it tells us,* CATI Publication no. 880106, Center for Irrigation Technology, p. 3.

Zoldoske, D.F., Genito, S. and Jorgensen, G.S. (1995) 'Subsurface Drip Irrigation: A University Experience', *Irrigation Notes,* Center for Irrigation Technology, Fresno, California, Jan, 35 pp.

CHAPTER 8

Turfgrass nutrition and fertilisers

S.T. COCKERHAM, *University of California, Riverside, California, and* **D.D. MINNER,** *Iowa State University, Ames, Iowa, United States of America*

Introduction

The sports field is a complex physical and biological system that supports the activities of sport and other events. Well managed turfgrass sports fields are safe, durable, and aesthetically attractive, but they are not low maintenance. Turfgrass under the pressure of intense traffic should be sustained at optimum vigour. Correct fertilisation of the turf is a key factor in that process. Research on sports turf fertilisation has been conducted in the programs of universities, government agencies and private institutes throughout the world. Much is site specific, sport dependent, and influenced by local climate. Still, there are basics that can vary little for turf managers wherever they might be.

Primary fertiliser elements

Nitrogen (N), phosphorus (P), and potassium (K) are the primary nutrient elements used in large quantities by plants. Sulfur (S) and iron (Fe) are secondary elements that sports turf managers often use in significant quantities.

Nitrogen is a major ingredient of the plant. As a component of chlorophyll, nitrogen deficiency first shows up as a yellowing of the turf. The deficiency is corrected by the application of nitrogen as a turf fertiliser. Phosphorus is used to make proteins and help transfer energy within the plant. It is especially important in the development of roots, rhizomes, stolons and tillers. The roots are vital to the sustenance of the grass plant. If there are no roots then there is no growth. Rhizomes, stolons and tillers are plant structures that have a spreading growth habit and allow thinly covered turf areas to quickly recover. Potassium is involved in several plant metabolic processes, many of which are related to water use. It is highly soluble and does not stay in the root zone very long. Sports turf on sandy soils and pure sands loses potassium rapidly and requires application almost as frequently as nitrogen. The turfgrass plant utilises potassium at about the same rate as nitrogen (Shearman, 1989; Catrice *et al.*, 1993). Potassium improves turfgrass wear tolerance, disease tolerance and aesthetic quality. Sulfur is important in protein synthesis and sulfur deficiency stunts the growth of the turf plant. Iron, essential for chlorophyll synthesis, is important to turf colour. Turf suffering from iron deficiency is chlorotic, and does not respond to nitrogen. Iron is usually present in the soil, but has a tendency to form insoluble compounds.

Nutrient testing in plants and soils

Soils provide the reservoir of nutrients required for turfgrass growth. For many situations where soils are very predictable, a test may be taken every three years and fertility adjusted accordingly. Traditional soil testing uses chemical extractants to simulate the release of elements and their eventual availability to the plant. This type of traditional testing provides an estimate of the potential pool of nutrients that may become available to the plant. Some laboratories also report the nutrients that are released from a water saturated paste extract. This test provides a reading on the immediate availability of nutrients at a given point in time, rather than the potential nutrients that may become available over time. Knowing the nutrient status by soil testing allows the turf manager to make timely decisions concerning fertiliser application. Testing the soil in this manner determines if sufficient nutrients are available for plant up-take by the roots, but it does not supply information about the actual nutrient status of the plant. Plant analysis provides the most accurate estimate of the tissue's elemental content. Using information from both soil and plant testing gives the most complete picture of nutrients supplied by the soil and taken up by the plant. Table 8.1 shows both the soil and plant nutrient requirements for sufficient growth by turfgrass.

Several events have changed the way we now approach fertility assessment of sports turf. Overnight express mail and laboratory techniques that supply rapid turn-around of samples have boosted the growth of private laboratories and consultants that specialise in nutrient testing and turfgrass recommendations. There is an immense supply of different products formulated for the turfgrass industry that can be related to solving nutritional problems. Some of those include granular, liquid, foliar fed, organic, inorganic, chelated, and slow release fertilisers; hormonal and biostimulant materials; peat, ceramic clay, and diatomaceous earth additives; and many others. Rapidly available information from testing has been combined with a supply of products in an effort to solve difficult growing conditions created on sports fields by intense traffic and artificial soils. Academic research has been slow to develop fertility guidelines based on testing strategies and new products. Nutrient testing is a valuable service that provides the necessary background to make informed decisions. Standardised testing procedures produce accurate and reproducible results, however the range of nutrient sufficiency and deficiency levels based on sports turf performance continues to be debated. From a strategic standpoint, start by adjusting pH and soil nutrients based on soil testing. Testing for plant nutrients can help determine if actual uptake by the plant is limited and if there is a need for specially formulated fertilisers. The final and most important determination is your evaluation of how the turf performs at certain nutrient test levels and how the turf responds to fertiliser application. Testing provides information that empowers your ability

Table 8.1 Suggested target range for soil and plant nutrients.

Nutrient	*Tissue target range (%)*	*Soil target range (ppm)*
N	2.8–5.5	
P	0.3–0.6	15–30
K	1.0–3.5	100–250
Ca	0.4–1.2	
Mg	0.2–0.6	
S	0.2–0.5	
	(ppm)	
Fe	35–400	20–60
Mn	25–150	5–20
Zn	20–60	3–15
Cu	5–20	0.5–3.0
B	10–60	0.5–1.5
Na	0.1–0.4	
Mo	not known	
Cl	not known	

to make changes in fertility strategy and a valuable service, especially when turf performance improves.

Collecting soil samples

Soil test results have little meaning if the sample does not accurately represent the area. The actual amount of soil used for laboratory nutrient analysis is only about 50 grams. If the sample represents an entire soccer field then the 50 gram sample will be used to provide fertiliser recommendations for about 2.5 million kilograms of soil. There is far greater chance for error by collecting a non-representative sample compared to the error that might occur in the laboratory phase of the nutrient extraction process. Start by assessing your facility and your fertility goals. Keep in mind that routine sampling in the same location can be used to develop a working history of how well your fertiliser program is supplying nutrients. Priority should be given to similar soil types when deciding which areas to combine into one sample, and thus one fertility program. Areas with similar topography, drainage, and turf species should also be considered. Sand-based sports fields should obviously be sampled separately from fields containing higher amounts of silt and clay. Areas where grass is actually growing on fill-soil or sub-soil should be sampled separately. Other situations that result in dissimilar soils include sodded areas (especially thick-cut sod), organic layers, and native soils overlaid by a layer of topdressing sand. Separate sampling of each distinct layer may provide useful information in addition to the average that results from mixing the layers.

Since sports fields have distinct traffic patterns some managers have developed specific fertility programs for specific high traffic areas of the field. For example, higher rates of phosphorus can help seedling plants in worn areas where constant reseeding is needed. In the same areas mature grass may benefit from additional potassium that is not necessarily needed on the rest of the field. Select a reputable soil testing company that is familiar with sports turf fertility testing and recommendations. There is nothing more frustrating than sending the same soil sample to two different laboratories and getting two different test results back. Follow the laboratory recommendations for sampling procedure and depth of sampling. Samples are generally taken from the top five centimetres of soil with the grass and thatch removed, some labs may specify eight centimetre or 15 centimetre depths. It is important to collect all samples from a uniform depth. Once you have selected a uniform area for sampling, the question then becomes how many samples are needed and where to actually probe the soil. A minimum of five samples should be taken from the five points as if an imaginary W were placed on the area (Adams and Gibbs, 1994). Not more than 20 samples will be needed to represent one area. Collect the samples in a clean plastic bucket. Crumble the sample plugs and mix the soil together. One subsample containing about 500 to 1000 cubic centimetres of soil should be sufficient for soil nutrient analysis, unless otherwise suggested by your laboratory. Small inexpensive paper bags with a waterproof lining and a sealable top can be purchased to submit soil samples. Use a permanent marker to date and identify each sample location. Develop a field location map that designates what areas make up a soil sample. Use the soil testing map so that plant samples can be tested from the same areas for comparison.

How often to sample and how many samples to submit may depend on your budget. Areas where soils seldom show a change in test results may only need sampling every three years. Sampling twice a year may be useful when your strategy involves application of lime or sulfur to adjust pH. Sand-based systems should be evaluated at least each year since they generally have a low cation exchange capacity and are prone to nutrient deficiencies. On intensely managed sand-based sport fields some managers test soil every month, plant analysis twice a month, and irrigation water twice a year (York, 1998). Where effluent or reclaimed water is used even more rigorous testing may be needed to deal with potential problems from sodium and bicarbonate (H_2CO_3).

Collecting plant samples

Tissue testing and plant analysis are two different approaches that have been used to assess the nutrient status of grasses. Tissue testing involves field analysis of extracted cellular sap using reagents and test kits. Plant analysis is a more accurate estimate of the tissue's elemental content and involves submitting plant samples to a laboratory (Jones *et al.*, 1992). Consult the plant analysis laboratory for sampling and sample submittal procedures. Unlike soil samples, plant samples must be collected from fresh living plants and shipped overnight express to the laboratory. The youngest grass leaves are desired when collecting plant material. A typical method of collecting and submitting samples involves collecting clippings from your normal mowing practice. Remove a handful of clippings from the catcher at several different locations on the test area and put them in a clean bucket or large plastic bag. Mix the clippings by hand and then fill a one litre plastic zip lock bag for submittal to the laboratory. Plastic zip lock bags for vegetables have small pin holes and are ideal for submitting plant samples. Label the bag with a permanent marker. Remember to sample the same areas that were used for soil testing so that you can determine if the available soil nutrients are actually taken up by the plant.

In developing your soil, water, and plant testing strategy it is important to remember that testing is simply another tool that should help you grow better grass. While very accurate values can be generated for nutrition levels, it may not have much meaning if there is no measurable response in turf performance. Add a section to your nutrient testing and fertiliser records that provides for some type of visual assessment of the turf's performance in response to your fertility program. Measure performance by visually assessing turf colour on a scale of 1 to 10 (10 = darkest green, 5 = lowest acceptable green colour, and 1 = brown turf); percent of area with living green grass cover (0 to 100 per cent); and percent of area with exposed soil showing (0 to 100 per cent). When expensive fertilisers are recommended to solve a problem or enhance growth, try a test application on a small area and compare it with the surrounding non-treated control area. Another approach is to keep a one square metre plywood board on hand. Simply lay the board on the turf before fertiliser application. Make your normal product application to the entire area and immediately pick up the board. The non-treated area under the board serves as a control plot that can be compared with the surrounding area that was treated. An untreated control plot or check is needed in order to determine if a grass response is associated with the product you have applied. Throughout the growing season your normal fertility program can become a series of treatments and controls that can help evaluate fertiliser effects for specific fields. Do not be overly dependent by letting a laboratory make your fertility decisions. Instead, determine your own fertility target values by keeping accurate records of test values, fertiliser applications, and your evaluation of the turf's performance.

Soil testing

Additional information found in a soil test report is often used when developing a fertility strategy. Organic matter content, cation exchange capacity

(CEC), and soil pH have a direct effect on plant available nutrients. CEC is an important value that does not actually measure soil fertility but does measure the potential for storing nutrients. Soil particles have a negative surface charge that attracts positively charged cations such as hydrogen H^+, calcium Ca^{++}, magnesium Mg^{++}, potassium K^+, sodium Na^+, and ammonium NH_4^+. Negatively charged anions such as nitrate NO_3^- do not attach to soil particles and are easily flushed from the soil. Nitrogen is taken up by grass mainly in the nitrate form, which is also the form of nitrogen that is most easily leached and often requires replenishing from nitrogen fertiliser. Soils have a wide range of CEC values. The higher the CEC value the greater the capacity to store nutrients. In general CEC, expressed as milliequivalents per 100 grams, increases with clay and organic matter content and decreases with sand content: sand 1–6, sand + peat 1–14, clay loam soil 25–30, clay 80–120, and organic matter 150–500 (Christians, 1998). Sand is often used in sports turf to improve drainage and reduce compaction. Soils high in sand content and with a CEC below 6 are more susceptible to nutrient deficiency and require greater attention to fertiliser scheduling. Sports turf managers who use sand rootzones are often faced with the decision of increasing CEC by adding either organic matter or soil containing clay. On a weight basis the same CEC can be obtained with five per cent organic matter or 30 per cent illitic clay (Adams and Gibbs, 1994). A stable form of organic matter that does not cause blocking of macropores is usually preferred to increase CEC without limiting water infiltration below a targeted value.

Soil organic matter associated with turfgrass comes primarily from leaf, stem and root residues of grasses. These plant fragments decompose rapidly at first and eventually become a more stable form of organic matter known as humus. Humus is colloidal material with a negative charge, finely dispersed particles, high CEC and properties similar to clay. Nitrogen is the only major plant nutrient that depends on organic matter to store soil nitrogen. As indicated earlier, nitrogen is not found on a soil test report because it easily fluctuates as organic matter decomposes. The organic matter content is, however, a good indicator of how much nitrogen is stored in the soil as organic matter for eventual release to the plant. Organic matter also contains other essential plant nutrients (phosphorus and sulfur) and trace elements (copper, zinc and boron).

In addition to supplying nutrients, organic matter affects soil physical properties and water holding. Generally organic matter has a positive effect by improving soil structure and increasing available water. Decaying roots can create large pores in the soil. Organic matter as humus consists of large molecules rather than plant tissues. Adams and Gibbs (1994) explains that with regard to sportsturf soils, plant fragments create pores that aid drainage but humus, which improves water and nutrient retention, may also block pores. An organic matter content of two to four per cent is usually sufficient to assist with nutrient retention on native soil fields.

Potential hydrogen or pH is a common soil test that measures the amount of H^+ in the soil solution. Typically an equal volume of both soil and deionised water are used to conduct the test. Technically the pH of a solution is the $-\log_{10}$ of the hydrogen ion concentration. Pure water (H_2O) dissociates into both H^+ (acidic) or OH^- (basic) and at equilibrium each has a concentration of 10^{-7} M, with the combined ion concentration equal to 10^{-14} M. The pH scale ranges from 0 to 14, with 7 being neutral. A pH of 14 indicates that no H^+ is present and all the ion concentration (10^{-14} M) is OH^-. At a neutral pH of 7 there are equal concentrations (10^{-7} M) of H^+ and OH^-. Soil pH generally ranges from a low of 3 to a high of 11, with 7 being neutral. Everything below 7 is acidic and everything above 7 is basic or alkaline.

Nutrient form and availability change in the soil as pH fluctuates. Most turfgrasses grow best under slightly acidic conditions (pH 6.2 to 6.5) because this is the pH range where most soil nutrients are available to the plant. Although not ideal, most turfgrasses can be grown and most nutrients are available under the broader pH range of 6 to 7. Cations that make up the CEC are constantly competing with each other to occupy the negative charges on the soil particles. It is the soil pH that measures which cations dominate the exchange sites and soil solution. Under low pH acidic conditions H^+, aluminium Al^{+++}, iron Fe^{++}, manganese Mn^{++}, and zinc Zn^{++} will be plentiful, while Ca^{++}, Mg^{++}, and K^+ may be deficient. In contrast, Ca^{++} usually dominates high pH soils causing Fe^{++} to be deficient. Adjustment of pH is a more permanent solution to pH induced nutrient deficiencies. In situations where pH adjustment is not possible, specific nutrients that are deficient should be supplied in a form that will be readily available for the plant.

Liming may be required to raise the pH of soils that are too acidic. It should not be applied without the recommendation from a qualified soil test. The buffer pH found on most soil reports is used by the laboratory to determine how much lime is required to raise the soil pH to a given level. Calcium carbonate ($CaCO_3$) is the standard liming material used to raise pH. As a general rule of thumb, lime should be applied in the spring or autumn and each application should not exceed 2500 kilograms per hectare on actively growing grass. Repeat applications over several years may be required to raise pH. When possible, higher rates of lime can be tilled into the soil to adjust pH.

Acidification is the process of reducing pH. It is important to understand what is causing a high pH before there is an attempt to effectively reduce it by application of sulfur. Soils that are irrigated with water high in calcium may have their soil pH elevated just above 7. In this case, sulfur applied to the soil will be converted to sulfuric acid (H_2SO_4) by soil microbes. The pH will decrease when sulfuric acid reaches a high enough concentration. In contrast, calcareous soils that have a pH greater than 8.2, because of excessive calcium carbonate ($CaCO_3$), are nearly impossible to acidify. As long as solid lime ($CaCO_3$) remains in the soil it is not possible to decrease pH. To dissolve one per cent calcium carbonate in the soil would cost approximately $US200 000 per hectare (Killorn and Miller, 1986). Many of the sands used to build athletic fields contain at least 20 per cent calcium carbonate. Christians (1998) calculates that it would take 1000 to 1500 years to lower the pH below 7 by applying the maximum rate of sulfur (480 kilograms of sulfur per hectare) to a soil containing twenty per cent $CaCO_3$. If the soil test indicates that the soil is highly buffered with calcium carbonate then do not expect to increase nutrient availability by reducing pH through acidulation with sulfur. Instead of making nutrients available by lowering pH, simply apply the nutrients in a form that will be quickly available to the plant. For example, iron chlorotic grass growing on calcareous soils can be quickly returned to a normal green colour by treatment with chelated iron. Turf colour of iron-chlorotic Kentucky bluegrass improved with applications of $FeSO_4$ up to a rate of 49 kilograms of iron per hectare. Compared to $FeSO_4$ lower rates of iron chelate can be used to achieve the same level of colour enhancement (Minner and Butler, 1984). Chelated nutrients are not affected by pH induced changes in solubility, and therefore remain more available to the plant. When nutrient deficiency in the plant is caused by pH imbalance and unavailability in the soil, then the goal should not be to supply fertilisers that store nutrients in the soil. The strategy should be to apply frequent, weekly or monthly, applications of readily available plant nutrients.

Fertilisers

Fertiliser labels contain the nutrient analysis indicated by the percentage by weight of N-P-K in the package. The nutrients are listed on the label as numbers always in the order N-P-K. As a fertiliser, phosphorus is referred to as phosphate and shown by the chemical notation P_2O_5. Potassium as a fertiliser is known as potash and is shown by the chemical notation K_2O. A fertiliser with 15-5-10 on the label would be 15 per cent nitrogen plus five per cent P_2O_5 plus ten per cent K_2O all by weight. For simplicity we will use phosphorus and potassium, meaning P_2O_5 and K_2O. Nitrogen is designated as N in fertiliser analysis. Fertiliser recommendations are often expressed as a ratio of N, P and K. A 15-5-10 product would be expressed as a 3:1:2 ratio.

Nitrogen fertiliser sources are separated into three groups: inorganic, natural organic, and synthetic organic. Inorganic nitrogen fertilisers produce quick plant response, are not very sensitive to temperature, and are low in cost per unit of nitrogen. They are highly soluble in water, so the response does not last much longer than four weeks. The most commonly used inorganic nitrogen fertilisers are ammonium nitrate, ammonium sulfate, and calcium nitrate.

The turfgrass plant generally absorbs nitrogen in the ammonium form and in the nitrate form. Ammonium is taken up by the roots, but most of it is absorbed by soil particles. Soil microbes convert it into the nitrate form, release it from the soil particle, and make it available again. Cool temperatures slow soil microbes, making the ammonium less available to the plant. Nitrate is easily absorbed by roots and the availability is not sensitive to temperature. All fertilisers have the potential to burn the turf foliage during high temperatures if not watered in immediately after application, with the soluble inorganics producing considerable risk (Figure 8.1).

Figure 8.1 Fertiliser burn on Kentucky bluegrass (right)

Ammonium nitrate (33.5-0-0) is a soluble, fast release inorganic fertiliser. The fertiliser is a good source of nitrogen for use in cool temperatures. The burn potential is high and care should be taken when using ammonium nitrate in warm to high temperatures. Ammonium sulfate (21-0-0), in addition to nitrogen in the ammonium form, contains 24 per cent sulfur. Because sulfur reacts to acidify soil, this fertiliser is recommended for use on alkaline soils. Ammonium sulfate is easy to use and produces quick turf response, but the risk of burning is high. Calcium nitrate (15-0-0) is a good cool weather turf fertiliser, but it absorbs moisture easily causing it to cake in the bag. The calcium content gives it an advantage in acid (low pH) soils. It should be stored in airtight containers.

Natural organic nitrogen turf fertilisers are derived mostly from animal wastes (manure and sewage sludge). These materials typically are not soluble. The plant growth response is slow and they are only effective for four to eight months. The low nitrogen analysis (two to 15 per cent) of natural organics makes their cost per unit of nitrogen higher than the inorganics. Even though they are expensive nitrogen sources, many turf managers use natural organics due to the presence of a number of other nutrients and the low burn potential. Since soil microbes break down natural organics to release nitrogen, effectiveness is poor in cool weather. Activated sewage sludge is a widely used natural organic nitrogen turf fertiliser. It contains four to seven per cent nitrogen and is produced by treating processed sewage. Manures are often applied as preplant fertiliser for turfgrass establishment, but those with high soluble salt content, such as poultry manure, are not suitable for salt sensitive grasses.

Synthetic organic nitrogen fertilisers are primarily urea and urea-based compounds, both soluble (quick-release) and slow-release. Soluble synthetic organic fertilisers give quick turf growth response that may last four to six weeks. Soluble urea (45-0-0) is by far the most popular soluble synthetic organic nitrogen source for use on turf. The prills are convenient to apply dry through a spreader, or they can be dissolved in a sprayer tank for liquid application. The very high nitrogen content means that less total quantity must be handled compared to other fertilisers. In warm weather the ammonium in the urea turns to free ammonia and evaporates, so a large percentage of nitrogen is lost to volatilisation, besides leaching due to the solubility. As much as 15 per cent of the urea can be lost to the atmosphere without irrigation (Sheard and Beauchamp, 1985). Cool-season use of urea minimises volatilisation. The burn potential is high and care should be taken when using urea in hot temperatures to avoid turf injury.

Slow-release insoluble synthetic organic nitrogen fertilisers have little turf burn risk, and plant growth response is slow, which on occasion may be a disadvantage. The release of nitrogen over an extended period makes the plant growth more consistent and reduces the number of applications. The latter is a particular benefit where sports fields with high use rates make it difficult to schedule fertilising the turf. The slow-release materials are more expensive per unit of nitrogen than other sources, therefore the decision to use them must be based on the management advantages and not product cost. Slow-release nitrogen fertilisers are made by either combining urea with other compounds to produce water-insoluble materials or by coating a readily soluble nitrogen source. The most commonly used water-insoluble urea compounds for turf are urea formaldehyde (UF) and isobutylidene diurea (IBDU). UF (38-0-0) fertiliser is often packaged with about 25 per cent of nitrogen in the soluble form to provide a quick turf response. Nitrogen is released from the insoluble fractions of UF through microbial breakdown and is dependent upon temperature. It is not very effective when temperatures fall below 10°C. The rate of nitrogen release from UF is controlled by the size of particles and the fertiliser chemistry. A smaller particle size increases the number of particles, thus increasing the surface area of the fertiliser that is exposed to microbial activity. IBDU (31-0-0) releases nitrogen as the particles slowly dissolve in soil moisture. Since release is not influenced by microbial breakdown or temperatures, IBDU is good for cool-season use. Like UF, the rate of nitrogen release depends on the size of the particle. Turf response may take as long as four weeks, but can last as long as sixteen weeks. Turf managers, often anxious for a quick response, sometimes add soluble nitrogen to IBDU for immediate impact.

Timing of the application of the slow release nitrogen fertilisers is important to the effectiveness. The use of UF and IBDU is preferred in cool climate for late spring and late autumn applications, with IBDU showing the better quality and reduced mowing (Szmcak and Lemaire, 1985). Slow-release sulfur coated urea (SCU) is produced by spraying molten sulfur onto granules of urea. The rate of nitrogen release from SCU is determined by the thickness of the sulfur coating and the particle size. The nitrogen analysis varies but 32-0-0 is a common formulation, and sulfur (12 to 22 per cent) is available to the plant. Response to common turf formulations of SCU lasts eight to ten weeks. SCU usually provides a better initial growth response than other slow-release fertilisers and the cool season activity of SCU is acceptable. It is the least expensive slow-release synthetic organic fertiliser per unit of nitrogen.

Resin-coated materials are produced by coating urea, or other compounds, with a water-permeable plastic resin. These products usually are 26 to 34 per cent nitrogen. The permeability of the coating is sensitive to temperature. Lower temperatures slow nitrogen release by causing the resin to contract, so resin-coated fertilisers are primarily for warm-season use. The coating is also subject to breaking during handling or when going through a spreader. The material should be handled carefully as the urea released from the broken pellets acts as a soluble fertiliser so the treatment pattern may be uneven. Resin-coated fertilisers do not cause turf to produce a flush of growth. Instead, the initial response is a gradual increase in growth without the rapid spurts common with other fertilisers. This growth can be sustained for up to six months. Resin-coated fertilisers are the most expensive nitrogen source.

Two common sources for phosphorus are superphosphate (0-15-0) and triple superphosphate (0-45-0). Monoammonium phosphate (11-48-0) and ammonium phosphate-sulfate (16-20-0 plus 15 per cent sulfur) are common nitrogen fertilisers that are high in potassium. Common potassium fertiliser sources are potassium chloride or muriate of potash (0-0-60) and potassium sulfate (0-0-50). Both are inexpensive. Often a nitrogen plus potash fertiliser product mixture is used to ensure that both nutrients are available to the plant.

The primary plant nutrient elements (N, P and K) are not always applied individually. When all three are in one fertiliser product, it is called a complete fertiliser. In a complete fertiliser, nutrients are balanced in ratios to each other depending upon the local climate, soils and grass needs. A starter or preplant fertiliser may have a ratio of about 1:2:2, which might be an analysis such as 5-10-10 or 10-20-20. The nitrogen is low, while the phosphorus and potassium are high to stimulate seedling development. A common maintenance ratio for complete fertilisers might be about 3:1:1, which might be an analysis such as 15-5-5. Plants generally contain nitrogen, phosphorus and potassium in a ratio of 3:1:2, and this ratio can be applied as a 15-5-10 fertiliser. Common sulfur sources are ammonium sulfate (24 per cent sulfur), potassium sulfate (18 per cent sulfur), sulfur-coated urea (12 to 22 per cent sulfur), ferrous sulfate (19 per cent sulfur), and elemental sulfur (99 per cent sulfur). Iron can be applied either as a salt or in a chelated form. Salts include ferrous sulfate (20 per cent iron) and ferrous ammonium sulfate (15 per cent iron with six per cent nitrogen and 16 per cent sulfur). Chelates are chemicals that bind iron to prevent insoluble compounds from forming, while still allowing uptake by plants. Modifying the pH of alkaline soil with repeated sulfur use or an acid soil with lime can cause a turf response from released iron.

Fertilising for turf performance

Maintenance practices, including turfgrass nutrition, are used to ensure the safety, playability, aesthetics and durability that is required of high use athletic facilities.

Sports turf nitrogen

A suggested tissue target range for nitrogen is 2.8 to 5.5 per cent. There are analytical tests for soil nitrogen, however they are seldom reported because of the dynamic nature of nitrogen in the soil. Nitrogen levels vary rapidly in the soil and a test value reported one day may be completely different 24 hours later. Soil nitrogen only accumulates as a part of organic compounds. Breakdown of organic compounds produces ammonium (NH_4^+) and nitrate (NO_3^-) which are plant available forms of mineral nitrogen. Ammonium and nitrate-nitrogen are quickly used by the plant or lost from the soil by leaching or through denitrification under anerobic conditions. Since mineral nitrogen is not stored in the soil in a plant available form, it does not lend itself to a testing procedure that predicts the potential availability of nitrogen. Application of nitrogen is based on the turf manager's skill and past experience with nitrogen application. Turf colour and rate of growth generally influence the decision to apply nitrogen fertiliser. Increasing nitrogen application, within limits, increases the above ground biomass on sports fields. This is beneficial in that it creates cushion and mat that add resilience to the surface and keep the athlete from contacting the soil below. When the grass biomass is sufficiently reduced and soils are exposed, turf growing conditions and field playing conditions rapidly deteriorate. When the protective mat of grass is lost, soil aggregation near the surface is greatly reduced by grinding foot traffic under dry conditions. When wet, soil structure further deteriorates by smearing of the soil surface. Loss of turf cover followed by lack of soil structure, increased compaction, and a wetter surface can lead to quagmire conditions and games played on slippery fields. Wear tolerance of cool-season grasses improves at 200 to 300 kilograms of nitrogen per hectare per year (Turner and Hummel, 1992). On both sand and soil sports fields, nitrogen application levels below 100 kilograms of nitrogen per hectare per year are insufficient to sustain enough turf biomass under high traffic conditions. The optimum application rate on perennial ryegrass for wear tolerance is 289 kilograms of nitrogen per hectare per year (Canaway, 1984a). Application rates of up to 48 kilograms of nitrogen per hectare per month increased root and rhizome growth in warm-season grasses (Turner and Hummel, 1992).

Increasing nitrogen rates increased the above ground biomass of cool-season grasses. At the upper limit, 625 kilograms of nitrogen per hectare per year, divided into eight equal applications three to four weeks apart, decreased total biomass particularly on soil fields compared to sand construction. Tiller numbers increase over the entire range of nitrogen applications with dry weight per tiller decreasing as nitrogen increases. Root biomass decreased with increasing nitrogen, with most roots found at zero to five centimetres. The root-to-shoot ratio decreased with increasing nitrogen, thus allowing for increased cushion while reducing the recuperative potential of the grass (Canaway, 1984b). The optimum nitrogen application for shear strength of perennial ryegrass is 225 kilograms of nitrogen per hectare per year (Canaway, 1984c). Nitrogen application not only increases biomass, it also promotes a greener appearance of the turf. Greener turf is not always better especially if the colour is from excessive or improperly timed nitrogen applications. Since nitrogen tends to increase top growth, under rapid growth conditions shoots take priority over roots and rhizomes. Excessive spring nitrogen causes rapid shoot growth and additional mowing. Plants enter the summer stress period with reduced root development and increased succulence, and are more disease susceptible (Street, 1986). Carrow and Petrovic (1992) suggest that turf managers should not attempt to use excessive nitrogen to compensate for slow growth and thin turf if the cause for these responses is compacted soil conditions.

Sports turf phosphorus

Phosphorus is used by plants in relatively high quantities. In most forms it is slowly soluble, resists leaching, and remains in the rootzone.

Phosphorus can become unavailable to the plant if pH gets too high or too low. Phosphorus deficient turf appears stunted and may show a red–purple colour beginning at the leaf tips. Test levels of 15 to 30 parts per million phosphorus in the soil and 0.3 to 0.6 per cent phosphorus in leaf tissues are sufficient. Even though phosphorus is relatively immobile in soils, it is seldom deficient in the plant under normal growing conditions because the extensive root system of actively growing grass does an excellent job of scavenging for phosphorus. Christians (1998) observed no measurable response of Kentucky bluegrass growing on soil with levels as low as seven parts per million phosphorus. Still, there is benefit to root development in applying phosphorus to new seedlings and new sod installations (under the sod) even if soil tests indicate the presence of adequate available phosphorus. Application of phosphorus immediately following core aerification will place phosphorus deeper in the soil profile as opposed to continually building phosphorus levels near the surface. Poor rooting can often occur in sports turf for reasons such as compaction, layering, excessive moisture, and others. Soils may have sufficient levels of phosphorus but poorly developed root systems may not take up adequate amounts for the plant. Analysis of plant phosphorus levels would help reveal this situation. Because of this, phosphorus is often applied as a starter fertiliser with seeding practices where root volume is initially limited. High traffic areas that routinely receive seeding and starter fertiliser may build higher than normal levels of soil phosphorus. Elevated levels of phosphorus may also benefit poorly rooted mature turf that is similarly affected in intense traffic areas. Sports turf managers should target the upper sufficiency level of soil phosphorus to ensure that this major nutrient is not limiting plant growth, especially on intense traffic areas. On cool- and warm-season grasses, phosphorus is applied at the rate of 20 to 40 kilograms per hectare per year. There is a good scientific basis for applying phosphorus once per year in conjunction with nitrogen and potassium to promote growth on sports fields (Adams and Gibbs, 1994).

Sports turf potassium

Our understanding of potassium as it relates to turfgrass has changed considerably in recent years, and consequently there has been increased use of potassium on sports turf. Specialised fertilisers containing higher ratios of potassium, and even slow release forms of potassium, have been developed. Test levels of 100 to 250 parts per million of potassium in the soil and 1.0 to 3.5 per cent potassium in leaf tissues are generally sufficient for turf growth. Yellowing of older leaves followed by tip dieback and necrosis along the leaf margin are signs of potassium deficiency. Visual observation of these deficiency symptoms is rare, but occasionally they may be observed on sand-based fields. Plants that do not show these classic signs of potassium deficiency can appear normal, but still have a mild potassium deficiency. It is assumed that there is a mild deficiency because application of potassium to apparently healthy turf has improved tolerance to various types of plant stress. Reduced wilting, ability to survive high temperature stress, and improved spring performance have been observed. The survivability of perennial ryegrass during cold winters is improved as is the ability of bermudagrass (couchgrass) to tolerate wear. Potassium increases the number and lifespan of new bermudagrass rhizomes. Maximum growth occurs at lower levels of potassium, however for sports turf and stress tolerance potassium can be applied a little heavier than for other turf. Potassium is applied at 10 to 20 kilograms per hectare per month during the growing season using the higher rate on sandy soils. Sand-based athletic fields present a special challenge for potassium availability because the extremely low CEC of sand does not let soil potassium build to an adequate test level. Sandy conditions with low CEC may require light and frequent application, 'spoon feeding', of potassium and other nutrients to ensure uptake by the plant. Slow release sources of potassium fertiliser are also beneficial for sandy conditions. Sands containing as little as five per cent clay by weight will hold substantially more potassium than sands with no clay present. Porous ceramic clays have been used as an additive to increase the water holding capacity of sands. Under normal fertility practices, topdressing of sand-based fields amended with calcined clay increases soil potassium compared to topdressing with sand only. Adding 10 per cent Profile® porous ceramic clay (by volume) to 85 per cent sand and five per cent Dakota peat caused CEC to increase by eight per cent, potassium by 104 per cent and magnesium by 41 per cent. Optimum wear tolerance occurs at 270 to 360 kilograms of potassium per hectare per month, if nitrogen is not deficient; nitrogen and potassium must be kept in balance (Turner and Hummel, 1992). The use of potassium irrespective of the nitrogen and phosphorus status reduces the water stress in turf and increases recovery from drought (Schmidt and Breuninger, 1980). Many fertilisers are specially formulated with equal amounts of nitrogen and potassium for sports fields.

Injury recovery

Turf density can be a function of the vigour and efficiency of the roots, rhizomes and tillers. Recovery from injury is most certainly a function of their vigour and efficiency. Excess phosphorus in the root zone allows the plant to utilise it. Low solubility makes phosphorus very immobile, and repeated application causes it to accumulate in the upper soil layers. This has not been a problem and may provide excess phosphorus on demand to the plant under stress. If the sports field is being resodded, putting one of the phosphate nitrogen fertilisers under the turf and on the top of the soil or sand enhances sod knitting.

Stress resistance

Turf does not usually respond visibly to added potassium. It is the increased stress resistance that is important. Drought, heat and cold tolerance are improved and disease resistance increased. Traffic is the most significant stress on sports fields, and potassium increases the turfgrass traffic tolerance (Cockerham *et al.*, 1994).

Occasional use of sulfur-containing nitrogen fertilisers (e.g. ammonium sulfate) usually takes care of the sulfur needs of sports turf. Many sports fields are built in marginal soils, including land fill, where sulfur applications may be beneficial. Sulfur fertilisers tend to lower pH due to their acidic reaction in the soil and are usually preferred for alkaline soils.

Disease incidence

High nitrogen levels necessary for sports turf may increase the incidence of some diseases, while offering resistance to others. Good internal drainage helps in disease resistance with the application of high nitrogen (Raikes *et al.*, 1997). Diseases that are less serious when a moderate and balanced level of nitrogen, phosphorus and potassium is maintained in the root zone include *Sclerotinia homeocarpa* (Dollar spot), *Laetisaria fuciformis* (red thread and pink patch), *Microdochium nivale* (pink snow mold or Fusarium patch), *Typhula incarnata* (gray snow mold or Typhula blight), *Pythium aphanidermatatum* (Pythium blight), *Erysiphe graminis* (powdery mildew), *Puccinia* spp. (rusts), *Rhizoctonia solani* (brown patch), *Magnaporthe poae* (summer patch), *Leptosphaeria korrae* (necrotic ring spot, spring dead spot), *Colletotrichum graminicola* (anthracnose), *Gaeumannomyces graminis* (take-all patch, spring dead spot), *Pyricularia grisea* (gray leaf spot), and *Bipolaris* or *Drechslera* spp. (leaf spot). When nitrogen is high in relation to phosphorus and potassium, there may be disease trouble particularly in hot weather. A high level of potassium helps reduce injury from *Rhizoctonia solani, Sclerotinia homeocarpa, Microdochium nivale, Typhula incarnata, Laetisaria fuciformis, Gaeumannomyces graminis, Phythium aphanidermatatum,* and *Bipolaris* or *Drechslera* spp. (Shurtleff *et al.*, 1987).

Turf colour

As any sports turf manager knows, nitrogen is not the only way to make turf a uniform dark green. Iron is a valuable tool to intensify green colour. If nitrogen is in adequate supply, an application of soluble iron (five to ten kilograms of iron per hectare) such as ferrous sulfate will almost always darken the colour of the turf within two to three days. Because ferrous sulfate will easily burn the turf, irrigation should follow application immediately. Application on a hot day or at a high rate will cause burn where the applicator tyres scuff the turf.

Spring start

In the early spring, it is often possible to 'jump start' *Cynodon dactylon* L. and the hybrids (bermudagrass, couchgrass) with a one-time application of 40 to 60 kilograms of nitrogen per hectare in soluble form (Cockerham *et al.*, 1988).

Late season fertilisation

Late season nitrogen applied before warm-season grasses go dormant improves autumn colour retention (Gibeault *et al.*, 1997). Nitrogen applied before the last mowing of cool-season grasses improves spring greenup. The roots of warm-season grasses are still capable of absorbing nitrogen while the plant is dormant. Increased late season nitrogen does increase freezing injury of the turf (DiPoala and Beard, 1992). Use of a complete (N-P-K) fertiliser for late season sets the turf up for better spring greenup (Cockerham, 1989).

Winter use of fertilisers depends upon the growth of the turf. As soil temperatures at five centimetres drop below 10°C, warm-season grasses such as bermudagrass and zoysiagrass stop needing fertilisers. For the cool-season grasses, the low soil temperature is around 5°C. When the soil temperatures begin to rise in the spring and pass these temperatures, it is time to begin fertilisation again.

Clippings

Leaving turfgrass clippings on the turf has a beneficial effect by returning nutrients to the soil, while contributing little to the formation of thatch. Wear on intensely used sports fields reduces the ability of the grass to take-up nutrients and thus their recycling back to the soil in plant residues. High traffic areas tend to need more nitrogen than areas of the field with less traffic (Adams and Gibbs, 1994).

As previously indicated, clippings usually breakdown quickly, especially when earthworm activity is good or when they are topdressed frequently. There have, however, been some observations on intensely managed sand-based fields where a 'slimy' layer develops at the upper most horizontal surface of the field. This is easily observed by placing all of your fingers in the turf surface and making a scratching motion. If present, there will be a slick or slimy feeling at your fingertips. Closer inspection below the turf canopy usually reveals a thin wet mat of clippings on the surface. In moderation this layer does not cause any adverse growing or playing conditions. When excessive, some managers feel that the slick layer seals the field from air movement and makes playing conditions more slippery. The contribution to this problem from clippings has not been documented, but this observation on intensely managed fields has been increasing. It seems to be more of a problem on cool-season grasses rather than warm-season grasses. Regardless of the cause, the following management practices reduce this problem: light frequent topdressing; temporary clipping collection; spiking, slicing or coring; and infrequent watering to dry the surface.

Gypsum

Gypsum ($CaSO_4$) is often applied but seldom needed in sports turf. The classic misunderstanding with gypsum arises from its association with improving water movement and soil structure on sodic (high sodium) soils. The CEC sites in sodic soils are dominated by Na^+. Other cations that help soil aggregation, such as Ca^{++} and Mg^{++}, are displaced by Na^+. The deflocculation that occurs in sodic soils results in a very tight arrangement of individually dispersed soil particles saturated with Na^+. Macroporosity is greatly reduced and water infiltration slows to near zero when wet sodic soils are slick, sticky, and have poor drainage. When dry they become quite hard. Gypsum is correctly used to remedy this situation. The Ca^{++} in gypsum ($CaSO_4$) displaces Na^+ on the exchange site. The Na^{++} reacts with sulfate (SO_4^-) to form sodium sulfate (Na_2SO_4); a highly water-soluble material that is leached from the soil. Removing Na^+ and replacing Ca^{++} on the exchange site reduces deflocculation and allows natural aggregation of particles that eventually restores soil structure. Gypsum is very useful when soil structure deteriorates because of high Na^+. The misconception arises when there is a belief that gypsum can improve structure and drainage in any heavy clay soil, even those not necessarily affected by Na^+. A Na^+ impact on soil structure that requires the application of gypsum only occurs on a small percentage of sports turf soils. A soil test will determine the need for gypsum application. The problematic symptoms of sodic soils are very similar to those of heavily trafficked clay soils that are not affected by Na^+; both are hard and have poor structure and drainage. To add confusion gypsum is often advertised as a 'soil softener' material. Most sports turf managers should not anticipate a reduction in compaction and improved drainage by using gypsum. Even with this misconception, there are situations where gypsum is useful in sports turf. Gypsum ($CaSO_4$) can be used to supply calcuim. When the pH is above 6.7 and calcium is deficient, gypsum instead of lime ($CaCO_3$) should be used to supply calcium. Lime applied to an already high pH would further increase pH and may lead to iron deficiency. Gypsum supplies calcium without increasing pH. A suggested target range for calcium in the plant is 0.4 to 1.2 per cent.

Many water supplies are often high in Na^+. Sand-based systems irrigated with high Na^+ water may have excessive Na^+ on the exchange complex. Since sands do not deflocculate, the high Na^+ in this case will not result in reduced drainage. Sands retain their macroporosity through particle size arrangement rather than by aggregation of particles. The high Na^+ irrigation water can easily displace Ca^{++} and make it deficient in sandy soils with low CEC. Gypsum can be used in this case as a source of Ca^{++}. Testing both soil and plants associated with sand-based sports turf has revealed that apparently adequate levels of Ca^{++} in the rootzone have produced apparently deficient levels of Ca^{++} in the plant. Application of gypsum in these situations increased plant calcium and improved turf growth (York, 1998). Calcium availability, uptake, and effect on turfgrass performance in athletic fields continues to be evaluated.

Fertilisers plus pesticides

Combining pesticide and granular fertiliser into a single application product has become especially popular for home lawn and golf turf. The concept is to be able to apply herbicides, fungicides, and insecticides for particular regional problems using the fertiliser as a carrier. These products are expected to remove the need to make decisions in turf care. Early mixtures of herbicides and fertilisers were effective for broadleaf weed control and developed an enthusiastic following among homeowners (Switzer, 1969). The germination of the grassy weeds, the activity of fungal diseases, and the injury from insects do not usually occur conveniently at the same time. In the combination products, some pesticides may be applied on occasions when they are not needed, increasing the cost and environmental risk.

The most effective combinations have been with herbicides and fertiliser. Post-emergent crabgrass herbicides work well in such products with some benefit over the herbicide alone (Neal *et al.*, 1995). Although selectivity may be decreased, it is not unusual to observe increased efficacy of the herbicide with the combination. Sports turf managers should be aware that combinations of fertiliser and most pre-emergent herbicides may completely inhibit germination of turfgrass seed. Pre-emergence herbicides usually remain active in the soil for six to 26 weeks. Siduron is the only pre-emergence crabgrass herbicide that can be used at time of seeding on cool-season turfgrasses. Corn gluten meal with ten per cent nitrogen has shown to have fertiliser as well as herbicidal properties in applications to turfgrasses. The use of a natural fertiliser with a natural weed inhibiting compound has considerable attraction to those restricted from using synthetic fertilisers and pesticides (Christians and Liu, 1992).

Turf managers motivated by the need to conserve energy, labour and perhaps, money, have been interested in the potential of mixing liquid fertilisers with herbicides and applying them together. Testing shows that some herbicides and fertilisers are quite compatible, some mixtures increased the phytotoxicity of the herbicide, and some decreased the phytotoxicity of the herbicide (Sander *et al.*, 1987). The turf manager is best served to understand the challenges of compatibility in mixing herbicides and fertilisers without close evaluation. Fungicides and insecticides are most effective if they have systemic properties. Systemic materials are absorbed by the plant and translocated throughout. The theory of systemic pesticides is for them to be persistent and active for a period of time even after irrigation and mowing. The activity is usually a few weeks.

Cost of fertilisers

Fertiliser is the lowest cost resource for the return-on-investment that can be put into a sports field. To determine the actual application cost, the nutrient content of the fertiliser product is used. Ammonium sulfate fertiliser has the analysis of 21-0-0 containing 0.21 kilograms of nitrogen per kilogram of product. To apply 50 kilograms of nitrogen per hectare, it is necessary to apply 238 kilograms ($50 \div 0.21 = 238$) of ammonium sulphate. Urea fertiliser has the analysis of 45-0-0 containing 0.45 kilogram of nitrogen per kilogram of product. To apply 50 kilograms of nitrogen per hectare, it is necessary to apply 111 kilograms of urea ($50 \div 0.45 = 111$). To get the per unit nitrogen cost for either product divide the price per kilogram by the percentage of nitrogen. Compare the price per unit of nitrogen for the relationship between the products.

Cost per unit nitrogen applied to turf is only one consideration as to resource investment. Urea has twice the amount of nitrogen as ammonium sulfate, therefore to apply the same rate of nitrogen it only takes half as much urea. Shipping and handling would be less. Slow release fertilisers are considerably more expensive than soluble materials. Yet the concept of using slow release fertilisers is to reduce the number of applications, an advantage in labour, time and scheduling, handling and storage.

Calibrating granular application equipment

Equipment manuals should be consulted for proper use and calibration of equipment. Fertiliser products may also list recommended settings for different models of popular applicators. These are helpful guides to start the calibration process. An excellent review of equipment calibration and mathematics used in turfgrass maintenance is presented by Christians and Agnew (1997). Keeping accurate records of calibration results will make it easier when recalibration of products is needed. Be sure to note the equipment model, operator and material used. Variables that affect granular application are meter opening, swath width and ground speed. The meter opening can be precisely adjusted, and be changed to allow more or less fertiliser through the opening. The swath width and ground speed should remain constant.

A constant ground speed can be maintained with motorised equipment by using a speedometer or fixed position governor. Ground speed will vary with hand-pushed applicators, and calibration should be done separately for each operator (Figure 8.2). Swath width depends on the type of spreader. Accurate calibration is not possible unless ground speed and swath width remain constant. The two basic types of granular fertiliser applicators are drop-type (gravity) spreaders and rotary-type (centrifugal) spreaders. With a gravity spreader, granular material drops vertically from a hopper and falls between the wheels of the spreader. The effective width of application for drop spreaders is fixed by the width of the hopper and will not change with different granular materials. Even with this, obvious fertiliser streaks may occur because of improper wheel overlap. Make several side by side passes on a hard surface just as if you were applying the material to turf. Observe the pattern of the dropped material to determine the effective width and how much wheel overlap is needed to produce uniform coverage. The distance between the wheels may not always be the effective width.

Figure 8.2 Accurate application of fertiliser with centrifugal-type applicator requires prior knowledge of meter opening, swath width and ground speed

With rotary-type spreaders, fertiliser falls from a hopper onto a spinning centrifugal disk that throws the fertiliser in a wider and more horizontal pattern. The dispersal width depends on the speed of the spinning disk. The centrifugal disk can be hand cranked, ground driven or motor driven. Rotary spreaders do not apply a constant amount of material across the entire width of application. More material is applied toward the centre and less at the edges. The effective width of application is about 65 per cent of the maximum width of the granular material spread. To help determine the effective width of rotary spreaders place a row of collection containers spaced 30 centimetres apart and perpendicular to the spreader's direction of travel. The effective width is needed to properly space each pass when operating the spreader and it will also be needed to calculate coverage area during calibration. Each container should have the same dimensions; a cigar box works well. To keep the containers level conduct this part of the test on a paved surface, not a grass area. Partially fill the hopper, set the gauge to mid range, and operate the spreader across the row of containers at normal speed. Several passes may be needed to collect enough material in each container. Visually observing the amount of material in each container may be sufficient to estimate the effective width, but weighing is more accurate. After determining the effective width, you are ready to calibrate the spreader by measuring the amount of material applied to a given area at a specific spreader setting. The approach used may vary at this point but the basic idea is to measure the amount of material delivered to a known area. Select a test coverage area that is large enough to operate the equipment in a normal manner and collect a measurable amount of fertiliser. For a push type rotary spreader 100 square metres works nicely. Next, determine the travel distance that when multiplied by the effective width will give you a coverage area of 100 square metres. For example, if the effective width of a rotary spreader is three metres then a travel distance of 33.3 metres will be required to cover an area of 100 square metres ($3 \text{ m} \times 33.3 \text{ m} = 100 \text{ m}^2$). Use a ten square metre coverage area for drop type spreaders since they have a smaller effective width. For example, the travel distance required for a 0.5, 0.75 and 1.0 metre spreader is 20, 13.3 and 10 metres respectively (coverage area = effective width x travel distance).

The next step is to operate the spreader over the coverage area and determine how much granular material was dispensed just on the coverage area. Here you have some options. Collection adaptors can be purchased or fabricated for both drop and rotary spreaders. Collection and measurement of fertiliser is simplified with these devices and they are highly recommended. Another method for drop spreaders is to catch the fertiliser on plastic or sweep it up from a hard surface. For example, a desired travel distance of twenty metres could be made with ten passes over a two metre length of plastic. Rotary spreaders cover too much area for collection of fertiliser. Simply load the rotary spreader with a known weight of fertiliser. Operate the spreader over the test coverage area. Weigh the remaining fertiliser in the spreader and subtract that amount from the starting weight to determine the amount used per 100 square metres. Regardless of the collection method used, the end result will be the weight of fertiliser applied to 100 square metres for a given spreader setting. Repeat the process at different spreader settings and record the material applied at each setting. Careful calibration is recommended for the entire spreader range. Spreader settings are not necessarily linear and therefore half of a particular application rate may not coincide with half of the original spreader setting. Calibration for a range of spreader settings is particularly useful as it is a common practice to apply half the recommended rate in two different directions. To minimise application skips apply half the recommended rate in one direction and then turn the spreader in a direction 90° from the initial application and make a second application.

Example calibration problem

Suppose you have a push-type rotary spreader and would like to apply 50 kilograms of nitrogen per hectare. In the fertiliser storage room you locate a 20-5-10 bag of fertiliser with no other markings on the bag. You decide to apply a half-rate in two different directions to reduce skips. How do you make the proper fertiliser application?

- Check the spreader to make sure all parts are working.
- Load the spreader about half full and practice your walking speed. Become familiar with the spreader by opening and closing the delivery gate as you continue to walk.
- Determine the effective width for calibration of the spreader as indicated above. After operating the spreader on a bare soil surface you determine that the spreader throws fertiliser a total distance of five metres and you estimate the effective width to be 3.25 metres ($5 \text{ m} \times 0.65 = 3.25 \text{ m}$). You place collection boxes as

described above and find that most of the material falls in a 3.3 metre swath. You settle on 3.3 metres as the effective width for this spreader and material.

- Determine the travel distance required for a 100 square metres coverage area. A 30.3 metre travel distance is required for a 3.3 metre effective width and a 100 square metres coverage area ($100\ m^2 \div 3.3\ m = 30.3\ m$).
- Operate the spreader over the coverage area at several different settings and record the amount of material collected each time. The following settings and application rates were determined.

Table 8.2 The weight of fertiliser spread on a coverage area and the corresponding application rate.

Spreader setting	*weight of 20-5-10 fertiliser collected on coverage area ($g/100\ m^2$)*	*corresponding application rate ($kg\ N\ ha^{-1}$)*
closed		
D	1000	20
F	1200	24
H	2000	40
J	3500	70
L	4200	84
N	5000	100
open		

- Determine how much 20-5-10 fertiliser is required to apply the recommended rate of 50 kilograms of nitrogen per hectare. To determine this, divide the rate desired (50 kilograms of nitrogen per hectare) by the percent nitrogen on the fertiliser bag expressed as a decimal (20 per cent or 0.2). Thus, $50 \div 0.2 = 250$ kilograms. Therefore 250 kilograms per hectare of the 20-5-10 fertiliser is required to supply 50 kilograms of nitrogen per hectare. Divide the recommended rate in half as you decided to apply a half-rate in two different directions to avoid fertiliser skips. Half of the recommended rate would be 125 kilograms per hectare (or 1250 grams per 100 square metres) of the 20-5-10 fertiliser (or 25 kilograms of nitrogen per hectare). To achieve the desired half-rate of 25 kilograms of nitrogen per hectare place the spreader setting slightly more open than 'F' on the adjustment gauge.
- Properly apply the fertiliser by walking at the same consistent speed used in the calibration process. Using the effective width of 3.3 metres, after each pass, move the spreader over 3.3 metres from the centre of the tyre tracks. After application of a half-rate to the entire area turn the spreader perpendicular to the direction of travel and apply the second application of a half-rate.

References

Adams, W.A. and Gibbs, R.J. (1994). 'Natural turf for sport and amenity: science and practice', *CAB International*, Wallingford, Oxon, UK, p. 27.

Canaway, P.M. (1984a). 'The response of *Lolium perenne* (perennial ryegrass) turf grown on sand and soil to fertiliser. I. Ground cover response as affected by football type wear', *J. of the Sports Turf Research Institute*, 60, 8–18.

Canaway, P.M. (1984b). 'The response of *Lolium perenne* (perennial ryegrass) turf grown on sand and soil to fertiliser. II. Above ground biomass, tiller numbers and root biomass', *J. of the Sports Turf Research Institute*, 60, 19–26.

Canaway, P.M. (1984c). 'The response of *Lolium perenne* (perennial ryegrass) turf grown on sand and soil to fertiliser. III. Aspects of playability—ball bounce resilience and shear strength', *J. of the Sports Turf Research Institute*, 60, 27–36.

Carrow, R.N. and Petrovic, M.A. (1992). 'Soils, soil mixtures and soil amendments' in *Turfgrass ASA Monograph No. 32* (eds. D.V. Waddington, R.N. Carrow and R.C. Shearman), American Society of Agronomy, Madison, WI., pp. 285–330.

Catrice, H., Grandchamp, G., Pierrang, M., Duc, P. and Bondeux, F. (1993). 'Turf for horse racing: turf nutrition by simultaneous analysis of soils, leaves, and roots', *J. of the International Turfgrass Society*, 7, 553–9.

Christians, N.E. (1998). *Fundamentals of turfgrass management*, Ann Arbor Press, Chelsea, MI, USA. In Press.

Christians, N.E. and Agnew, M.L. (1997). *The mathematics of turfgrass maintenance*, Ann Arbor Press, Chelsea, MI, USA, 149 pp.

Christians, N.E. and Liu, D.L. (1992). 'The use of corn gluten meal as a natural "weed and feed" material for turf', *Hortscience*, 27, 587.

Cockerham, S.T. (1989). 'Late season sports field care', *Grounds Maintenance*, 24, 11:30–2.

Cockerham, S.T., Gibeault, V.A. and Borgonovo, M. (1994). 'Effects of nitrogen and potassium on high-trafficked sand rootzone turfgrass', *California Turfgrass Culture*, 44, 1, 2:4–6.

Cockerham, S.T., Gibeault, V.A. and Leonard, M.K. (1988). 'Fertilizing high-traffic turf', *Sportsturf*, 4, 10:21–4.

DiPaola, J.M. and Beard, J.B. (1992). 'Physiological effects of temperature stress', in D.V. Waddington *et al.* (eds.), *Turfgrass Agronomy*, 32, 231–67.

Gibeault, V.A., Cockerham, S.T., Autio, R. and Ries, S.B. (1997). 'The enhancement of zoysia winter colour', *J. of the International Turfgrass Society*, 8, 445–53.

Jones, B.J., Wolf, B. and Mills, H.A. (1992). Plant analysis handbook II. Micromacro Publishing, Inc., Athens, GA. pp. 6–10, 112–18, 112-115.

Killorn, R. and Miller, G. (1986). *Management of calcareous soils*, Iowa State Univ. Ext. Bul. Pm-1231.

Minner, D.D. and Butler, J.D. (1984). Correcting iron defi'ciency of Kentucky bluegrass, *Hort Sci.*, 19:109–10.

Neal, J.C., Morse, C.C. and Macksel, M.T. (1995). 'Postemergent crabgrass control with granular herbicides', *Proceedings of the 49th Annual Meeting of the Northeastern Weed Science Society*, 49, 75.

Raikes, C., Lepp, N.W. and Canaway, P.M. (1997). 'The effect of different nitrogen application levels on disease incidence on five different soccer pitch constructions', *J. of the International Turfgrass Society*, 8, 517–31.

Sander, K.I., Burnside, O.C. and Bucy, J.I. (1987). 'Herbicide compatibility and phytotoxicity when mixed with liquid fertilisers', *Agronomy J.*, 79, 48–52.

Schmidt, R.E. and Breuninger, J.M. (1980). 'The effects of fertilisation on recovery of Kentucky bluegrass turf from summer drought', *Proceedings of the 4th International Turfgrass Research Conference*, 4, 333–40.

Sheard, R.W. and Beauchamp, E.G. (1985). 'Aerodynamic measurement of ammonia volatilisation from urea applied to bluegrass-fescue turf', *Proceedings of the 5th International Turfgrass Research Conference*, 5, 549–55.

Shearman, R.C. (1989) 'Improving wear tolerance of sports turf', *Grounds Maintenance*, 24, 2:84–106.

Shurtleff, M.C., Fermanian, T.W. and Randell, R. (1987). *Controlling turfgrass pests*, Reston, Prentice-Hall, Englewood Cliffs, N.J.

Street, J.R. (1986). 'Athletic field fertilisation', *Proceedings of the 56th Annual Michigan Turf Conference*, 15, 120–6.

Switzer, C.M. (1969), 'Control of broadleaf weeds in turf', *Proceedings of the 1st International Turfgrass Research Conference, Harrogate*, 1, 412–20.

Szmczak, E. and Lemaire, F. (1985). 'Effect of four ternary fertilisers containing slow-release nitrogen and timing of application on *Lolium perenne* L. turf', *Proceedings of the 5th International Turfgrass Research Conference*, 5, 533–47.

Turner, T.R. and Hummel, N.W. Jr. (1992). 'Nutritional requirements and fertilisation', in D.V. Waddington *et al.* (eds.), *Turfgrass. Agronomy*, 32, 231–67.

York, D. (1998). Personal communication.

CHAPTER 9

Turfgrass machinery and equipment operation

R.G. LETTNER, *Agriculture Department, Fairview College, Fairview, Alberta, Canada*

Introduction

The scope of machinery and equipment presented in this chapter is limited to units that are required in the maintenance of turfgrass on a regular basis. Operation and service manuals are the preferred reference source for detailed and specific information. They should be accessible at all times for equipment operators and service technicians. Integrating the features of specialised equipment with turfgrass cultural requirements has been undertaken. The outcome is a practical discussion to assist in evaluating turfgrass cultural practices and the machinery and equipment available to perform particular tasks.

Sport and recreation facilities include intensively managed turfgrass areas. A smooth, uniform turfgrass surface of high density is important for playability. Ball roll, ball control and athlete manoeuverability are influenced by the surface condition of the turf. The high expectations of near flawless performance by athletes cannot be jeopardised by a less-than-ideal playing surface. Professional turfgrass managers are responsible for professional, amateur and recreational sports turf facilities. Competitiveness drives athletes and turfgrass managers alike in their pursuit of perfection. The human and physical resources available to the turfgrass manager directly influence turfgrass quality. The selection, operation and servicing of turfgrass equipment are integral requirements of successful turfgrass management.

Evolution of turfgrass equipment

The evolution of professional turfgrass equipment has been influenced by a quest for superior turfgrass playing conditions and aesthetics, the environmental impact of turfgrass equipment, available financial resources, and sources of new technology.

The ability of turfgrass mowers to cut at lower mowing heights has dramatically increased the challenge of playing many sports. Lightweight mowing units with precise manoeuverability are effective in establishing visually appealing turfgrass striping patterns. Turfgrass density, uniformity and smoothness have been improved by incorporating the use of specialised rollers, groomers, vertical mowers and topdressers into maintenance programs. Core cultivation and aeration equipment reduces soil compaction, improves air exchange within the soil, and assists in rectifying drainage problems. Turfgrass root depth and root mass are increased. Turfgrass plants are healthier and stronger; consequently they are better able to survive the stress to which intensively maintained sports turf is subjected.

Turfgrass equipment is required to meet increasingly stringent environmental standards. Exhaust emission and noise restrictions have prompted the development of electric battery powered machinery. Cleaner burning and more fuel-efficient internal combustion engines are used, and alternate fuels such as natural gas are being promoted. Hydraulic and lubricating oils can be formulated to be environmentally acceptable. Sprayers and spreaders that accurately apply and monitor fertiliser and pest control product applications are available. Spot treatments can be used as an alternative to general applications.

Professional turfgrass equipment is expensive in a competitive market. It is being sold into an industry that is overall financially well resourced. Value for equipment investment is dependent on safety and operation features, reliability, serviceability and longevity. Research and development expenses must be factored into the equipment cost. Microchip computer technology has been introduced into the operation, troubleshooting and servicing of turfgrass equipment. Operator safety and comfort are attended to with interlock switches, the design and layout of control panels and roll-over protection to name a few. Lightweight yet durable machinery that can be easily serviced and is well supported through an efficient parts distribution network is preferred.

Mowing equipment

The most basic turfgrass cultural practice is mowing. Regular mowing helps maintain a healthy, dense sward and provides good playability. The maximum range of 30 to 50 per cent turfgrass shoot growth should be removed at one time. For example, a turfgrass stand maintained at a height of 25 millimetres (1 inch) would be cut when it reached 38 to 50 millimetres (1.5 to 2.0 inches) respectively. Lowering the mowing height increases the mowing frequency and reduces the photosynthetic leaf area necessary for growth. Turfgrass plant health and vigour are improved when the cutting height is kept to the maximum acceptable height for sports turf playability. Frequent mowing removes smaller amounts of leaf tissue and is desirable when clippings are not collected. The short leaf fragments drop to the soil surface. Leaf tissue decomposes quickly in a moist environment (Beard, 1992). The rate of decomposition exceeds accumulation at temperatures above 120°C. Clipping removal can improve turfgrass quality by preventing the temperature of the turfgrass canopy from becoming elevated. Temperature control through clipping removal should be considered during periods of turfgrass heat stress. Clippings left on the turf contain nutrients that are recycled and composting or disposal is not required. The time that would have been spent managing clippings becomes available for other activities.

Selection of mowing equipment requires thoughtful examination of the intended use(s). Three common cutting techniques are reel, rotary and flail mowing (Figure 9.1). Reel mower design and maintenance is more advanced and generates greater discussion than either rotary or flail mower technology. Hydraulic or electric motors drive the cutting heads of most reel mowers. This is a tremendous improvement from the ground driven reel mower technology where ground speed has a direct affect on the revolutions per minute (rpm) of the reel. Hydraulic and electric motors provide constant reel speed independent of ground speed. When properly set, the blades of the reel come into near contact with the bedknife (ability to cut paper with minimal reel/bedknife contact). Manufacturers specify the recommended angle of relief to be maintained on the reel blades. The relief angle is often greater when grasses with tough vascular tissue are being cut, such as perennial ryegrass (*Lolium perenne*) or zoysiagrass (*Zoysia* spp). Water contained in turfgrass leaves is the lubricant used to reduce heat buildup and excessive wear on the cutting surfaces. Reel mowers should not be operated over dry dormant grass and reels should be disengaged when the mower is being transported to prevent extensive reel damage from friction and overheating. The contact edges of reel blades and bedknives can be maintained by using a motorised backlapper. The reel is spun in the reverse direction. Steel grit (lapping compound) is brushed onto the reel blades prior to starting the backlapping. Additional lapping compound is applied until the desired cutting edge is achieved. The reel and bedknife settings require adjustment during backlapping to maintain minimal contact. Residual backlapping compound is removed once the process is complete. Reel and bedknife settings are checked and adjustments are made.

Sharpening reel mower cutting units can be undertaken using manual grinders or automated spin grinders. In either case it is essential that the cutting unit is accurately prepared for grinding. Worn bearings must be replaced prior to grinding to eliminate any deviation in the positioning of the central shaft of the cutting unit on the grinder. Properly sharpened blades (which are helical in design) must contact the bedknife along its full length after grinding. The reel may be lowered to come into contact with a stationary bedknife (reel to bedknife) or the reel may remain stationary and the bedknife adjusted to come into contact with the reel (bedknife to reel). Keep in mind that the height of cut is changed whenever bedknife to reel adjustments are made. A reel that does not spin freely is in too close contact with the bedknife. This will result in increased wear on the reel and bedknife and increased energy consumption (Beard, 1982).

Setting the height of cut for reel mowers must be undertaken in a precise fashion. It is especially important when there is more than one reel on a mower (e.g. triplex greens mower). Each cutting unit must be accurately set to the same height. When this adjustment is completed in the shop it is referred to as the bench setting. The bench setting will be slightly higher than the actual height of cut. The reason for this can be understood when consideration is given to the fact that the weight of the mower and the thatch condition of the turf will result in the mower settling down lower in the turf than it would on a hard shop surface. The height of cut is referenced in fractions of an inch (3/16 inch), thousandths of an inch (187.5 thousandths) or millimetres (4.69 millimetres). Each leaf blade is cut several times as the reel moves forward. Clippings are dispersed over the cut turf when baskets are not

Figure 9.1 Examples of (a) reel and (b) rotary cutting techniques

used. Desired cutting height and quality of cut influence the reel cylinder diameter and the number of cutting blades. Small diameter reels with closely spaced blades are used on low cut greens. Larger diameter reels with fewer blades are used for higher cut turf.

Clip of the reel is defined as the forward distance travelled between successive clips. Three factors influence the clip of the reel: number of blades on the reel cylinder, rotational velocity of the reel, and speed of forward movement. It is desirable to have the clip of the reel as close as possible to equal to the cutting height. When ground speed is excessive and the clip of the reel is greater, then the mowing height of the cut turf may appear wavy. This is referred to as marcelling. Alternatively, if the clip of the reel is substantially less than the height of cut the quality of cut will be reduced. Uncut grass leaves are not effectively pulled in by the reel blades prior to being cut. The maximum cutting height of a reel cylinder is the distance from the cutting edge of the bedknife to the horizontal centre shaft of the reel. Grass extending above the central shaft of the reel will be pushed away rather than being pulled in by the reel blades (Beard, 1982; Turgeon, 1996).

Attachments are available for reel mowers to improve the quality of cut. Rollers, groomers and brushes are used to vertically orient grass leaf blades and shoots prior to cutting. Extreme care is required when setting groomers and brushes to make sure they are not going to damage the turf as a result of being set too low. Turfgrass damage is most likely to occur when the mowing unit is turning. The side of the cutting unit opposite the direction of turn will be lower and there is potential for it to scalp the turf or more severely cut into the turf (Turgeon, 1996).

Rotary mower technology is based on a horizontal cutting blade rotating at a high speed. There is only one contact surface, the cutting blade. The impact of the cutting blade against the turfgrass leaf blade provides the cutting action. Dull blades and low rpm are the primary causes of an undesirable quality of cut. Turfgrass species with tough vascular tissue in their leaves do not cut well using a rotary mower. Leaf blades tear rather than cut cleanly. Ragged cut leaf blades form protective callus tissue slowly. Callus tissue, once formed, results in a brown discolouration to the surface of the turf.

Mulching technology has been introduced for use with rotary mowers. The design of the cutting blade and the mower housing keep the cut grass blades suspended. The cutting blade repeatedly cuts the grass blades. Small pieces of grass drop to the ground where they decompose. Mulching technology is best suited to mowing dry turf. Wet turf adheres to the interior housing of the mower. This changes the aerodynamics of the mower and reduces the capacity of the mower to mulch the cut leaf blades. Wet clippings fall to the ground in unsightly clumps that must be promptly removed to avoid discolouration of the turf below. It is imperative that mulching mower blades maintain their design integrity after being sharpened and that the mower housing be kept clean for this technology to be effective.

The rotary mower deck of larger units is best located in front of the power train. Manoeuverability is improved and the turf is cut in front of the wheels, which prevents it from being flattened prior to mowing. Mid mount (belly mount) and rear mount rotary mowers are available. They improve mower diversity when an alternate attachment (e.g. scraper blade) is attached in front of the unit without removing the mower. The height of cut is adjusted by raising or lowering wheels and rollers attached to the mower deck. Rotary mowers are not designed to cut turf lower than 12 millimetres (½ inch). The width of cut, number of blades, and deck configuration vary with rotary mowers. Scalping and gouging can occur on uneven terrain. A floating deck design reduces the risk of turfgrass damage. Belts are required to drive all but direct drive single blade rotary mowers. Belt accessibility, availability, cost and ease of installation can be influential in the selection process of a rotary mower.

Frequent cleaning of rotary mower decks is required. Wet grass clippings can clog the discharge shoot and adhere to the inside of the deck. Reducing ground speed and increasing mower revolutions per minute improve performance in wet or heavy turf. A sharp cutting edge on rotary blades must be maintained. Failure to do so will have noticeable adverse effects on turf appearance. A sharp blade will cleanly cut turfgrass and result in minimal callus formation on the cut leaf blade ends. In contrast, a dull blade will tear and shred leaf tissues. Rotary mower blades should be sharpened frequently so that the cutting edge remains free of nicks. Sharpened blades should maintain the manufactured relief angle. Ensure the blade is properly balanced after sharpening. Magnetic blade balancers are convenient. They hold the centred blade in place while allowing the blade to freely rotate vertically to determine if one side of the blade is heavier than the other. Unbalanced blades cause serious vibrations that will fatigue mower parts and cause premature wear and/or breakdown of the mower (Emmons, 1995).

Flail mowers at one time were utility mowers that provided a low quality of cut. There are commercial flail mowers designed specifically for clearing and brushing operations. Turfgrass flail mowers utilise the same principles, but on a refined basis. Improvements in the design and spacing of the cutting blades that pivot on a horizontal shaft have resulted in substantial improvements in the quality of cut that is now possible. The high-speed rotation of the horizontal shaft causes the cutting blades to swing out due to centrifugal force. Leaf clippings are repeatedly cut by the rotating blades and fall to the turf as mulch. Cutting blades coming in contact with an obstacle, such as a rock, will swing back on their pivot without causing serious damage to the blade or more importantly to the mower shaft. Sharpening flail mower blades can be weighed against installing a complete new set when the cost of labour is considered. All blades must be of the same length to achieve a uniformly cut turf. There is no minimum height of cut since the rotating knives can be set to come into contact with the soil surface. Effective turfgrass striping patterns are substantial secondary benefits of flail mowers. Single or multiple flail units are operated by the power take-off from a power train.

Accumulation of grass clippings on mowing equipment occurs due to moisture within the clippings and the manner in which they are dispersed. A pressure washer preferably, or other water supply, should be used to clean mowers after each use before clippings dry. Thorough cleaning of the surface and undersurface of mowers prevents a build-up of clippings. The cutting performance of mowers can be substantially reduced when routine cleaning is not practised.

Core cultivation and aeration equipment

Conventional aeration, which is also referred to as core cultivation, is one of the most intrusive turfgrass cultural practices. Hollow tynes or spoons remove cores from the turf. The cores left scattered on the turf surface are frequently picked up. Manual shovelling of cores remains a common practice for intensively maintained greens although mechanised removal is an option. Larger areas are more likely to have cores mechanically removed or broken up using a steel mat. Aeration, as the name implies, is intended to reduce the surface compaction of turfgrass soils and allow oxygen-depleted air from the soil to be exchanged for atmospheric air higher in oxygen.

It is very common to see increased turfgrass root depth and activity in old aeration holes. Reducing compaction and creating surface holes in the turf has the added advantage of improving water infiltration. Other reasons for core cultivation are modification of the root zone growing media by adding an amended material in place of the soil that was removed, and supplying nutrients such as phosphorus (which has very limited movement in the soil) into the growing media. Until recently the most common aeration depth was 75 millimetres (3.0 inches). It is important to realise the limited amount of turfgrass surface area affected by a single aeration. Take for example a turfgrass stand aerated using 12 millimetre (0.5 inch) diameter hollow tynes at a spacing of 50 millimetres (2 inches) on centre. The total surface area to be affected is only five per cent (Emmons, 1995). A single aeration cannot rectify serious turfgrass compaction or other related problems.

Turfgrass managers are no longer limited to using hollow tyne core aerators with fixed spacing. Features available with aerators offer a selection of spacings as a result of changing the ground speed of the aerator. A 50 millimetre (2 inch) spacing between tynes remains standard, however the distance between holes will vary with the speed the aerator travels. A selection of tyne diameters is available from 6 millimetres (0.25 inch) and up. Tynes must be monitored for wear and replaced when the depth of penetration is considered unacceptable. Hardened carbide tips on hollow tynes may be available but the substantially higher cost may not be economical. Core aerating heavy clay soils can result in cores becoming lodged within the tynes. Normally cores within the tynes are pushed out as the tynes re-enter the soil. Physical removal of lodged cores must be undertaken during aeration and also once aeration is complete. Leaving cores within tynes after aeration can result in a very long and difficult task of removing the dried cores prior to using the aerator again.

Long-term use of hollow tyne aeration has raised concern of a possible plough pan compacted layer at the soil depth of full tyne penetration. This is normally considered the 75 millimetre (3 inch) depth. Longer tynes were seen as a method of breaking through the compacted plough pan layer. Hollow tynes are not well suited to deep tyne aeration; as a result solid tynes were developed. Turfgrass managers have often been reluctant to support and adopt the philosophy of solid tyne aeration. How can a solid object being pushed into the soil (without the removal of a plug) reduce the level of soil compaction? It is important to realise that when a solid tyne enters the soil, displacement of the soil occurs in all directions. The direction of least resistance is upward at the turfgrass soil surface. Upward movement of the soil diminishes the potential for compaction along the walls of the hole. Consideration must also be given to the moisture content in the soil. Water acts as a lubricant and as such assists in tyne penetration. Too high a soil water content at the time of aeration will allow movement of soil aggregates into macropore spaces and thus can cause compaction along the walls of the holes. Soil moisture should be below field capacity at the time of aeration, in particular for solid tyne aeration. Soil moisture provides the needed lubricant for tyne penetration, yet the soil should be dry enough to fracture or produce fissures due to the force exerted by the tyne entering the soil. These horizontal and vertical fissures act as additional channels for improved gaseous exchange within the soil (Emmons, 1995).

Soil texture will dictate the approximate depth to which tynes penetrate. The force needed for tyne penetration limits the depth of deep aeration. Compacted heavy clay soils are the most difficult for deep aeration. Solid tynes 300 millimetres (12 inches) and longer are available. Knowledge of underground services is required prior to deep aeration. Irrigation lines and drain lines are the most prone to being punctured as a result of the tynes extending beyond the installation depth of these services. Water injection has provided an alternative means of unobtrusive aeration for smaller, intensive use locations such as greens. Pressurised water moving at a high velocity is dispersed through specialised injector nozzles. The water coming in contact with the turfgrass surface forms a small opening. The high velocity water is forced through the soil. Soil particles shift and leave holes. These openings become rectangular rather than circular in shape deeper in the soil. Impenetrable objects such as stones within the soil will restrict the depth to which the water will penetrate. Water will penetrate 200 to 300 millimetres (8 to 12 inches) under normal operation. Concern has been expressed that clay and silt soil fines dispersed in the water float up through the open channels to the soil surface. These fines are a factor in soil layering and can partially seal the soil surface restricting water infiltration rates.

The quantity of water applied will cause the turfgrass surface to temporarily rise. Rollers are an integral part of a water injection unit and are used to smooth the turfgrass surface and provide a playable surface immediately after the turf has been aerated. The typical water supply for water injection aeration is through an irrigation system. A hose can be attached to a quick coupler valve, or some alternative modification to the irrigation system is undertaken to provide water. Clean water is essential to minimise wear on the injection components. Filters are installed at some point along the hose between the irrigation supply and the water injection aerator. Additional water filters are located on the aerator. All water filters must be routinely inspected. The frequency of filter changes will be dictated by water quality.

Adequate water flow at an acceptable pressure must be maintained for the water injection aerator to perform within design specifications. Consideration has been given to diversify the use of water injection aeration to include application of nutrients and/or pesticides directly into the soil. The formulation of these products for use in water does not guarantee the absence of abrasive materials. Abrasives will prematurely wear the injection components. Caution is advised. Read the warranty information to determine the extent of acceptable use.

Topdressing equipment

Topdressing improves the turfgrass surface condition in addition to assisting in the control of thatch. The topdressing material applied combines with the thatch (organic dead and decaying turfgrass plant parts). The integrity of a uniform soil texture throughout the surface soil layer is necessary. Soil layering causes a disruption in water infiltration and movement through the soil profile, and may result in more serious problems such as black plug layer. Uniform, light and frequent topdressing is desirable in response to improved turfgrass playability and thatch control. Topdressing is also used to fill aeration holes and bring the turfgrass back into a playable condition (Turgeon, 1996). At one time the only method of applying topdressing was manually using a shovel. This technique is time consuming and requires considerable skill to spread the topdressing material uniformly. The desire for frequent topdressing of large turfgrass areas has resulted in mechanised topdressing. Topdressers can be dedicated units or attachments to utility vehicles. Smaller units are self-propelled with operator control provided while walking beside the topdresser.

Mechanised topdressers consist of a storage hopper for the topdressing material, and a platform (conveyor belt) at the base of the hopper that moves horizontally to direct the topdressing to the back of the hopper where it is directed downward to the turf by a rotating brush. Topdressing material, often containing a high percentage of sand, is abrasive and causes wear to the mechanical drive mechanism. It is important to clean accumulated topdressing away from mechanical parts such as chains, belts, gears and bearings.

The application rate of topdressing is dependent on the ground speed and the setting (opening between the conveyor and the brush). Calibrating a topdresser can be accomplished by placing a predetermined volume of topdressing into the hopper, setting a reasonable ground speed and brush opening, and travelling until the hopper is empty. The distance travelled and the width of topdressing application are measured, and the area calculated. For example, a predetermined volume of 0.5 cubic metres of topdressing was placed in the hopper and a distance of 25 metres travelled. If the width of topdressing was 1.5 metres, then an area of 30 square metres (1.5 m × 20 m) was covered. The area was covered to a depth of 0.0167 metres or 16.7 millimetres (0.5 m^3/30 m^2). So an area of 30 square metres was covered to a depth of 16.7 millimetres.

The ground speed would be recorded and remain constant. Adjustments to the brush opening would be undertaken to change the application rate, that is the thickness of the topdressing being applied. Continue to calibrate until the desired thickness of topdressing is being applied.

Features of topdressers to consider are:

- volume of the hopper—larger volumes of topdressing held in the hopper will reduce the frequency of filling but also increase the weight
- dimensions of the hopper—the hopper should be either longer or wider than the loader used to fill it to eliminate spillage
- compacting pressure—it is important that the weight of the topdresser when full does not result in serious compaction to the turfgrass soil (the number, width and design of the tyres will influence the pressure)
- design of conveyor belt—a ribbed conveyor belt (chevron™) surface will pick up the topdressing more effectively than a smooth surface. This feature is especially important when the topdressing is damp and has a tendency to hold together.

Only high quality topdressing material should be used. It should be free of pebbles and other debris. Weed seed contaminated topdressing is to be avoided. It is normally preferred to use topdressing that has the same physical properties as the original growing media. Maintaining a consistent rootzone profile is necessary to avoid layering problems. Topdressing has also been used to amend and improve the existing growing media without taking the turf out of play. It is a gradual process that may take several years. The speed of soil modification can be increased when incorporated with a program of core aeration. The desired growing media replaces the existing material removed in the cores.

Topdressing is matted or brushed in after application. The material distributed over the turfgrass surface must be moved into the soil–thatch layer below the turfgrass leaves. The thickness of the topdressing application will dictate how aggressively it must be worked in. Metal drag mats can be pulled behind a lightweight utility vehicle. Alternatively, brushes or brooms are available for incorporating the topdressing into the soil surface. Mechanised oscillating brush attachments are very effective when used to fill aeration holes with topdressing. Topdressing with material containing a high percentage of sand can be very abrasive to turfgrass leaves. Dragging angular topdressing sand across the turf during the process of working it in is similar to rubbing sandpaper over the turf. Turfgrass leaf tissues become bruised and will be predisposed to stress factors. Quarry and manufactured sand are likely to be angular in shape. In contrast, lake and river sand is rounded due to the tumbling action of sand against sand in water.

Vertical mowing

Vertical mowing is a turfgrass cultural practice that vertically slices through the turfgrass foliar material and deeper into the thatch and soil. A series of thin cutting blades are attached along a horizontal shaft. The distance between cutting blades should be adjustable by adding or removing spacer washers that fit on the shaft between the blades. The horizontal shaft is powered to spin at high revolutions per minute. It is possible for the shaft to rotate in a forward direction (the direction of travel) or in a reverse direction (opposite to the direction of travel). When the shaft rotates in the direction of travel it assists in the forward movement of the mower. Reversing the direction of horizontal shaft rotation results in more aggressive mowing and consequently increased thatch removal from the turf. Blade replacement is required when there is no longer adequate height adjustment possible within the normal adjustment settings or when the manufactured cutting edge of the blades has worn out.

The vertical orientation of the blades provides a result similar to raking turf (Figure 9.2). Power raking is an alternate name used for vertical mowing. Mechanised vertical mowing is more effective and efficient than hand raking in cutting through stolons, lifting thatch from the turfgrass surface, and slicing through thatch into the soil. Vertical mowing shaves off raised turfgrass surface imperfections whereas topdressing fills small depressions in the turf. The result is a truer, smoother turfgrass surface. Vertical mowing can be used to break up aeration cores left on the surface after hollow tyne aeration, reducing the need for core removal (Emmons, 1995).

Groomers are a type of vertical mower that lifts up the turf to a more vertical orientation immediately prior to cutting. The beneficial result is turfgrass that is more uniformly cut and the elimination, or at least reduction, of grain (procumbent turfgrass leaves, stems and stolons). They are not intended to penetrate into thatch and thus do not lift out thatch material. Grooming attachments are installed in front of the cutting heads of reel mowers. This practice is used most often on closely cut turf and is unlikely to be undertaken at each cutting when the turfgrass is being cut on a daily basis.

Figure 9.2 Vertical mowing lifts thatch from the turfgrass surface and shaves off raised imperfections

Slicing and spiking are turfgrass cultivation practices with similarities to both vertical mowing and aeration. The similarity to vertical mowing is the type of knives used to penetrate the soil. Vertical knives cut intermittent slits through the turf, the thatch and into the soil. The advantage is that no thatch or cores are brought to the surface. Surface playability is not substantially affected. They are used for short-term improvements to soil aeration and water infiltration. Mechanical slicing incorporates a series of v-shaped knives attached to a horizontal cylinder or shaft. The length of these knives is 100 to 150 millimetres (4 to 6 inches) which allows penetration through the turf and into the soil. The rotating cylinder may be weighted or have a positive downward pressure introduced to ensure full penetration of the knives. Spiking is similar to slicing, however the knives are shorter and penetrate to a 25 millimetre (1 inch) depth. Spiking may also refer to the practice of using short solid tynes to punch holes into the surface of the turf. The spacing and orientation of knives on the rotating cylinder dictate the number of perforated openings in the turf as a result of slicing or spiking. Spiking attachments may be used on reel mowers by replacing the cutting heads. Greens and other highly manicured turf areas can be spiked during periods of heavy play and/or environmental stress. At these times it is unacceptable to perform more disruptive cultural practices. It is common for slicing and some spiking implements to be ground driven. They are hitched to and pulled by a tractor. Service and maintenance requirements consist of lubricating moving parts, inspecting for worn, broken or bent knives and replacing knives when necessary.

Rolling equipment

Rolling has long been considered a desirable practice used to smooth the turfgrass surface. Minor imperfections that negatively influence playability are removed. The roller weight and pressure exerted by it on the turfgrass soil directly influence how much the surface contour will be affected (Turgeon, 1996). The effects of soil moisture must also be considered. Water acts as a lubricant in soil. Rolling wet soil causes the soil separates and aggregates to slide together and into existing air spaces. The result is a more compacted soil. Vibrating rollers have an even greater impact on compacting soil. Rolling when the soil moisture level is between one-half field capacity and field capacity will reduce the degree of surface compaction. A compacted soil reduces the rate of water infiltration and the percentage of macropores within the soil. Macropores are important in supplying essential oxygen to turfgrass roots.

Heavy rollers are now primarily limited for use on sod production fields prior to harvest. It is critical that the surface is smooth for uniform thickness of harvested sod. Lightweight rollers are preferred for sports turf and are in common use on sand-based greens. Their use has improved the trueness of the turfgrass surface. Electric rollers can be considered when a power supply is available (Figure 9.3). Self-propelled rollers provide a similar specialised function. Triplex-attachment rollers are mounted in place of cutting heads on mowers. Caution must be exercised. The rollers are much heavier than the cutting heads. Possible damage can result to the frame of the Triplex and/or to the hydraulics that raise and lower the rollers. Drum rollers may be used manually or hitched to a utility vehicle. Roller weight can most easily be adjusted when water is used as ballast. A plug located on the side of the roller can be removed for either addition or removal of water. Water should always be drained from rollers after use and before they are stored. Steel rollers rust when water is left in them for an extended time.

Rolling has other potential applications. Intimate seed–soil contact is achieved by rolling after seeding. Soil is wetted when water forms a film that surrounds individual soil particles. Turfgrass

Figure 9.3 Electrically-powered roller used in the maintenance of bowling greens (Photo courtesy of G.W. Beehag, Globe Australia Pty Ltd, NSW)

seed that is in contact with wetted soil particles benefits from the moisture that is essential for germination. Sodded areas are also rolled to eliminate air pockets under the sod, ensure roots are in contact with the soil, and provide a more uniform turfgrass surface.

Application equipment

The application of fertiliser and pest control products is required on sports turf. Correct application rates and accurately calibrated equipment are essential in managing high quality turf. Applying an appropriate product at the correct rate using application equipment that has been accurately calibrated will minimise environmental concerns. Hand application of dry products is no longer recommended. On occasion seed and fertiliser can be applied by hand to small areas or irregular shaped areas where equipment use is impractical. A constant walking speed and a steady, rhythmic distribution technique is necessary for uniform application of the product. Most often turfgrass fertiliser and pest control products are applied using either a sprayer or a spreader. Sprayers are used for liquid applications, whereas spreaders are used for the application of dry products.

Sprayers are used for foliar and drench applications of liquid fertilisers, herbicides, fungicides and insecticides. Dependent on product formulation, the prepared solution can either consist of a product dissolved or suspended in water. Suspensions may precipitate out of solution if not agitated repeatedly or continuously during application. The efficacy of a solution can depend on water quality. Organic and mineral particles present in a water supply deactivate the active ingredients of some products thus leaving them partially or totally ineffective. The pH and mineral content of the water supply also can reduce the efficacy of some products. It is preferred to use water with a near neutral pH (7.0) and a low mineral concentration. Combining two or more products should only be considered when their compatibility is known. Knowing the compatibility of the respective active ingredients is not enough. The carrier materials may not be compatible and could render the mixture ineffective, or worse produce a precipitate in the tank. It is important that the spray volume specified on the product label be followed for maximum product efficacy. This may require the use of high volume nozzles for some applications (Vargas, 1994).

Sprayers consist of the following basic components: tank, pump, spray boom, transport and power supply. Plastic tanks with a low profile design have many advantages over metal tanks. Plastic is lightweight, does not rust or corrode, and is durable. Low profile tanks molded to fit onto utility vehicles offer improved stability and safety when operated on undulating terrain. The sprayer tank capacity should be adequate to keep the number of fills to a minimum during a normal spray program. The solution level can easily be monitored when plastic tanks are not opaque or when there is a view window on the tank.

Pumps are required to provide constant pressure. Fluctuating pressure can result from a worn pump or a sprayer system that is not capable of maintaining a constant pump rpm as a result of travelling over hilly terrain. Sprayer nozzles deliver the required flow rate, litres per minute (gallons per minute), when the specified pressure is maintained. A change in the flow rate can occur with a change in pressure, when the nozzle size changes, when worn nozzles have not been replaced, or clogged nozzles have not been cleaned (Emmons, 1995). Increasing the flow rate (larger nozzles) while maintaining a constant pressure when the diameter of the hoses remains constant causes the solution to travel at a higher velocity through the system. Increasing the velocity of a liquid in a confined vessel (hose) results in increased turbulence and a resultant reduction in pressure due to friction loss. This provides a brief explanation as to why the gauge pressure at the pump can be higher than the pressure at the furthest nozzle on a boom sprayer when product is being sprayed. Spray products can be corrosive and it is important that pumps are thoroughly flushed with clean water after each use. Water must be drained from pumps that are exposed to freezing temperatures. Cast iron and plastic pumps alike can crack when confined water expands as it turns to ice.

Sprayers are noted for providing a uniform distribution of product. Unfortunately there is also the potential for spray drift. A substantial reduction in spray drift can be realised by increasing the droplet size of the solution coming from the nozzles. Lowering pressure at the pump will increase the spray droplet size but also affect the discharge rate. The use of low-pressure nozzles can provide an acceptable alternative. Shrouds are available that attach over the booms to further reduce the amount of product drift caused by wind. Maintaining the specified boom height above ground level is necessary for accurate distribution of product. It is a challenging and sometimes impossible feat when a large sprayer is used over undulating terrain. The width of sprayer booms is variable, from a single nozzle backpack sprayer to three section booms with the two outer wings designed to break-away (swing back) when they contact an obstruction such as a tree trunk or a fence post.

Nozzles must be checked frequently for wear. Worn nozzles change the flow rate and the distribution pattern of product being sprayed. Product formulations that suspend particles in solution are most abrasive and their use will require more frequent monitoring and changing of nozzles. A second reason for frequent monitoring of nozzles is to check for partial or full blockage of nozzle orifices. When foreign debris becomes trapped in the screen or orifice of a nozzle less product will be delivered than is required, and the distribution of the product that does pass through the nozzle is unknown. Removal of nozzles should be undertaken in an area where drainage of product from the boom will not damage the turf. Caution must be used to prevent direct bodily contact with pesticides. Never blow through a contaminated nozzle in an attempt to dislodge debris causing the blockage. Nozzle orifices are easily damaged when wire or some other device is forced through to remove the foreign material. Soaking clogged nozzles in water and using fine brass wire to gently remove foreign material is preferred.

Dedicated sprayers and those installed on utility vehicles can be equipped with computerised control systems. Sprayer precision is improved when spray volume is automatically adjusted for changes in ground speed. Sprayer calibration cannot be overlooked even with sophisticated sprayer systems (Vargas, 1994). The purpose of calibration is to physically determine how much of the spray solution is being applied to a known area, such as 100 square metres (1000 square feet). It is safest to calibrate using water. Calibration must be repeated for each change in ground speed, pressure and nozzle size. Changing any one will result in a difference in the volume of product applied.

The simplest method of calibration is to collect the liquid from one nozzle while the sprayer covers the specified area of 100 square metres (1000 square feet). Multiplying the volume of liquid collected by the number of nozzles on the boom represents the total volume of liquid applied per 100 square metres (1000 square feet). The calibration should be repeated four to six times to ensure a consistent output of liquid. Any one or combination of the following can be used to adjust rates that are not satisfactory. Pressure can be increased (increase rate) or lowered (decrease rate), ground speed can be slowed (increase rate) or sped up (decrease rate), and nozzles can be changed. All nozzles used at any one time must be the same model number and have the same discharge rate. All nozzles should be checked prior to each spray application and replaced at least once per year. Consideration should be given to frequency and duration of use. Wettable powder formulations are particularly abrasive (Emmons, 1995).

Spreaders, which include seeders, are used for the application of dry products. The most commonly used spreaders are categorised as either drop (gravity) spreaders or centrifugal (rotary, broadcast or cyclone) spreaders. Drop spreaders consist of a hopper to hold the product, a mechanical rotating agitator that keeps the dry material from clumping, a series of adjustable openings at the bottom of the hopper through which the product passes, and a gate to control the release rate of the product. At least one of the two or more wheels drives the agitator. Wheels are located outside the hopper and it is imperative that wheel marks overlap on adjacent passes a drop spreader makes over the turf. Failure to do so results in narrow strips of turf that do not receive an application of the product. Too much overlap can result in an overapplication of product and visual streaking, if not phytotoxic damage to the turf. Dividing the material into two equal quantities and applying it at half rate in two directions minimises inconsistency and phytotoxicity. The distance material falls from the hopper to the turf is limited and wind does not normally alter the application rate. Drop spreaders are well suited for the application of small, light material such as grass seed and fine fertilisers or amending agents, and should be used when it is important to confine the product, such as a fertiliser or herbicide, to a specified area. Small drop spreaders can have a width of 60 to 100 centimetres (2 to 3 feet). Larger spreaders pulled by a tractor are available. To avoid excess product being applied when starting, begin forward movement before opening the hopper. When operated correctly, drop spreaders apply material uniformly (Turgeon, 1996).

Centrifugal spreaders have an impeller disc that is located beneath the hopper. Product passes through adjustable openings in the bottom of the hopper and onto a spinning disc where it is discharged outward from the spreader in a semicircular pattern. Speed of impeller rotation and the mass of the product influence the distance that the dry material travels before dropping to the ground. On push-type units, one of the spreader wheels drives the impeller disc. The speed of impeller rotation is directly related to the ground speed of the spreader. Increasing the speed of impeller rotation increases the velocity of the material as it leaves the disc. An increase in velocity causes the product to travel further out from the spreader and consequently decreases the application rate. Application of material having various particle sizes further reduces the uniformity of coverage. The majority of larger, heavier granules will travel further and be distributed farthest out from the spreader. Smaller, lighter material is first to drop to the ground. Tractor mounted centrifugal spreaders can have the impeller rotation driven by a power take-off which is independent of ground speed. Wind has a substantially greater influence on product distribution from a rotary spreader because of the velocity of the product as it leaves the spreader and the greater distance above the turf that it is released. Overlap of product is necessary when using a rotary spreader. Adjacent passes of a spreader should be overlapped 30 to 50 per cent of the diameter of product application. For example, a rotary spreader that applies product a width of 3 metres (10 feet) should have a distance of 2.1 metres (7 feet) to 1.5 metres (5 feet) between passes to provide 30 per cent and 50 per cent overlap of the diameter respectively (Turgeon, 1996).

Calibration of spreaders can be accomplished using several techniques. In all cases it is important to use care and precision. The calibration process must be repeated until settings are determined that will apply the required amount of product over a known area. Record for future reference the product name (include particle size information), application rate, walking speed (push spreaders) and spreader setting. Dry material that is not free flowing cannot be calibrated accurately. This can occur when too much moisture (humidity) is present in the product and it sticks together.

Calibrating drop spreaders using the catch pan method is convenient. It is necessary to fabricate a catch pan to be temporarily attached beneath the hopper. Cardboard and plastic materials are commonly used to construct the catch pan because they are easy to modify. The catch pan is used as a trough to catch and contain the material released from the spreader. It is placed below the full length of the hopper opening making sure there is no interference with the spreader controls. The hopper is filled with the dry material requiring calibration and operated over a specified area, such as 10 square metres (100 square feet). The material collected in the catch pan is weighed and the application rate calculated. Ensure that the capacity of the catch pan is substantially greater than the quantity of material applied (Emmons, 1995).

Application rate
= mass of material collected / 10 sq. m. (100 sq. ft.) × 100 sq.m. (1000 sq. ft.)
= g of material applied per 100 sq.m. or lbs of material applied per 1000 sq. ft.

Repeat the calibration process until the spreader setting is determined that applies the product at the required rate.

Rotary spreaders can be calibrated by placing a known weight of product in the hopper (i.e. 25 kilograms, 50 pounds) and spreading it over a specific area (100 square metres or 1000 square feet). Remove the unused material from the hopper and weigh. The difference in weight between the starting weight of material put in the hopper and the weight of the unused material is the amount of material applied by the spreader.

Application rate
= mass of material applied/100 sq.m. (1000 sq. ft.)
= kg of material applied per 100 sq. m. or lbs of material applied per 1000 sq. ft

A variation of this technique is to place a known weight of product in the hopper and spread it until the hopper is empty. Determine the area covered. The application rate will be over an undesirable reference area, for example 25 kilograms (50 pounds) per 125 square metres (1250 square feet). Convert to a standard area, such as 100 square metres (1000 square feet). So the application rate would be 20 kilograms of material per 100 square metres or 40 pounds of material per 1000 square feet.

Repeat the calibration process until the spreader setting is determined that applies the desired rate of product over a specified area.

It is important to thoroughly clean spreaders after each use. Particular attention must be given to the gate that controls the release of material located at the bottom of the hopper. The slide mechanism collects material that, if not removed, will cause the gate to stick or even corrode. Accurate calibration of the spreader relies on consistent openings for material to pass through as it leaves the hopper. Spray the spreader with water to clean and lightly lubricating the moving gate components after cleaning.

References

Beard, J.B. 1982. *Turf Management for Golf Courses*, Burgess, Minneapolis, MN, 642 pp.

Emmons, R.D. 1995. *Turfgrass Science and Management*, 2nd edn, Delmar, Albany, NY, 512 pp.

Turgeon, A.J. 1996. *Turfgrass Management*, 4th edn, Prentice-Hall, Upper Saddle River, NJ, 406 pp.

Vargas, J.M. 1994. *Management of Turfgrass Diseases*, 2nd edn, Lewis, Ann Arbor, MI, 294 pp.

CHAPTER 10

Turfgrass plant health and protection

D.E. ALDOUS and **J.S. BRERETON,** Institute of Land and Food Resources, University of Melbourne, Burnley College, Victoria, Australia

Introduction

Today plant health and protection professionals present an integrated approach to the management and control of weeds, diseases and insect pests of turf. Weeds, diseases and pests interfere with the production of high quality turf surfaces and can be very expensive to control. Sometimes the approach to control is simplified, and reliant on one method of control. Such an approach is not sustainable in the long term and fails to recognise that there are often other sanitary, environmental, biological and cultural management strategies that should be implemented to prevent or manage pest populations to an acceptable level.

Turfgrass weeds

A weed is defined as a plant growing in a place where it is not wanted (Shurtleff *et al.*, 1987). Weeds detract from the overall appearance of the turfgrass surface, and also compete with the more desirable grasses for moisture, light, space and nutrients. A plant species may also be designated as a weed when it causes interference to play, unacceptable differences in colour and growth, and durability problems within the sward. Turfgrass weeds may be classified as exotic narrow-leaf species and broad-leaf weed species. Some grass species in turf, such as Bermudagrass (*Cynodon dactylon*), may be considered a weed in some turfgrass communities and a desired turfgrass in others. Weeds may be categorised by their life cycle and growth habit. Summer annuals are weeds that grow each spring or summer from seed and generally overwinter before germinating the following spring. Winter annual weeds germinate in the autumn and late winter from seed, mature, produce seed during the following spring, and die in early summer. Seeds of most of these species are dormant during the spring. Indeterminate annual weeds include weeds such as chickweed (*Stellaria media*) and annual bluegrass or winter grass (*Poa annua*), that germinate and grow during most seasons in certain regions. Biennial weeds have a two-year life span, developing to partial maturity during the first growing season. Perennial weed species persist over longer periods, and usually overwinter in a dormant state. Many weedy perennials spread primarily by seed, while others such as yellow wood sorrel (*Oxalis corniculata*) and nutgrass (*Cyperus rotundis*) can spread both by seed and vegetative means.

The degree to which a weed adapts to its local environment determines its persistence in the turfgrass community. Several factors, such as climate, soil conditions and maintenance operations, will influence this persistence. Weeds, such as annual bluegrass and crabgrass or summer grass (*Digitaria sanguinualis*), establish where soil has been exposed by cultivation or is compacted or poorly drained. Cultivation may also stimulate weed seed in the soil to germinate as well as activate underground vegetative structures. Some annual weeds are prolific seed producers, ensuring future generations. Weed seed numbers of 45 000 to 200 000 seeds per square metre have been recorded in *Poa annua* dominated turfs (Lush, 1993; Beehag, 1994). Still others, such as plantain (*Plantago lanceolata*), annual bluegrass and bentgrass (*Agrostis tenuis*), are known for their persistence in the soil seed bank (Grime, 1980). Turfgrass seed lots may also be contaminated by weed seeds. Weed seed is commonly disseminated by the wind, water, propagating materials, topdressing materials, turfgrass equipment, animals, birds and humans. The timing and intensity of many cultural operations, such as mowing, verticutting and fertilisation, can influence the germination and growth of weed seed and vegetative plant parts. For example, excessive phosphate in particular is likely to encourage annual bluegrass and clover. Thirty-five broad-leaf and ten grassy weeds, their seasonally, and control strategies are presented in Tables 10.1 and 10.2.

Turfgrass weed management and control

Weed control has been defined as any practice designed to prevent weed emergence or cause a shift from an undesirable vegetation towards a more desirable turfgrass situation (Turgeon, 1991). Successful weed control programs begin with the use of strategic cultural practices that provide the desired species with a competitive advantage. Annuals reproduce only by seed. Control methods must therefore aim at preventing the setting of seed. With many biennials the first year of growth is purely vegetative with the flower and seed forming in the second year. Control methods should be aimed largely at the vegetative growth or seedling period. Perennials live for many years. Therefore control methods should aim at limiting seed production and then at reducing the infestation.

Weed control is primarily by chemical and non-chemical means. Herbicides are classified into two different groups based on the nature of their activity: pre-emergence and post-emergence (Vengris *et al.*, 1982). Pre-emergence herbicides are applied to established turf before the weeds have emerged. Some herbicides recommended for preplant use in turf areas include the fumigant methyl bromide. Methyl bromide is a non-selective soil fumigant that requires special application techniques using a gasproof covering like polyethylene, but which can be effective against bermudagrass, nutgrass and other persistent perennials as well as existing annual weeds and many germinating seeds. Methyl bromide however is recognised as an ozone depleting agent and is therefore likely to have restrictions placed on its usage in the future. Limited restrictions on the use of methyl bromide in Victoria, Australia, will progressively extend in scope over the next few years to eliminate the use of this chemical altogether.

In established turf, herbicides may also be applied before (pre-emergence) or after (post-emergence) weeds emerge. Pre-emergence herbicides are used to control weeds in turf and are applied to the soil before the weed seeds germinate. Often they are best incorporated into the top 25 to 60 millimetres of soil by irrigation where the chemical is taken up by the roots and shoots of the emerging weeds. Post-emergence herbicides are either translocated systemically or act as a contact herbicide on the turf plant. Such herbicides are translocated in the vascular system and eventually reach a site of action where they interfere with a key plant process and ultimately kill the weed. Before using any herbicide, contact your department of agriculture representative, and read and heed the label. Herbicide applications may be broadcast or spot applied. Larger more uniform infestations should be controlled with broadcast spray applications, smaller infestations by spot application. Chemical controls can be discussed with your local supplier but make sure that these comply with local regulations concerning usage and that due care is taken to assess any potential risks to the environment. Recently a range of combination (pesticide on fertiliser carrier) products, which address pre-emergent grass and broad-leaf weed control, have been released onto the market with formulations currently being assessed for efficacy, stability of formulation, and occupational health and safety (Cooper, 1997). Corn gluten hydrolysate has also been shown to have an inhibitory effect on some annual grasses when applied pre-emergent (Liu and Christians, 1997).

Non-chemical means may include mechanical

Table 10.1 Common and scientific names, season and method of control of common broad-leaf weeds in turfgrass.

Common name	*Scientific name*	*Season*	*Control strategy*
Birdseye pearlwort (1,4)	*Sagina procumbens*	prostrate annual	a,b,d
Buckhorn weed (1,4,5,6)	*Plantago lanceolata*	prostrate annual	a,b,d
Catsear (4)	*Hypochaeris radicata*	cool-season perennial	a,b,d
Centella (5)	*Centella uniflora*	low growing perennial	a,b,d
Common hawksbeard (4)	*Crepis capillaris*	low growing perennial	a,b,d,e
Cotula spp. (2)	*Cotula australis*	cool-season, low growing annual	a,d,e
Creeping oxalis (1,3,4)	*Oxalis corniculata*	low growing perennial	a,b,e
Creeping buttercup (1,4)	*Ranunculus repens*	cool-season perennial	b,d
Clover, white (1, 3, 4)	*Trifolium repens*	cool-season perennial	a,b,d,e
Common chickweed (1,3,4,5)	*Stellaria media*	cool-season, winter annual	b,d,e
Cudweeds (2)	*Gnaphalium involucratum*	low growing perennial	a,b,d
Curly dock (2)	*Rumex crispis*	low growing perennial	a,b,d,e
Dandelion (1,2,3,6)	*Taraxacum officinale*	cool-season, low growing, perennial	a,b,d
English daisy (1,4,3,5,6)	*Bellis perennis*	low growing perennial	a,b,d
Field bindweed (1,3)	*Convolvulus arvensis*	cool-season, low growing perennial	a,b,e
Field pennycress (3)	*Thlaspi arvense*	low growing, winter annual	a,b,d,e
Healall (1,3)	*Prunella vulgaris*	cool-season, low growing perennial	a,b,d
Hydrocotyle (1)	*Hydrocotyle moschata*	low growing perennial	b,d
Juncus spp., toadrush (1,4)	*Juncus bufonius*	perennial	a,b,d
Knotweed (1,3,4)	*Polygonium aviculare*	summer annual	c,b,d
Lawn burweed, onehunga (1,6)	*Soliva valdiviana*	spring annual	b,d
Lawn pearlwort (1,6)	*Sagina procumbens*	low growing annual	b,d
Mayweed chamomile (2)	*Anthemis cotula*	winter annual	a,b,d
Mouse-ear chickweed (1,3)	*Cerastium glomeratum*	low growing perennial	a,b,d,e,f
Mullumbimby couch (2)	*Cyperus brevifolius*	perennial	a,b,d
Parsley peart (1,3)	*Aphnes spp.*	perennial	b,d
Pennyroyal (1,2,3)	*Mentha pulegium*	low growing perennial	b,d
Plantain (3)	*Plantago major*	cool-season perennial	a,b,d
Purple nutsedge (1)	*Cyperus rotundus*	grass-like perennial	a,b,d
Red sorrel (1,3)	*Rumex acetosella*	cool-season, perennial	b,e
Selfheal (4)	*Prunella vulgaris*	creeping perennial	a,b,d,e
Speedwell (1,3,4,6)	*Veronica serpyllifolia, filiformis*	cool-season, annual/perennial	a, b,d,e
Twin, swinecress (1,6)	*Coropus didymus*	winter annual	a,b,d
Yarrow (3)	*Achillea millefolium*	cool-season, rhizomatous perennial	a,b,d
Wild carrot (1,3)	*Daucus carota*	cool-season, biennial	a,b

Adapted from: 1. United States of America, Shurtleff et al. (1987); 2. Australia, Australian Turfgrass Research Institute (1988); 3. United States of America, Watschke *et al.* (1995); 4. Great Britain, Adams and Gibbs (1994); 5. New Zealand, New Zealand Institute for Turf Culture, Walker, (1960).

Control strategies: (a) physically or manually remove weed before seeding; (b) use a pre or post-emergent herbicide under labelled directions, that is safe to dominant turfgrass; (c) eleviate compaction; (d) avoid nitrogen sources when weed is most competitive; (e) spot treatment with non-selective pre or post-emergent herbicides under labelled directions that is safe to dominant turfgrass; (f) decrease shade and improve drainage.

Table 10.2 Common and scientific names, cool- or warm-season, season, and control strategies of foreign grass weeds in turfgrass.

Common name	*Scientific name*	*Cool or warm season*	*Control strategies*
Annual blue, winter grass (1,2,3,4)	*Poa annua*	cool-season, winter annual	b,g,d
Bentgrass (1, 2, 3, 5)	*Agrostis spp.*	cool-season perennial	c,h
Bermudagrass, common couch (1,3)	*Cynodon dactylon*	warm-season perennial	b,c,h
Carpetgrass (1)	*Axonopus affinis*	warm-season, summer annual	a,b,c,d
Crab, summergrass (1,2,3)	*Digitara sanguinalis*	warm-season, summer annual	a,b,c,d
Goose, crowsfoot grass (1, 2, 3)	*Eleusine indica*	warm-season, summer annual	a,b,e,d
Kikuyugrass (1)	*Pennisetium clandestinum*	warm-season perennial	b,c,h
Lovegrass, stinkgrass (1,2)	*Eragrostis cilianensis*	warm-season, summer annual	a,b,c,d
Paspalum, dallisgrass (1,2,3)	*Paspalum dilatatum*	warm-season perennial	f,b,d
Velvetgrass, Yorkshire fog (1,2,3,4,5)	*Holcus lanatus*	cool-season perennial	c,h

Adapted from: 1. United States of America, Shurtleff *et al.* (1987); 2. Australia, Australian Turfgrass Research Institute (1988); 3. United States of America, Watschke *et al.* (1995); 4. Great Britain, Adams and Gibbs (1994); 5. New Zealand, New Zealand Institute for Turf Culture, Walker, (1960).

Control strategies: (a) avoid establishment when weed germinating; (b) adjust mowing height to collect weed seed or compete with weed; (c) avoid nitrogen sources when weed is most competitive; (d) use appropriate pre or post-emergence herbicide under labelled directions, that is safe to dominant turfgrass; (e) reduce compacted surface to aid desirable grass recovery; (f) avoid coring when weed is germinating; (g) renovate and establish with more competitive desirable grass; (h) spot treatment with pre or post herbicide, under labelled directions, that is safe to dominant turfgrass, or fumigation.

methods, the use of cultural processes, and biological control. Mechanical methods include hand pulling, hand hoeing, scarifying, rotary hoeing, ploughing, burning and smothering, and the use of weed-free propagation materials. Mechanical methods are particularly effective with annuals where the seed supply in the soil is limited and short-lived. Biennials and perennials are discouraged by cutting at the soil surface, but many grow back from their underground plant parts.

Cultural controls are based on the fact that plants differ in their ability to compete with one another, and may include selection of a more suitable turf species, renovation to introduce resistant species and/or cultivars, competitive mowing heights and frequencies, changes in nitrogen amounts and source, improving the surface and sub-surface drainage, more favourable soil conditioners and pH, reducing compaction, and proper adjustments to equipment and machinery (Watschke *et al.*, 1995). On mature turfgrass surfaces, weed problems are often the result of overwatering or underwatering, mowing too close or too high, low fertility, excessive wear, disease or insect damage, soil compaction, and low light levels. Excessive fertiliser use or watering beyond the needs of a turfgrass can frequently be the cause of serious weed problems. Similarly, frequent mowing will prevent or reduce seed production in some weed species. Most erect growing weeds can be controlled by frequent mowing at the correct heights so that the seedheads are removed before viable seed is produced (Beard, 1982). Persistent removal of top growth will eventually reduce biennial and perennial weed populations by depleting their reserves of carbohydrates.

Biological weed control is an approach that uses other natural living organisms to control or reduce the population of the undesirable weed species (TeBeest, 1991). Some examples of living organisms that have been used in biological weed control programs are insects, mites, fungi, nematodes and aquatic and terrestrial vertebrate herbivores. However only limited success has been achieved.

Figure 10.1 lists the more common turfgrass weeds found in natural grass surfaces.

Semi-aquatic weed management and control methods

Many water courses and irrigation dams associated with racetracks, golf courses and parkland

(a) Capeweed (*Cryptostemma calendula*)

(b) Creeping Speedwell (*Veronica filiformis*)

(c) Curly dock (*Rumex crispis*)

(d) English daisy (*Bellis perennis*)

Figure 10.1 Common weeds of natural grass surfaces

may be sources of algae and aquatic weeds. Weed populations may lead to eutrophication, block filter and pump outlets, and produce dangerous toxins that can be poisonous to grazing animals and native wildlife. Twenty-six aquatic weeds, their characteristics and control strategies are presented in Table 10.3. The major methods of aquatic weed prevention include manipulation of the habitat, active removal of aquatic plant material, biological control, and chemical treatment. Manipulating the habitat may mean regular draining of the water body, the use of liners, dyes and colourants that limit the amount of light available for plant growth, or the use of aerators and ozone generators that increase the water's oxygen level. In larger water bodies, the most appropriate option may be physical or mechanical harvesting of the aquatic weed or algae source. This material may also be a useful source of compost. In the US, mechanical harvesters are sometimes used several days before a chemical treatment is applied to increase the efficacy of the aquatic herbicide (Nus, 1993). In smaller water bodies, where the degree of weed infestation is slight, biological control may be possible. For example, in the US, the exotic weevil (*Neochetina eichhorniae*) has been used to control water hyacinth (*Eichhornia crassipes*), and certain fish, particularly the triploid grass carp, feed on aquatic vegetation. In enclosed waterways, such as dams and lakes within private clubs and racecourses, algaecides and aquatic herbicides are often used. Straw has successfully controlled algal blooms on enclosed waterways and is a relatively inexpensive and safe means of keeping these waterways clean (Newman, 1994). The control of aquatic plants using herbicides is similar to terrestrial plants, with the exception that many aquatic herbicides have a post application withholding period before treated lakes can be used for irrigation. All chemicals should be checked with your state water authority and any other appropriate authorities before use over water or in close proximity to waterways.

Turfgrass diseases

A disease is defined by Smiley (1983) as an abnormal alteration in the physiological processes or morphological development of a plant by a pathogenic organism or environmental factor. Such agents of disease can either be infectious (biotic) agents such as plant pathogenic species of bacteria, fungi, mycoplasmas, nematodes and viruses, or non-infectious (abiotic) agents which

Table 10.3 Common and scientific names, plant types, and control methods for common algal and aquatic weeds.

Common name	*Scientific name*	*Plant type*	*Control methods*
Alligator weed	*Altmanthera philoxeroides*	Emergent plant	a,b,c,d
Arrowhead	*Sagittaria graminea*	Emergent plant	a,b,d
Bladderwort	*Utricularia spp.*	Submersed plant	a,b,d
Blue-green algae	*Anacystis cyanea, Anabaena crinalis*	Plantonic algae	a,b,d
Bullrush	*Scripus spp.*	Emergent plant	a,b,d
Cat's tail	*Typha latifolia*	Emergent plant	a,b,d
Chlorella	*Chlorella spp.*	Planktonic algae	a,b,d
Cladophora	*Cladophora spp.*	Filamentous algae	a,b,d
Duckweed	*Lemna minor, L.trisulca*	Free floating plant	a,b,d
Elodea	*Anacharis spp.*	Submersed plant	a,b,d
Euglena	*Euglena spp.*	Euglenoid	a,b,d
Horse hair clump	*Pithophora spp.*	Filamentous algae	a,b,d
Mougeotia	*Mougeotis spp.*	Filamentous algae	a,b,d
Muskgrass	*Chara spp.*	Weedlike algae	a,b,d
Navicula	*Navicula spp.*	Diatoms	a,b,d
Parrot's feather	*Myriophyllum brasiliense*	Submerged plant	a,b,d
Pondweed	*Potomogeton spp.*	Submerged plant	a,b,d
Red azolla	*Azolla filiculoides*	Free floating plant	a,b,d
Salvinia	*Salvinia rotundifolia, S. molesta*	Free floating plant	a,b,d
Water hyacinth	*Eichhornia crassipes*	Free floating plant	a,b,d
Water lettuce	*Pistia stratiotes*	Free floating plant	a,b,d
Water meal	*Wolfia spp.*	Free floating plant	a,b,d
Water milfoil	*Myriophyllum spp.*	Submerged plant	a,b,d
Water net	*Hydrodictyon spp.*	Filamentous algae	a,b,d
Water silk	*Spirogyra spp.*	Filamentous algae	a,b,d
Zygnema	*Zygnema spp.*	Filamentous algae	a,b,d

Adapted from: 1. United States of America, Nus, (1993); 2. Australia, Beehag, (1996), Rolfe *et al.* (1994), Richard and Shepherd, 1992, Aldous, (1991).

Control strategies: (a) drainage; (b) harvesting; (c) biological control; (d) chemical treatment.

are physiological disorders brought about by unfavourable environmental conditions or physical injury. Such injury can be induced by chemical (pesticides, animal urine or salts, or chemical spills), physical (extremes of temperature, lightning and soil compaction), or mechanical (scalping with the mower or abrasion injury) agents. A pathogen is a disease-causing agent. Most pathogens are also parasites. Parasites are organisms that obtain part or all of their nutrients from a living turfgrass host. Disease organisms such as rust (*Puccinia spp.),* or powdery mildew or yellow tuft (*Sclerophthora macrospora*) are called obligate parasites as they are dependent on that living host for nutrients. Other organisms, such as slime mould (*Mucilo crustacea*), live on dead organic matter only, such as thatch and mat, and are called saprophytes. Still other organisms can live as parasites but under certain conditions are capable of saprophytic growth; these are called facultative saprophytes. Finally there are organisms, such as fusarium patch (*Microdochium nivale*) and damping off (*Pythium spp.*), that usually live as saprophytes but under certain conditions may become parasitic. Such organisms are called facultative parasites (Vargas, 1981). As a result of different injuries, turf exhibits various responses known as symptoms which show themselves as changes in leaf colour, leaf, stem and root rots, and swellings. Symptoms are often unique to a particular pathogen and its interaction with a particular host plant which assists in their identification.

For any disease to become active, three interactive conditions are necessary: the existence of the pathogen, a susceptible host or grass, and a favourable environment. These three conditions comprise the plant disease triangle. If any one of these components is missing, the disease will not develop. Therefore correct identification of the disease organism is an important clue in identifying the probable cause. Procedures for collecting and submitting turfgrass samples to a diagnostic laboratory for identification should include correctly identifying the affected plants, recording the overall symptoms, cultural conditions, specific symptoms, and any abnormalities, adopting a correct sampling technique and enabling rapid delivery to the diagnostic laboratory (Woodcock, 1983). Samples should contain healthy as well as unhealthy turf. Turf samples 30 x 30 centimetres and 5 centimetres deep are ideal for investigation. Where such sampling is not possible, several cylinders of turf 10 centimetres in diameter and 10 centimetres deep should be collected. All samples should be placed in a plastic bag and correctly labelled. Twenty-four of the more important turfgrass diseases, their common and scientific names, causal agents and control methods are presented in Table 10.4. For a more comprehensive coverage of turfgrass diseases, work by American (Smiley *et al.*, 1992; Watchke *et al.*, 1995), English (Smith *et al.*, 1989) and Australian (Woodcock, 1983; Australian Turfgrass Research Institute, 1994), researchers is recommended.

Turfgrass disease management and control

Smiley (1983) proposed four important disease control strategies: prevent the pathogen from becoming established in a new turfgrass stand, change the genetic composition of the turfgrass plant so that they will resist attack by pathogens, change the surrounding environment so that susceptible plants can overcome an attack by the pathogen, and protect plants from being infected by pathogens. These outcomes translate into making better use of sanitation procedures, greater use of disease resistant plant cultivars, and implement operations leading to best practice in cultural management.

Sanitary practices include the careful inspection for, and removal of, inoculum found in topdressing and contaminated vegetative material

(a) Fairy ring (*Marasmius spp.*) with stimulated bluegrass displaying typical 'frog eye' effect grass and fruiting bodies (mushrooms) (photo courtesy Rhone-Poulene Rural Australia)

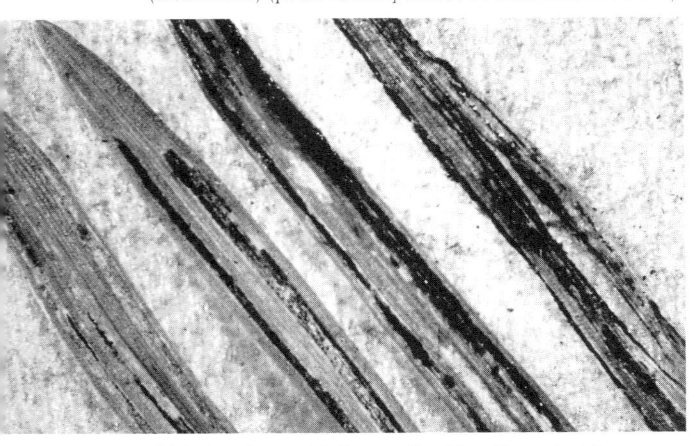

(b) Stripped smut (*Ustilago striiformis*) on Kentucky bluegrass

(c) Slime mold fungus (*Mucilo crustacea*) on bluegrass and bermudagrass leaves

Figure 10.2 Common turfgrass diseases

and equipment that may enter a healthy, disease-free turf. Diseases caused by *Drechslera* and *Fusarium spp*, sporulate profusely and therefore can be more difficult to contain. Leaf spot (*Bipolaris sorokiniana*) spores are easily dispersed by the wind. Other fungi, such as *Lanzia* and *Sclerotium*, are limited to an existence in soil, so they can be suppressed by fumigating the soil before new turf establishment. Turfgrass cultivars vary in their susceptibility to pathogens and to adverse environmental conditions. For example, brown patch is favoured by excess thatch and mat along with high temperatures (23 to 35°C (75 to 95°F)) and high humidity, whereas red thread (*Laetisaria fuciformis*) usually appears on turf deficient in nitrogen during periods of prolonged cool, wet weather. Therefore there is a need to establish a turfgrass that is more likely not to succumb to the most favourable pathogen that exists in the area. Turf surfaces comprised of different grass species fare better against rust than turf surfaces composed of a single species. Although there are many degrees of disease resistance in turfgrass, an immune plant is one that is not attacked by a particular pathogen. A susceptible plant is one that is severely damaged by disease. Resistant plants are intermediate in reaction. A highly resistant plant may exhibit slight development of the disease under ideal conditions for development; a slightly resistant plant may be almost as severely damaged as the fully susceptible ones. Varieties of plants that are considered highly resistant or immune in certain areas because no disease attacks have been observed, may prove to be more susceptible in other areas. Determining the degree of resistance of a new variety should be accomplished with greenhouse inoculations under ideal conditions for the pathogen, and should be supplemented by field testing at many locations throughout the country.

Management practices and environmental factors necessary to maintain turf have a decided effect on disease incidence and development.

Table 10.4 Common name, causal agent, primary hosts, predisposing conditions and management controls of twenty-four fungal diseases of turfgrass.

Common Name	*Causal Agents*	*Primary hosts*	*Predisposing Conditions*	*Management Controls*
Anthracnose (1, 2, 5, 6)	*Colletotrichum graminicola, Microdochium bolleyi*	a,f	g	Increase height of cut, renovate, fungicides.
Rhizoctonia (Brown patch) (1, 2, 3, 4, 5)	*Rhizoctonia solani, R. ceralis, R. zeae*	i,j	b	Reduce N levels & thatch, improve drainage, fungicides.
Curvularia diseases (Curvularia blight) (1, 3, 5)	*Curvularia geniculata*	a,b	h	Alleviate stress, fungicides.
Dollar spot (1, 5, 6)	*Lanzia homeocarpa, Moellerodiscus spp.*	i	e	Balanced N&K fertility, remove dew, fungicides.
Downy mildew (Yellow tuft) (1, 2, 5)	*Sclerophthora macrospora*	i	d	Improve drainage, fungicides.
Fairy rings (1, 2, 4, 5)	*Clitocybe, Lepiota, Lycoperdon, Marasmius.*	i	g	Drench, wetting agents, soil fumigation.
Fusarium blight & F. blight syndrome (1, 3, 5)	*Fusarium equiseti, F. acuminatum, F. poae*	a,b,f	h	Increase height of cut, acidifying N fertilisers, fungicides.
Fusarium patch (Pink snow mold) (1, 3, 4, 5, 6)	*Gerlachia (Microdochium) nivalis*	a,i	b	Balanced NPK fertiliser prior to dormancy, fungicides.
Kikuyu yellows (2)	*Verrucalvus flavofaciens*	h	d	Not yet available.
Helminthosporium leaf spot (1, 2, 5)	*Bipolaris sorokiniana, B. cynodontis, B. tetramera*	a,,b,c,f	a	Increase mowing height, resistant cultivars, fungicides.
Leopard spot (2)	Several fungi, yet to be identified.	e	d	Not yet available.
Pythium blight, (Damping off) (1, 2, 4, 5)	*Pythium aphanidermatum, P. graminicola*	a,i	d	Improve drainage, water & thatch control, fungicides.
Powdery mildew (1, 5)	*Erysiphe graminis*	b,f	f	Improve sun & air circulation, tolerant grasses, fungicides.
Rust (1, 2, 5, 6)	*Puccinia and Uromyces spp.*	b,c,i,j	c	Balanced NPK, avoid stress, resistant cultivars, fungicides.
Drechslera diseases (Melting out) (1, 3, 5, 6)	*Drechslera poae, D. dictyoides, D. erythrospila*	a,c,d,f	c,d	Increase mowing, resistant varieties, fungicides.
Red thread (1, 2, 3, 4, 5, 6)	*Laetisaria fuciformis*	a,b,c,d,f	a	Balanced N&K, pH 6.5-7.0, fungicides.
Southern blight, (Rolf's' disease) (1, 3, 4, 5)	*Sclerotium rolfsii*	a,b,d,k,l	h	Reduce thatch, core, raise pH >8.0, fungicides.
Slime mold (1, 5)	*Mucilo crustacea, Physarium spp. Fuligo spp.*	i	d	Forcefully wash leaves, fungicides seldom used.
Spring dead spot (1, 2, 3, 5)	*Leptosphaeria namari*	d	c	Cease use of N 6 weeks prior to dormancy, fungicides.
Stripe, flag and loose smuts (1, 5)	*Ustilago striiformis* (Stripe smut), *Urocystis agropyri* (Flag smut), *Entyloma spp.* (Blister smuts)	a,c,b,f	b	Improve drainage, balanced NPK, fungicides.
Summer patch (1, 5)	*Pyricularia (Magnaporthe) grisea, P. oryzae*	b,c,d,f,g	b	Increase height of mowing, acidifying N sources, fungicides.
Take all patch (1, 3, 5, 6)	*Gaumannomyces graminis*	a,b,f	c	Acidifying N sources, improve drainage, fungicides.
Typhula blight,(Snow scald) (1, 5)	*Typhula incarnata and T. ishikariensis*	i	b	Avoid heavy N applications prior to dormancy, fungicides.
White helminthosporium (2)	Several fungi, yet to be identified.	e	d	Not yet available.

Key to authorities: 1, United States, Smiley, (1983); 2, Australia, Australian Turf Research Institute (1994), Wong (1996); 3, Australia, Woodcock, (1983); 4, New Zealand, Latch, (1960); 5. United States, Watschke *et al.* (1995), 6, United Kingdom, Adams and Gibbs, 1994.

Primary hosts, a = bentgrass, b = Kentucky bluegrass, c = perennial ryegrass, d = bermudagrass, e = bermudagrass hybrid, f = fescues, g = centipede, h = kikuyugrass, i = many turfgrass species, j = zoysiagrass, k = cotula, l = dichondra, n = St. Augustine grass.

Predisposing factors; a = overcast, rainy weather; b = prolonged cold, wet weather; c = cool wet autumn followed by hot summer; d. = cool, wet, humid, spring; e = warm days, cool nights, f = shade; g = prolonged rain or hot/dry days; h = hot, dry summer.

Most cultural practices exert an indirect effect on disease development by modifying the micro-environment. Others are directly involved with disease development or with the level of damage caused by disease. Temperature has a direct effect on the performance of fungi, other disease organisms, and the growth of turf. For example, Couch and Bedford (1966) reported that the bentgrasses, Kentucky bluegrass and red fescue, in that order, are most susceptible to fusarium blight (*Fusarium roseum*) under hot dry weather with daytime temperatures of 29 to 35°C (85 to 95°F) and night temperatures of 21°C (70°F) or above. Spring dead spot (*Leptosphaeria namari*) in Bermudagrass is favoured by cool conditions in spring and early autumn, as well as drought stress and compacted soils (Woodcock, 1997).

The level of humidity at the turf surface is considered important in disease development as it provides the free water for germination and mycelial growth. Fusarium patch requires 30 to 50 hours of wet leaf surface to penetrate uninjured tissue, whereas rust (*Puccinia graminis*) requires moisture in the form of dew for 10 to 12 hours to infect plants. Recommendations to control the *Helminthosporium* leaf spots require the removal of water droplets by syringing the surface to reduce the incidence of this disease on bentgrass putting greens during the summer months. Frequent cycles of wetting and drying can greatly favour the severity of diseases such as powdery mildew, and rust in heavily shaded areas. Extreme drought stress appears to enhance the turf diseases take-all (*Gaumannomyes graminis*), strip smut (*Ustilago striiformis*), powdery mildew (*Erysiphe graminis*), and fairy ring (*Marasmius spp.*) (Watschke *et al.*, 1995).

Similarly, proper soil fertility can improve turfgrass performance and its tolerance to disease. However, Couch and Bloom (1960) reported that turfgrass susceptibility to disease is probably not due to a single nutritional factor, but rather to a combination of environmental factors including soil fertility. High levels of nitrogen (N) nutrition can increase the outbreaks of many turfgrass diseases such as snow mold and brown patch. Turf deficient in nitrogen tends to develop more dollar sport (*Lanzia homeoecarpa*) and red thread (*Laetisaria fuciformis*) than turf adequately fertilised. High nitrogen, potassium and sulfate levels and alkaline soil conditions favour fusarium patch (*Microdochium nivale*). There is little evidence that the other major nutrients, phosphorus and potassium, or the trace elements, singularly have a great impact on disease incidence although in New Zealand research has demonstrated that the fertiliser potassium carbonate applied at 0.9 kilograms per 100 square metres has been used as a treatment in eliminating fairy rings from Leptinella bowling greens (Walmsley, 1996). Fertiliser combinations of N+K and N+P+K have been shown to reduce the incidence of diseases such as red thread, spring dead spot, take-all, and Fusarium patch (Watschke *et al.*, 1995).

Most diseases are favoured by mowing practices which create wounded tissue. Fungi, such as *Rhizoctonia* and *Fusarium spp.*, may be transported by mowers and clippings and readily colonise cut leaf tips. Close mowing is known to exacerbate Helminthosporium diseases, brown patch, dollar spot and rust. Thatch provides fungi with moisture, as well as being a repository for the resting structures of pathogens. *Helminthosporium spp.* and Rolf's disease (*Sclerotium rolfsii*), among other diseases, produce spores on the thatch layer (Healy and Britton, 1968).

A number of fungicides and fungistats have been formulated to eliminate fungal pathogens or to render them incapable of further growth and development. Most fungicides are designed to prevent the pathogen from infecting the host plant through destruction of the spores, mycelia and sclerotia in the soil or thatch, applying a protective spray onto the turf leaf surface that eliminates the pathogens present, or applying systemic fungicides which prevent direct infection. Contact or non-systemic fungicides need to be applied more frequently during times of high disease activity, as tissue is constantly growing and/or being removed. Systemic fungicides are those which are absorbed and translocated throughout the plant and are not subject to weathering or breakdown by sunlight or irrigation (Beard, 1973; Smith *et al.*, 1989; Turgeon, 1991). Fungicides are available in sprayable (wettable powder, flowable or soluble) formulations and some are available as granular formulations. With wettable powders, the active ingredient is carried on finely ground clay materials and treated with adjuvants so they are easily suspended in water. Emulsifiable concentrates are formulated with various liquid organic solvents that contain special adjuvants that make them miscible in water. Most systemic fungicides are actually fungistats, or substances that prevent the growth of a fungus without killing it. Instructions for the use of such fungicides are found on the label. These should be read carefully and adhered to.

Turfgrass insect pests

Six major classes of Arthropods attack turf: the Arachnida (spiders, mites and ticks), Crustacea (shrimp, sowbugs and billbugs), Chilopoda (centipedes), Diplopoda (millipedes), Symphyla (garden centipedes) and Insecta (Watschke *et al.*, 1995). The largest of the classes, the Insecta, are characterised by three pairs of legs, one pair of antennae, often the presence of wings, and three separate body divisions of the head, thorax and abdomen. An exoskeleton surrounds the insect body and supports the internal muscles and organs. Insects respire through a set of tubes, trachea, which run the length of the body. Insects belong to one of two categories based on life cycle or metamorphosis. Insects can be divided into two broad groups that display either gradual or complete metamorphosis. The young of insects, such as the chinch bugs, grasshoppers and leafhoppers, complete a gradual metamorphosis in which the nymphal and adult stages look similar. Insects such as beetles, moths and butterflies undergo complete metamorphosis from an egg to larvae to pupa through to adult. The larva of a beetle is known as a grub, the larva of a moth, a caterpillar, and that of a fly, a maggot. In most cases, only certain insect stages attack turf or annoy people or cause economic damage.

Insect pests of turf can be divided into three groups: those that feed below the surface of the soil, those that eat the leaves and stems, and those that suck plant juices (App and Kerr, 1969). Grubs are the larvae of many species of beetles belonging to the family *Scarabacidae*, which include the black beetle (*Heteronychus arator*) and the pasture scarab (*Aphortius howitti*) in Australia, and the oriental beetle (*Anomala orientalis*) and Japanese beetle (*Popillia japonica*) in the United States. All live in the soil and feed on turfgrass roots. Although the adult beetles differ considerably in colour, structural markings and habits, the grubs themselves and the injury they cause to turf are quite similar. The typical c-shaped grub remains underground and feeds on the rootmat, roots and/or organic matter. Grub damage is often most evident during the spring and autumn. When populations average more than 30 grubs per square metre, chemical treatment may be necessary. Other grubs, such as the green june beetle (*Cotinis nitida*) develop mounds of earth which disfigure the turf and may smother the grass. Grub species can be identified by examining the grub's raster or terminal segment under a hand lens (Watschke *et al.*, 1995). The soil-dwelling larvae that chew the turfgrass root system include the white-fringed weevil (*Pantomorus leucoloma*), the amnemus or clover root weevil (*Amnemus quardrituberculatus*) and the cotula weevil (*Steriphus variabilis*). Billbugs or la plata weevil (*Sphenophorus brunnipennis*) can cause damage to both Australian and American cool- and warm-season turfgrasses (Juska, 1965). Weevils often feed on the inside of the grass stem and crown, then move into the soil where they feed on the roots. Fine, whitish, sawdust-like larval excrement (frass) may be observed on the soil surface. Field crickets (family *Gryllidae*), mole crickets (family *Gryllotalpidae*), and a recently introduced insect pest to Australia, the Changa mole cricket (*Scapteriscus didactylus*), also chew the stems and leaves of turf causing injury (Kappro, 1996). Earwigs are beetle-like insects that can occasionally cause damage in turfgrass. The Australian sod fly (*Inopus rubriceps*) can affect all turf species, feeding on the sap of roots and shoots. Earthworms are generally known to aerate the soil and some turf authorities believe they are beneficial to turf soil. However others consider the earthworm to be a nuisance because of castings left on the surface of the turf.

Grubs of several kinds of caterpillars, such as the sod webworm, armyworm and cutworm, eat the leaves and stems of turf. The sod webworm, the bluegrass webworm, (*C. teterrellus*) and tropical sod webworm (*Herpetogramma phaeopteralis, H. licarsisalis*) (Brogden and Kerr, 1964; Ward, 1995) are particularly damaging under heavy populations. The sod webworm is widely distributed having been identifed in the United States, Australia, India, China, Japan and Hawaii. The fully-grown larvae of the fall armyworm (*Spodoptera maurtia*) (Hassan, 1997) or the pasture webworm (*Psara Iicarsisalis*) are about 30 millimetres long, vary from light green to almost black, and have several stripes along the sides of the body. Cutworm and armyworm caterpillars, from the Family *Noctuidae* are the larvae of the common brownish, buff-coloured or greyish black moths which fly about lights on warm nights. The black (*Agrotis ipsilon*) and variegated cutworm (*Peridroma saucia*) are common in the United States, whereas several species of common (*Agrotis infusa*) and brown cutworm (*Agrotis munda*) are found in Australia. When food is scarce, armyworms will move as a group, feeding indiscriminately on plants in their path. In New Zealand, the porina (*Wiseana cervinata*) and army caterpillar (*Pseudaletia separata*) are common pests of sports turf. Other foliage and stem feeders include the Australian plague locust (*Chortoicetes terminifera*), the wingless grasshopper (*Phaulacridium vittatum*) (Whittet, 1965), and the maggots of the frit fly (*Oscinella frit*) (Schread, 1964).

Sampling techniques to detect soil dwelling insects include the use of chemical irritants, such as water (submerge area for 10 minutes), or the use of pyrethrin solutions (4 litres of 0.0015 per cent pyrethrin solution), soap detergent (0.25 per cent concentration in 4 litres of water), potassium permanganate (4.5 litres solution containing 7 grams of potassium permanganate), formalin (4.5 litre solution of 10 millilitres of formaldehyde per 0.36 square metre), or the use of boards and hessian bags (leave overnight to cover the area)(Ward, 1995).

Many insects pests suck the juices from roots, stems and leaves of turf and include the lawn chinch bugs (*Blissus spp.*), Rhodesgrass scale (*Antonina graminis*), and Bermudagrass scale (*Odonaspis ruthae*). They feed by inserting a slender piercing beak into the grass and suck the plant juices. Several species of ground pearl (*Margarodes spp*) attack the roots of grasses and are difficult to control. The leafhoppers or Jassids (*O. Hemiptera*), meadow spittlebug (*Philaenus spumarius*), and the mite family; the grass webbing mite (*Oligonychus digitatus*), the two spotted mite (*Tetranychus urticae*), the pasture (*Bryobia repensi*) and couchgrass mite (*Dolichotetranychus australianus*), retard growth by sucking the sap from the stems and leaves. Damage often appears as whitened areas which can be mistaken for drought or disease damage. Slugs

(a) Billbug (*Sphenophorus brunneipennis*) (Photo courtesy Bayer Australia)

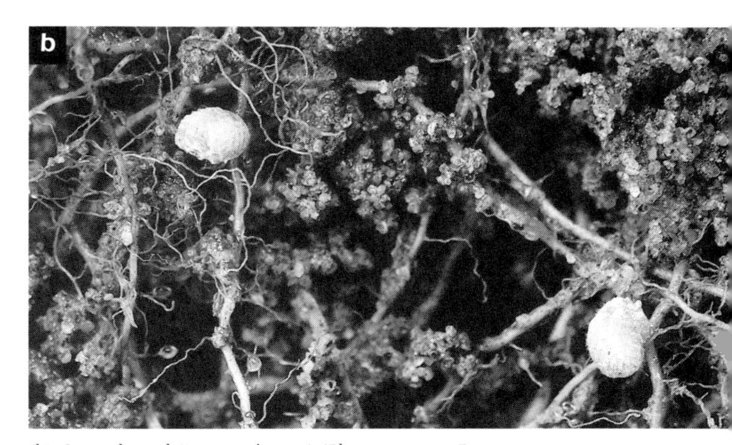

(b) Ground pearl (*Margarodes spp.*) (Photo courtesy Bayer Australia)

(c) Argentine stem weevil (*Listronotus bonariensis*) (Photo courtesy Bayer Australia)

(d) African black beetle (*Herterenychus arator*) (Photo courtesy Bayer Australia)

Figure 10.3 Common turfgrass insect pests

(phylum *Mollusca*) can become serious pests of pastures by feeding on the foliage, especially tender, young shoots. Similarly with sowbugs and pillbugs, which belong to the class Crustacea, the millipedes (class *Diplopoda*) and centipedes (class *Chilopoda*), all of which feed on decaying vegetable matter but occasionally attack living plants. Other pests of turf usually cited include the earthworm (phylum *Annelida*), spiders, ticks, fleas, ants and the nuisance vertebrate pests, the starling and magpie, moles, and the American raccoon. Figure 10.3 illustrates the more common insect pests of turfgrass.

Fifty-one of the major insect pests, their characteristics and control are presented in Table 10.5.

Turfgrass insect management and control

The basic steps in controlling insect pests include identification of the insect and its life cycle, active monitoring and sampling to determine the insects' potential for damage, deciding what strategies to use to suppress the population if taking action against the insect is required, and sound evaluation of the program (Shurtlett *et al.*, 1987). In selecting the appropriate control mechanism, the

Table 10.5 Common and scientific names, type of feeding pattern and damage, and control options of common insect pests of turfgrass.

Common Name	*Scientific Name*	*Type of Feeding **	*Type of Damage*	*Control Options*
Ant, Argentine (2, 3)	*Irodomyrmex humilis*	—	Moulds, smother, covers	Modify habitat, chemical.
Armyworm, Fall (2,4)	*Spodoptera frugiperda*	S	Larvae defoliate leaf	Biological control, chemical,
Armyworm, Common (1,4)	*Pseudaletia convecta*	S	Larvae defoliate leaf	weed & aeration management
Aphid, Cherry (2)	*Rhodalosiphum padi*	J	Feed actively on leaf	Chemical
Beetle, Black (1,3)	*Heteronychus arator*	R	Adults, larvae feed on roots	Chemical
Beetle, Asiatic (2,4)	*Maladera castanea*	R	"	Chemical, biological control
Beetle, Japanese (2,4)	*Popillia japonica*	R	"	"
Beetle, May (2,4)	*Phyllophaga spp.*	R	"	"
Beetle, Green June (2,4)	*Cotinis nitida*	R	"	"
Beetle, Flea (1)	*Chaetocema australica*	S	Feed actively on leaf	Chemical control
Beetle, Dichondra flea (3)	*Chaetocema repens*	S	"	"
Beetle, Sugar cane (5)	*Rhypardia morosa*	R	Larvae feed on roots	"
Billbug, Common (1,2)	*Spheno. brunnipennis*	R (S)	Burrow & feed in stem	Chemical, biological control
Billbug, Bluegrass (2,4)	*Spheno. parvulus*	R (S)	"	"
Billbug, Hunting (2,4)	*S. venatus vestitus*	R (S)	"	"
Chinchbug, Common (4)	*Blissus leucopterus*	J	Feed actively on leaf	"
Chinchbug, Southern (4)	*B. insularis*	J	"	"
Centipede (2,4)	*Chilopoda spp.*	-	—	Crushing, Chemical control
Chiggers, Redbugs (2)	*Eutrombicula spp.*	-	—	"
Caterpillar, Army (3)	*Pseudaletia separata*	S	Feed actively on leaf	Chemical control
Cockchafer (1,4)	*Aphodius tasmaniae*	R	Larvae feed on roots	Chemical, biological control
Chafer, European (2,4)	*Rhizotrogus majalis*	R	Larvae feed on roots	"
Cockchafer, Red-headed (5)	*Adoryphorus couloni*	R (S)	"	"
Corbia, Ataenius (1,2,4)	*Ataenius imparalis*	R	"	"
Cricket, Black field (3)	*Tellogryllus commodus*	S	Burrow, leaf feeders	Chemical control
Cricket, Northern mole (5)	*Gryllotalpa hexadactyla*	S	"	Chemical, biological control
Cricket, Changa mole (1,4)	*Scapteriscus didactylus*	S	"	"
Cricket, Common (1,5)	*Gryllus commodus*	S	"	"
Cutworm, Black (2,4)	*Agrotis ipsilon*	R (S)	Leaves, stems removed	"
Cutworm, Common (1)	*Agrotis infusa*	R (S)	"	"
Earwig, European (2,4)	*Forficula auricularia*	—	—	Chemical control, barriers
Fly, Australian Sod (2)	*Inopus rubiceps*	J	Turf stunted & thin	Chemical control
Fly, Frit (2,4)	*Oscinella frit*	J	"	"
Fly, Couch (1)	*Delia urbana*	J	"	"
Grasshopper, small (5)	*Austroicetes cruciata*	S	Feed actively on leaf	Chemical control
Grasshopper, Wingless (5)	*Phaulacridium vittatum*	S	"	"
Grasshopper, Common (5)	*Austracris guttalosa*	S	"	"
Hoppers, Leaf (2,4)	*Empoasca fabae*	J		"
Mite, Grass (5)	*Oligonychus stickenyi*	J	Blothing of leaves	Chemical control

continued

Table 10.5 Common and scientific names, type of feeding pattern and damage, and control options of common insect pests of turfgrass.—*continued*

Common Name	*Scientific Name*	*Type of Feeding **	*Type of Damage*	*Control Options*
Mite, Two-spotted (5)	*Tetranychus urticae*	J	Blothing of leaves	"
Mite, Pasture (5)	*Bryobia repensi*	J	"	"
Mite, False Spider (5)	*Eriophyes cynodoniensis*	J	"	"
Mite, Bermudagrass (2)	*Aceria neocynodomis*	J	—	"
Pearl, Ground (2,4)	*Margarodes spp.*	J	—	Fumigation
Scale, Rhodesgrass (2,4)	*Antonia graminis*	J	—	Chemical, biological control
Scale, Couchgrass (2,4)	*Odonaspis ruthae*	J	—	"
Weevil, Argentine stem (4)	*Listronotus bonariensis*	S	Adults, larvae feed on	Destroy sites, Chemical
Weevil, white fringed (1)	*Pantomorus leucoloma*	S	leaf and stem base	"
Weevil, Cotula (3)	*Steriphus variabilis*	S	"	"
Weevil, Hyperode (2)	*Hyperodes spp.*	S	"	"
Webworm, Larger sod (2,4)	*Pediasia trisectus*	S	Shoots defoliated, larvae	Chemical, biological control

Key to authorities: 1. Australia, Australian Turf Research Institute (1988); 2. United States of America, Shurtleff et al. (1987); 3, New Zealand, New Zealand Institute of Turf Culture, Helson, (1960); 4. United States of America, Watschke et al. (1995); 5. Australia, Whittet (1965).

* Type of feeding: R= root, stolon and rhizome feeders, S =leaf feeders, J= suck cell contents.

aim is not eradication of the insect pest but rather the suppression of the pest to an acceptable level. This has been termed the aesthetic damage level or the level at which the insect population causes observable, unacceptable damage (Watschke *et al.*, 1995). Control measures must be implemented before this point is reached, therefore an action threshold level must be established. Preventative measures that suppress the occurrence of destructive turfgrass pests include sanitation, such as the selection of clean seed, sod or topsoil, and reducing sources of inoculum, crop rotation and cultivation, the cleaning up of areas where insects may overwinter or build up resistance, and making greater use of turfgrass host resistance cultivars. For example, resistant cultivars of Kentucky bluegrass to sod webworm (Pass *et al.*, 1965), resistant bermudagrass cultivars to bermudagrass mite (Butler and Kneebone, 1965), and the resistance to the fall armyworm in *Zoysia japonica* cv. Cavalier, *Buchloe dactyloides* cv. Prairie, *Poa pratensis* cv. Baron and three Poa hybrids (*P. pratensis x P. arachnifera*) and *F. arundinaceae* cvs. Rebel 11 and Rebel Jr. (Reinhert *et al.*, 1997). Biological control uses natural parasites, predators and diseases to control pest populations. For example laboratory trials with organophosphorus insecticides, and the entomogenous nematode (*Steinernema carpocapsae*) (Bio-Safe®), have shown a high efficacy towards the larvae of the Hunting billbug (*Sphenophorus venatus vestitus*) (Hatsukade, 1997). Chemical control has often meant only the use of pesticides, but it should also include the use of chemicals such as attractants (providing alternative food sources) and sex pheromones (attractiveness to another of the species).

There are four general groups of insecticides used in the control of turf insects: inorganics such as elemental sulfur, bioinsecticides such as *Bacillus thuringiensis*, botanicals or plant extracts such as pyrethrum or rotenone, and the synthetic organics. This last group comprises the chlorinated hydrocarbons such as aldrin, chlordane, dieldrin and D.D.T., all of which have now been replaced as they left undesirable residues or caused other environmental problems; organophosphates, such as tetrapropyl, carbophenothion, diazinon, chlorpyrifos, malathion and isofenphos; the carbamates, which include carbaryl and bendiocarb, and the artificial insect growth regulators, which act like natural insect hormones.

Insecticides are marketed in several formulations: dusts, wettable powders, emulsifiable concentrates, granules and oil solutions, the last of which are not recommended for use on turf as they may burn the grass. In general, spray formulations are preferred over dusts. Granular insecticides do not usually kill leaf-feeding insects as quickly as liquid sprays, however their residual activity is longer. Insecticide sprays work quickly and will kill the majority of insects within the first 48 hours. Soil-borne insect pests are the most difficult to control with some materials being tied up with the mat and thatch. Irrigation following the insecticide treatment can accelerate their effectiveness. Insecticides may be applied in all kinds of equipment, but the equipment must be designed to handle the material being applied. All equipment must be properly calibrated to dispense the correct amounts. Dusts and wettable powders can be applied by power equipment or hand-operated equipment. Granules can be applied through most fertiliser applicators, either power or hand-operated. With all insecticides read the label very carefully and check to make sure that the product is registered for use in your locality and complies with any other regulation concerning usage.

Nematodes

Nematodes are microscopic animals that can live saprophytically in films of water and soil, or as parasites of plants and animals (Beard, 1973; Neylan, 1996). Parasitic plant nematodes are round in cross-section, usually eel-shaped, but some, such as the root-knot nematode (*Meloidogyne spp.*), can become swollen to a kidney or pear shape. Nematodes range from 350 to 2500 microns in length and 15 to 35 microns in diameter. All inhabit the soil as eggs and may either hatch immediately or remain inactive for several months or even years, as with cyst nematode (*Heterodera spp.*). Overwintering may occur at any stage of the life cycle but more commonly as larvae within an egg-like shell. Generally there are two hatchings a year, autumn and spring, with 3 to 4 weeks required to complete the life cycle. All parasitic nematodes possess a spear-like stylet which pierces the plant cell wall. Secretions are then injected into the cells which degrade the protoplasmic contents of the cell, facilitating withdrawal through the stylet. The dagger (*Xiphinema spp.*), pin (*Pratylenchus spp.*) and stubby root (*Paratrichodorus spp.*) nematodes are known to function as transferors of virus organisms in this way. The location and feeding sites of different nematodes are shown in Figure 10.4.

Nematode activity is favoured by well aerated soils, mild to warm soil temperatures (>15°C) and readily available soil moisture. Dissemination of nematodes may include equipment, irrigation water, topsoil, vegetative material, irrigation and drainage waters. Movement is in the film of water surrounding the soil particle. Lateral movement seldom exceeds 30 centimetres (12 inches) per year. Activity is restricted in dry soils and those saturated long enough to be deficient in oxygen. Damage appears to be more extensive in warm temperate and sub-tropical regions than in cooler regions, and they are most active in cool-season grasses in mid to late spring and autumn, and on warm-season grasses during summer and autumn. Nematode injury to turfgrasses may appear as slight to severe chlorosis and declining growth usually in circular to irregular patches on the foliage, plants may exhibit 'turfiness', and in *Leptinella* managed surfaces a well defined 'tide mark' appears on the surface. Typical root symptoms may include a sparseness of roots, discolouration, a stubby root condition, galled or slightly swollen rootlets, or a combination of these. Symptoms are generally

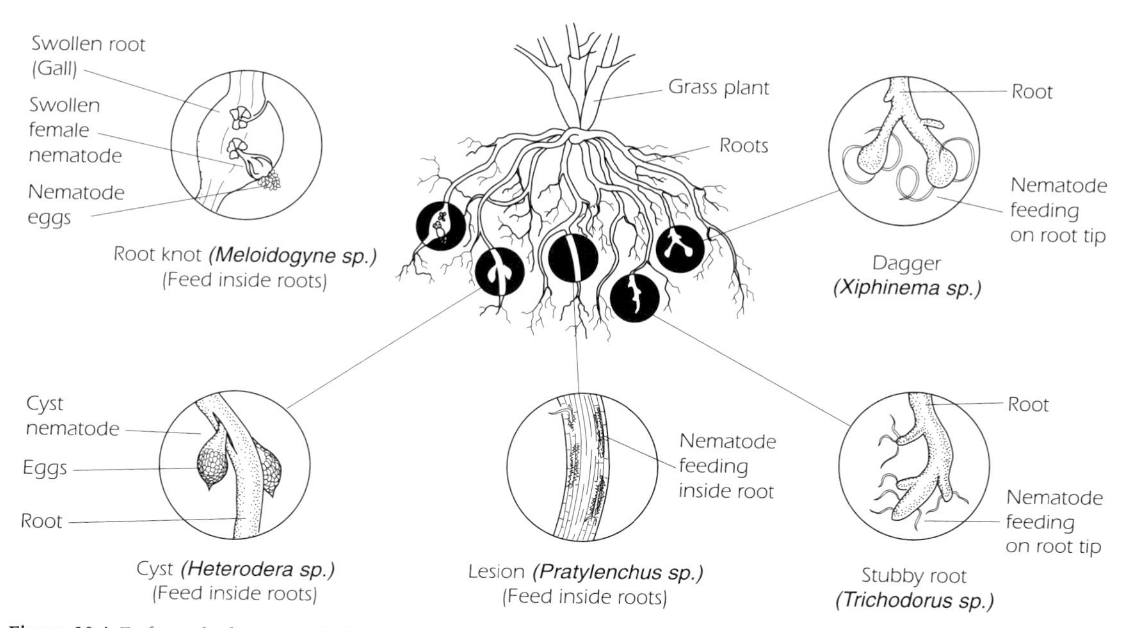

Figure 10.4 Turfgrass feeding sites of different nematode populations (courtesy of Bayer Australia)

more evident under periods of hot weather, moisture stress and poor fertility. At least ten species of parasitic nematodes are now associated with unthrifty turfgrass (Colbran, 1964; Heald and Perry, 1969; Beehag, 1995). Fourteen of the more important nematodes, their characteristics and associated turfgrass species, are presented in Table 10.6.

Nematode control can be achieved through effective sanitation and preplant preventation, and the use of resistant turfgrass cultivars (Beard, 1973). Sanitary practices include the purchase of vegetative material for plugging or sprigging from reputable sources, and the effective cleaning of equipment that may be used in turf areas other than the infected area. Heald and Wells (1967) reported that nematode populations were reduced in hybrid bermudagrass cv. Tifgreen and Tifdwarf by a hot water treatment 48.8 to 51.6°C (120 to 125°F) for 30 to 45 minutes. Preplant prevention methods include the use of heat treatments or fumigation products to be used at recommended rates. Early work by Heald and Burton (1968) and Nutter and Christie (1958) have shown that fertilisers containing organic nitrogen in the form of processed sewage sludge will suppress nematode populations. Other chemical alternatives currently being tested for nematode control include a mustard seed meal product that has demonstrated bio-fumigation properties, the ground stalk of the sesame plant (Westall, 1997) and a liquid extract from the seaweed *Ecklonia maxima*, which is known to inhibit nematode larvae penetration and retard development (Hodgson, 1995). Resistance of grasses to nematodes has been reported by several workers (Riggs *et al.*, 1962; Gaskin, 1965). As early as 1970 Johnson reported that bermudagrass cv. Tifdwarf was more tolerant of ring and stunt nematodes than other bermudagrass hybrids. Varying degrees of resistance have also been reported for St. Augustine grass and zoysiagrass cultivars.

Chemical control is warranted if monitoring levels indicate greater than 50 per cent shrivelled, brown roots, turf quality has deteriorated, and if nematode numbers detected are significantly above established threshold guidelines.

Dibromochloropropane (DBCP) has been used to reduce nematode populations in centipedegrass (Walker *et al.*, 1975), ethoprop in *Leptinella* greens (Yeates, 1977), and fenamiphos in bermudagrass, creeping bent and perennial ryegrass turf (Siviour, 1978). Fumigants, such as methyl bromide, are also useful if applied preplanting. In all states and countries check with state authorities before using a nematicide. Overuse may also lead to biodegradation problems.

Integrated pest management

In an attempt to overcome the reliance on pesticides and also to address environmental and occupational health and safety concerns, there has been a philosophical shift towards the implementation of integrated pest management (IPM) or best management practices (BMP) systems. IPM has been described as a pest management system that incorporates various control tactics to keep pests from reaching injurious levels, while minimising effects on humans and the environment (Bruneau *et al.*, 1992). The process aims at optimising natural mortality factors through the systematic and balanced use of all environmentally, economically and socially compatible control methods while attempting to keep potential pests from reaching unacceptable levels. IPM does not exclude the use of pesticides, but insists on their rationalisation. Pests of turfgrasses develop only when several specific criteria occur simultaneously over a given period of time. The pest or extremes in environmental conditions, a susceptible host, and environmental conditions conducive to pest development must be present for a certain period of time and need to be monitored as part of an IPM program. IPM is reliant on the successful monitoring of pests and diseases and the provision of data so that informed decisions can be made with

Table 10.6 Common and scientific names of common cyst, endo- and ectoparastic nematodes, and associated turfgrass species.

Common name	*Scientific name*	*Associated turfgrass species*
CYST		
Cyst (1,2)	*Heterodera spp.*	Tall fescue, fine-leaf fescue, bermudagrass, rough bluegrass, ryegrass
ENDOPARASITIC		
Root-knot (1,3,4)	*Meloidogyne spp.*	Bermudagrass, Kentucky bluegrass, rough bluegrass, bentgrass, cotula, St. Augustinegrass
Root lesion (1,2,4)	*Pratylenchus spp.*	Prairiegrass, bermudagrass, Kentucky bluegrass, Queensland blue couch, perennial ryegrass
Burrowing (1)	*Radopholus spp.*	Bermudagrass, St. Augustinegrass
Foliar and floral (1,2,3)	*Anguina spp.*	Browntop, chewings fescue, bentgrass
Stem and bulb (1)	*Ditylenchus spp.*	Annual bluegrass
ECTOPARASITIC		
Spiral (1,2,3,4)	*Helicotylenchus spp.*	Bladygrass, paspalum, creeping bentgrass, bermudagrass, Kentucky bluegrass, cotula, zoysiagrass.
Sting (1,4)	*Belonolaimus spp.*	St. Augustinegrass, zoysiagrass, bermudagrass,
Stunt or stylet (1,4)	*Tylenchorhynchus spp.*	Bermudagrass, Kentucky bluegrass, annual bluegrass, creeping bentgrass
Ring or root (1,2,4)	*Criconemoides spp.*	Queensland blue couch, zoysiagrass, kangaroo grass, Kentucky bluegrass, St. Augustinegrass
Pin (1,4)	*Paratylenchus spp.*	Kentucky bluegrass, tall fescue
Stubby root (1,2,4)	*Paratrichodorus spp.*	Rhodesgrass, bermudagrass, St. Augustinegrass
Dagger (1,2,3,4)	*Xiphinema spp.*	Rhodesgrass, cotula, zoysiagrass, St. Augustinegrass
Lance (1,4)	*Hoplolaimus spp.*	Bermudagrass, annual bluegrass, centipedegrass

Key to authorities: 1. United States of America, Couch (1993); 2. Australia, Colbran (1964), Aldous (1983); 3. New Zealand, Yeates (1977); 4.US: Watschke *et al.* (1995).

respect to the selection and timing of control measures.

Four significant pest control goals need to be considered: prevent pests from becoming established in the first place, change the genetic composition of plants so these plants can resist attack, alter the environment so that susceptible turfgrass plants can overcome an pest attack, and offer protection of the plant attacked by the pest. The tactics used to achieve these four goals could include seed and seedbed sanitation, use of pest-resistant plant cultivars, use of best practice cultural management operations, and the application of recommended chemicals. Sanitation includes planting of pest-free seed and vegetative material, purchasing this material from a high health nursery, or using fumigation of the soil as a means to eliminate the pest population in the seedbed. For example seedling diseases caused by damping off (*Pythium spp.*) and brown patch (*Rhizoctonia spp.*), are especially well controlled by seedbed fumigation.

Many organisms have natural enemies, such as fungi, bacteria and nematodes, that keep the turf system in balance and help to prevent or minimise the establishment of a pathogen and the development of disease problem. *Rhizoctonia solani* can be controlled naturally by the common soil fungus *Trichodermis viride* which produces an effective toxin against the disease. The choice of organic material is important because it must not only provide a supply of organic matter to support microorganism activity, but also support the relevant microorganisms that can control the disease (McGeary, 1996). Leslie (1994) reviewed the literature on organic materials as they relate to turf, and reported that some of the best sources of both organic matter and populations of antagonistic microorganisms are composted materials. These can be either incorporated with the sand/soil rootzone mix after fumigation and before establishment, or part of the topdressing material. Monthly applications of topdressing material containing one kilogram of compost per 100 square metres have been shown to be effective in suppressing diseases such as dollar spot, brown patch, pythium root rot and red thread (Leslie, 1994). *Pythium* populations can be reduced on putting green turf that receives continuous applications of compost in the absence of any fungicide application. The Chiba Prefectural Government in Japan has examined several chemical-free pest control methods and found that the grass cutworm and bluegrass webworm can be satisfactorily controlled with combinations of entomopathogenic nematodes and bacteria, sex pheromones, and/or by covering the turf with nets (Anon, 1997).

Turfgrass cultivars vary in their susceptibility to pests and adverse environmental conditions. Selection of turfgrass species and cultivars tolerant to pest attack, or adapted to special environmental conditions, help turfgrass establishment on specific sites and their tolerance of pests. It is now commonplace to plant blends or mixtures on new turf surfaces that will have a maximum chance for survival with minimal maintenance costs. Bio-insecticides have already had promising results in controlling pests and diseases. The bacterium *Bacillus spp.*, and nematodes have been shown to control mole crickets (Bruneau *et al.*, 1993; Kaya *et al.*, 1993) and the Japanese beetle (Wong, 1996), and *Bacillus thuringiensis* has shown good control against many Lepidopterous pests. *Metarhizium ansopliae*, a strain of fungus, has been shown to specifically attack the red headed cockchafer (*Adorphorus couloni*) (Rath *et al.*, 1995), and a bacterial strain of *Enterobacter clocae* has been shown to effectively control Pythium blight in pots of bentgrass cv. Penncross, and perennial ryegrass cv. All star (Palmer 1996). Successful development of microbial control agents depends on their compatibility with the other control agents such as herbicides, insecticides and fungicides. An insecticide derived from organic plant extracts has been shown to be effective in controlling lawn armyworm and African black beetle (Hassan, 1997). This product, and presumably others of like kind, are now achieving recognition with national registration authorities as insecticides. This type of approach recognises organic oils offer potential as environmentally sound means of controlling many insect pests. Over the past decade, symbiotic fungal endophytes have also been found to enhance insect resistance; billbugs (*Spehnophorus parverlus*) (Johnson-Cicalese and White, 1990), and sod webworm (*Parapediasia spp.*) in cool-season grasses, enhance fungal disease resistance; leaf spot (*Rhizoctonia zeae*) and brown patch (*R. solani*), in tall fescue (*Festuca arundinaceae*), also improve the drought tolerance and nitrogen utilisation of the same grass (Bacon *et al.*, 1997).

The best way to prevent pest problems is to manage the turf environment and so discourage pests. Problems generally occur when the turf is weakened by poor management, excessive use or through environmental stress. Pest populations will be governed mainly by temperature, humidity, light intensity and quality, and moisture, as well as cultural practices. Integrated pest control involves acknowledging the turfgrass surface as an ecological system, and therefore requires the turf manager to have a greater understanding about the turfgrass, its environment, the pests involved, the alternative control measures, as well as the various interactions that will occur in the system (Martin, 1994).

References

Adams, W.A. and R.J. Gibbs. 1994. 'Natural turf for sports and amenity: science and practice', *CAB International*, Cambridge, 404 pp.

Aldous, D.E. 1983. 'Nematode problems in turfgrass', in D.E. Aldous (ed.), *Turfgrass Disease Proceedings*, Victorian College of Agriculture and Horticulture and the Turf Research Institute, Department of Agriculture, Victoria, 12–13 July, 118 pp.

Aldous, D.E. 1991. *Weed control for the home garden*, Lothian Press, South Melbourne, 64 pp.

Anon. 1997. *Guidelines for non-chemical pest control in golf courses*, Chiba Prefectural Government, Japan, Agriculture and Forestry Department, Chiba Prefectural Government, Japan, 43 pp.

App, B.A. and S.H. Kerr 1969. 'Harmful insects', in A.A. Hanson and F.V. Juska (eds), *Turfgrass Science*, American Society of Agronomy Inc., Madison, pp 336–57.

ATRI. 1988. *Weeds and their control*, Australian Turfgrass Research Institute, Concord West, NSW.

ATRI. 1994. *Turf disease manual*, 3rd edn, Australian Turfgrass Research Institute Ltd, Concord West, NSW, 51 pp.

Bacon, C.W., M.D. Richardson and J.F. White. 1997. 'Modification and uses of endophyte-enhanced turfgrasses:a role for molecular technology', *Crop Science*, 37(5):1415–25.

Beard, J.B. 1973. *Turfgrass: science and culture*, Prentice-Hall, Englewood Cliffs, N.J., 658 pp.

Beard, J.B. 1982. *Turf Management for Golf Courses*, McMillian N.Y., 642 pp.

Beehag, G.W. 1994. 'A review of *Poa annua* (L.) management', *Proceedings 1st ATRI Research Conference, Sydney*, Australian Turfgrass Research Institute, vol. 1:61–5.

Beehag, G.W. 1995. 'Nemacur®—Could the turfgrass industry manage without it?', *TurfCraft*, July, pp. 74–7.

Beehag, G.W. 1996. 'Management of aquatic plants in a sportsturf environment: Part 1', *A.T.R.I. Turf Notes*, Summer, p. 3.

Brogden, J.E. and S.H. Kerr. 1964. *Home gardners lawn insect control guide*, Florida Agri. Exp. Sta. Circ. 213A.

Bruneau, A.H., J.E. Watkins and R.L. Brandenburg. 1992. 'Integrated pest management', in D.V. Waddington, R.N. Butler, G.D. and W.R. Kneebone, 'Variations in response to bermudagrass varieties to bermudagrass mite infestations with and without chemical control', *Arizona Turfgrass Research Report*, 230:7–16

Colbran, R.C. 1964. Studies on plant and soil nematodes. 7. *Queensland Records of the Order* Tylenchida *and the Genera* Trichodorus *and* Xiphinema, Bulletin No. 269, Queensland Dept. of Primary Industries. pp. 77–123.

Cooper, R. 1997. 'Combination products for the Australian turf industry', *The 1997 Green Pages Annual*, Strategic Publications, New Gisbourne, Victoria. pp. 62–4.

Couch, H.B. 1993. *Diseases of turfgrass*, 2nd edn., R. Kreiger, Malabar, Fla.

Couch, H.B. and E.R. Bedford. 1966. 'Fusarium blight of turfgrasses', *Phytopathology*, 56:781–6.

Couch, H.B. and J.R. Bloom. 1960. 'Influence of environment on diseases of turfgrasses. II. Effects of nutrition, pH and soil moisture on *Sclerotinia* dollar spot', *Phytopathology*, 50: 761–3.

Funk, C.R., F.C. Belanger and J.A. Murphy. 1994. 'Role of endophytes used in grasses for turf and soil conservation', in C.W. Bacon, and J.F. White, Jr. (ed.), *Biotechnology of endophyte fungi*, CRC Press, Boca Ration, Fl, p. 201-2-9

Gaskin, T.A. 1965. 'Susceptability of bluegrass to root-knot nematodes', *Plant. Dis.Reptr.*, 49:89–90.

Grime, J.P. 1980. 'An ecological approach to management', in I.H. Rorison and R. Hunt (ed.), *Amenity grasslands—an ecological perspective*, University of Sheffield, John Wiley & Sons Ltd, pp. 13–55.

Hassan, E. 1997. 'Natural oils as insecticides', *International Turfgrass Society, University of Sydney, NSW, poster presentation*, 2 pp.

Hatsukade, M. 1997. 'Biology and control of the hunting billbug, *Sphenophorus venatus vestitus* Chittenden, on golf courses in Japan', *International Turfgrass Society, Research Journal*, vol. 8:987–96.

Heald, C.M. and V.G. Perry. 1969. 'Nematodes and other pests' in A.A. Hanson and F.V. Juska (ed.), *Turfgrass Science*, American Society of Agronomy, Number 14, Agronomy series, Madison, Wisconsin, US, chap. 12:358–69.

Heald, C.M. and H.D. Wells. 1967. 'Control of endo- and ecto-parasitic nematodes in turf by hot water treatments', *Plnt. Dis. Reptr.*, 51(11):905–7.

Healy, M.J. and M.P. Britton. 1968. 'Infection and development of *Helminthosporium sorokinianum* in *Agrostis palustris*', *Phytopathology* 58:273–6.

Helson, G.A. 1960. 'Insect pests of turf' in G. Robinson (ed.), *Turf Culture*, New Zealand Institute for Turf Culture, pp.153–66.

Johnson, A.W. 1970. Pathogenicity and interaction of three nematodes species on six bermudagrasses, *J. of Nemat.*, 2:36–41.

Johnson-Cicalese, J.M. and R.H. White. 1990. 'Effect of *Acromonium* endophytes on four species of billbugs found in New Jersey turfgrasses', *J. Am. Soc. Hort.Sci.*, 115:602–4.

Juska, F.V. 1965. 'Billbug injury in zoysiagrass turf', *Park maintenance*, 18(5):38.

Kappro, J. 1996. 'The changa mole cricket', *A.T.R.I. Turf Notes*, Autumn, p. 5,7.

Kaya, H.K., M.G. Klein and T.M. Burando. 1993. Impact of *Bacilluspopilliae, Rickettsiella popilliae* and entomopathogenic nematodes on a population of the scaraeid, *Cyelocephala hirta. Biocontrol Science and Technology*, 3(4): 443–53.

Latch, G.C.M. 1960. 'Diseases of turf and disease control', in G.S. Robinson (ed.), *New Zealand Institute for Turf Culture*, 362 pp.

Leslie, A.R. 1989 'Development of an IPM program for turfgrass', in A.R. Leslie and R.C. Metcalf (ed.), *Integrated Pest Management for Turfgrass and Ornamentals*, US Environmental Protection Agency, Washington, US, pp. 315–18.

Liu, D.L. and N.E. Christians. 1997. 'The use of hydrolyzed corn gluten meal as a natural preemergence weed control in turf', in J.Hall, (ed.), *International Turfgrass Society Research Journal*, University of Sydney, NSW, vol. 8:1043–50.

Martin, P. 1994. 'Integrated pest management: is it working?', *13th Australian Turfgrass Conference, Australia Golf Course Superintendents Association, Adelaide*, 4 pp.

McGeary, D. 1996 'Disease control and prevention', *TurfCraft*, no. 49, July/August, pp.25–30.

Newman, J. 1994. 'Control of algae with straw', *Information Sheet 3*, Aquatic Weeds Research Unit, Long Ashton Res. Stn, Berkshire, UK, pp. 1–9.

Neylan, J. 1996. 'Nematodes', *Golf and Sports Turf Australia*, October, pp 11–15.

Nus, J. 1993. 'A knowledge based approach: managing ponds and lakes', *Golf Course Superintendent*, June, pp. 6, 7, 10.

Nutter, G.C. and J.R. Christie. 1958. 'Nematode investigation on putting grass turf', *Proc. Fla. State Hort. Soc.*, pp. 445–9.

Palmer D. 1996. 'Biological control agents for turf pests not far way', *TurfCraft*, no. 49, July/August, pp. 35.

Pass, B.C., R.C. Buckner and P.R. Burrus.1965. 'Differential reaction of Kentucky bluegrass strains to sod webworms', *Agron. J.*, 57:510–11.

Reinert, J.A., M.C. Engelke, J.C. Read, J. Maranz and B.R. Wiseman. 1997. 'Susceptibility of cool and warm season turfgrasses to Fall Armyworm, *Spodoptera frugiperda', International Turfgrass Society Research Journal,* vol. 8:1003–11.

Richardson, R.G. and R.C.H. Shepherd. 1992. 'Recommendations for weed control in temperate Australia, vol. 1', *Weed Science Society of Victoria Inc.,* Melbourne.

Riggs, R.D., J.R. Dale, and M.L. Hamblen. 1962. 'Reaction of bermudagrass varieties and lines to root-knot nemtatodes', *Phytopathology,* 52:587–8.

Rolfe, C., A. Currey and I. Atkinson. 1994. *Managing water in plant nurseries,* H.R.D.C./N.I.A.A./N.S.W. Agriculture, 64 pp.

Schread, J.C. 1964. 'Insect pests of Connecticut lawns', *Connecticut Agr. Exp. Sta.Circ.* 212 (Rev.).

Shurtleff, R.C., T.W. Fermanian and R. Randell. 1987. *Controlling turfgrass pests,* Reston Book, Prentice-Hall Inc., Englewood Cliffs, N.J.

Siviour, T.R. 1978. 'Biology and control of *Belonolainus lolii n.sp.'* in *Turf. Proc 3rd Nat. Plant Path. Conf. A.P.P.S. Melbourne, May 14–17,* Abstract no. 63., Aust. Plant Path. Soc. Newsletter 7(1 sup.):37.

Smiley, R.W. 1983. *Compendium of turfgrass diseases,* American Phytopathological Society, St. Paul, MN, 102 pp.

Smiley, R.W., P.H. Dernoeden and B.B. Clarke. 1992. *Compendium of turfgrass diseases, 2nd edn,* American Phytopathological Society, St. Paul, MN, 98 pp.

Smith, J.D., N. Jackson and A.R. Woolhouse. 1989. *Fungal diseases of amenity turf grasses,* E & F.N. Spon., London, New York, 401 pp.

TeBeest, D.O. 1991. *Microbial control of weeds,* Chapman and Hall, Great Britain.

Turgeon, A.J. 1991. *Turfgrass management,* 3rd edn, Prentice Hall, New Jersey.

Vargas, J.M. 1981. *Management of turfgrass diseases,* Burgess Publishing Co., Minneapolis, MN, 204 pp.

Vengris J. and W.A. Torello. 1982. *Lawns basics factors, construction and maintenance of fine turf areas,* Thomson Publications, California.

Walker, C. 1960. *Turf culture,* New Zealand Institute for Turf Culture, Palmerston North, pp. 352.

Walker, J.T., R. Molsmeer and J. Melin. 1975. 'Effects of repeated annual and semiannual nematicide application to centipede grass', *J. of Nemat.,* 7(4):331.

Ward, A. 1995. 'Non-destructive sampling techniques for the lawn armyworm and the African black beetle in turf', *Golf & Sports Turf Australia,* April, pp.12–18.

Watschke, T. L., P.H. Dernoeden and D.J. Shetlar. 1995. 'Managing turfgrass pests', *Advances in Turfgrass Science,* Lewis Publishers, 361 pp.

Westall, D. 1997. 'Nematode management-using chemical alternatives', *The 1997 Green Pages Annual,* pp. 13–14

Whittet, J.N. 1965. *Pastures of NSW,* Department of Agriculture, Sydney, NSW, 632 pp.

Wong, P. 1996. 'Soil-borne patch diseases of turfgrass and their management', *A.T.R.I. Turf Notes,* Autumn, p. 2, 10.

Woodcock, T. 1983. 'Patch diseases of turf', in D.E. Aldous, (ed.), *Turfgrass Disease Proceedings,* Victorian College of Agriculture and Horticulture and the Turf Research Institute, Department of Agriculture, Victoria, July 12–13, 118 pp.

Woodcock, T. 1997. 'Couchgrass diseases', *Golf & Sports Turf Australia,* February, p. 41–3.

Woodcock, T. 1983. 'Sample preparation and disease diagnosis', in D.E. Aldous (ed.), *Turfgrass Disease Proceedings,* Victorian College of Agriculture and Horticulture and the Turf Research Institute, Department of Agriculture, Victoria, July 12–13, 118 pp.

Yeates, G.W. 1977. 'Incidence, effects and control of nematodes in *cotula* greens', *Sports Turf Review,* N.Z. Turf Culture Institute Inc., No. 111, October, pp. 123–6.

CHAPTER 11

Turfgrass facility business and administration

W.R. WITHERSPOON, Guelph Turfgrass Institute, Ontario Agricultural College, University of Guelph, Guelph, Ontario, Canada

Introduction

One of the greatest challenges for a turf manager is the management of finances and people. The vast majority of problems that arise on a day-to-day basis are either financially based or, most often, related to some aspect of human resource management or dealing with people. A basic understanding of financial principles and effective human resource management techniques can go a long way to ensuring that your facility is run effectively and efficiently. This chapter is intended as a general overview of basic business and human resource management principles and techniques. You are encouraged to seek out more detailed information or support.

Financial management

Accounting

Accounting is simply keeping track of expenses and revenues. In small businesses, this may be accomplished using a ledger book or a computer and a good basic accounting program. However, the best single tool to manage your business finances is to have a good accountant. Professional advice is invaluable to establish the accounting method that best suits your operation. A good accountant can also provide assistance with respect to taxation and other aspects of law that affect the way you financially manage your operation.

Developing a budget

A budget is more than just how much revenue you expect and how much money you have to spend in a given period. It is a statistical and financial expression of the goals and objectives of an operation. The budget should be closely aligned with the basic goals and objectives of your organisation. For example, if your primary goal is to maximise profit, the budget would be developed to optimise revenue through minimising expenses. In most businesses, quality of product or service has a major impact on revenue so that expenses are effectively targeted to providing a product or service that will generate the maximum revenue.

A budget is developed by determining the goals you wish to accomplish in a given period, developing a list of tasks that must be undertaken to achieve the goals and then assigning a dollar value to the tasks. In most operations, labour (salaries and benefits) is the primary expense. Other costs include the purchase and maintenance of capital items such as machinery (trucks, mowers, tractors, etc) and buildings and the various supplies required for your business to operate (seed, soil, fertiliser, pesticides, etc).

Table 11.1 provides a sample maintenance budget for an 18-hole golf course (provided for illustrative purposes only, amounts will vary significantly from region to region and course to course).

Table 11.1 White Pine Golf Club: Maintenance budget

	$
Personnel	
Salaries, wages, bonuses	152 500
Payroll taxes and benefits	12 500
Professional development	2 750
Travel	1 000
Hospitality	300
Contracted services	500
Emergency services	500
Sub-total	170 050
Utilities and services	
Water	7 500
Electricity	5 000
Heating fuel	700
Telephone	700
Sub-total	13 900
Supplies	
Fertiliser	8 000
Pest control products	10 000
Seed and sod	7 000
Fuel, lubricants, cleaners	5 000
Expendable supplies	3 500
Plant material	1 000
Printing	250
Miscellaneous	500
Sub-total	35 250
Repair and maintenance	
Parts and repairs	7 500
Tree care	5 000
Irrigation and drainage repairs	3 500
Licences and permits	250
Rentals	500
Depreciation (equipment)	12 500
Miscellaneous	1 550
Sub-total	30 800
Total	250 000

(adapted from Daniel and Freeborg, 1979)

Estimating costs and making a profit

An essential part of budgeting for existing operations and new projects is the development of good cost estimates. Companies that bid on maintenance or construction contracts require the ability to make accurate estimates of costs for each job and include a margin for profit.

Each operation or job may have all of some of the following costs: materials, labour (including benefits, taxes, insurance, etc), equipment and subcontractors. In addition, each of these costs may have additional costs that must also be added in as part of the estimate; items like supervision, start-up and clean-up time each day, storage containers, etc. Overhead for each job should also be recovered. Overhead is the fixed cost of operating your business whether it is a lawn care company, golf course maintenance operation or city parks department. It includes many things including rent, advertising, depreciation, professional dues, insurance, telephone, utilities and so on—anything that costs you money but can not be attributed to one specific job or task. Each year you should forecast your overhead costs based on historical records as well as your forecast of what you are planning to do in the next year and what you will need to do it. On each job you are estimating, apply this overhead to each of the four costs: materials, labour, equipment and subcontractors, based on the portion of your overhead that you can attribute to each of those items. A general rule-of-thumb in allocating overhead is to add 10 per cent for overhead to material, 25 per cent to equipment and 5 per cent to subcontractors. When you apply these percentages as your overhead costs, the remaining overhead must be recovered from the labour portion of your estimate. Keeping good records of your costs is critical to developing a good overhead recovery system in any business that must provide accurate cost estimates or bid for work.

If you are operating a business, profit is a return on your investment in your business. In pricing your services or developing a bid to provide services, you should include a profit percentage that reflects the return you desire from your investment in your business. Your profit should take into account a variety of considerations including how badly you need the work, the risk associated with taking the work, the size of the job, and the economic and competitive conditions of the marketplace (Vander Kooi, 1997).

Purchasing system

Every operation requires some form of purchasing system to track and control purchases. The most basic system is to have one person (or their designate) responsible for approving all purchases above a certain value. Larger operations control purchases through a purchase order system where purchases above a certain value must be approved prior to ordering. Larger purchases may be let out to tender—the product or service is described in detail and sent to several suppliers for a price quote. In most instances the lowest price quote is accepted. For record keeping, all purchases should be attributed to a specific job or task so that the funds can be tracked against a budget or job estimate (Ostmeyer, 1994).

Record keeping

Effective management of budgets and job costing involves accurate records. The detail of your record keeping system will be dictated by the resources available for record keeping. Computer systems offer the most efficient method of keeping financial records such as payroll, accounts payable and accounts receivable. As previously mentioned, accurate records are critical in developing an effective overhead recovery system in any operation. This is another area where professional assistance in the form of a good accountant can save you significant time and money in managing your business or operation.

Time sheets

Time sheets are a specific type of record used to determine payment for employee time. Larger companies may use time clocks where employees 'punch in' at the start of the work day and 'punch out' at the end of the work day. The time cards are collected at the end of each pay period and pay cheques are made out. More detailed time sheets are also valuable in tracking labour costs for specific tasks and jobs. Card systems are now available that allow employees to 'swipe' their employee card when they start and end work, with the information being transferred directly to a time sheet program for calculation of pay. More detailed time sheets providing hours spent on particular activities can be used for future planning as well as developing methods to more effectively allocate staff to various tasks.

Human resource management

If you ask an experienced turf manager what single task takes the most of his or her time, the answer will invariably be managing people. People are any business' greatest asset. However, they can also be your worst nightmare. Equipment can break down but in most situations it can be repaired or replaced in a short period of time. Equipment doesn't have children, drink excessive amounts of alcohol, get sick, or leave at your busiest time to pursue other interests. As you gain more responsibility within an organisation, or develop and expand your business, human resource management will quickly become your greatest challenge. The following is a very basic introduction to the intricacies of managing people for optimum efficiency. A significant portion of this section is adapted from a series of publications developed for horticultural and agricultural employers in Ontario (McEwan and Owen, 1995).

Hiring

Determining labour needs

Labour needs are basically determined by the job you have to do. On a golf course, park or sports field, your requirements are usually fairly straightforward. You know the area you have to maintain and probably have a good idea of the labour time required for the various routine maintenance tasks. Take into account the abilities of your current staff as well as the abilities of seasonal or part-time staff based on your past experience. It is a good idea to develop an annual schedule of tasks and estimate your labour needs over the course of the year. At times where labour needs are decreased you may be able to schedule new projects to help balance your labour requirements. Utilisation of part-time labour and sub-contracting are two ways you can gain additional flexibility in managing your labour needs.

Table 11.2 provides an example of one method of determining annual labour needs and the number of employees required. You should use past records or best estimates of labour needs to complete the charts for each task you are required to carry out over the course of the year. It is helpful to chart your labour requirements for each month to determine your seasonal staffing needs (Hannebaum, 1995).

Table 11.2 Sample task analysis (adapted from McEwan and Owen, 1995).

NAME OF TASK (e.g. greens mowing, fertilising, pest control, etc.)

Type of work and when
Length of time
Total hours
Staff required

TOTAL LABOUR REQUIREMENTS FOR THE BUSINESS (sum of all tasks)

Type of work and when
Total hours per year
Total hours per week
Who does it
Number of hours per week per person
Employee schedule / hourly analysis

Employee name
Number of normal weekly hours
Number of normal daily hours
Number of weekly hours including overtime
Number of daily hours including overtime
Number of possible overtime weeks

Job descriptions

Job descriptions are valuable for a number of reasons. They help you to find and select applicants for jobs, orientate new staff, develop performance standards for jobs, evaluate performance, reward employees and in some cases dismiss employees. Most importantly, they clearly communicate to employees your expectations for their performance. A job description should include the following: job title, name/title of supervisor, principal duties and responsibilities, minor duties, physical requirements, productivity expectations, wages and benefits, work hours and vacation time, work environment, safety responsibilities, job hazards and job qualifications. As always, it is wise to check with local regulations to determine specific detail related to legislative requirements that should be included in job descriptions (employment standards, pay equity, etc).

Recruiting

The sources of labour vary from region to region. Some areas have government operated job recruitment centres. High schools, colleges and universities are good sources for summer employees and new graduates. Retired persons in good health are excellent candidates for part-time work (Graham, 1997). Certain jobs may be appropriate for persons with disabilities. Determining the most appropriate sources will depend upon the type of position you are trying to fill. Job advertisements may be placed in local newspapers or in regional trade publications. The advertisement should be clearly worded to inform potential candidates about the position (the job description is useful for this information), promote your business in a fashion that will attract high quality applicants, and help screen people who do not have appropriate skills to perform the job. It may be helpful to design an application form for applicants to complete which will help you gather the information that is most important to you.

A sample job advertisement could read:

White Pine Golf Club, a nationally ranked public golf course built in 1923 and operated with the highest standards of maintenance and customer service, requires the services of a

TURF EQUIPMENT OPERATOR

We require an equipment operator with a practical knowledge of:

- Current golf course maintenance techniques
- Basic turf equipment maintenance practices
- Safe equipment operation

Qualifications: Related experience and preferably technical training or certificate in turf equipment operation from a recognised college or university and a valid driver's licence. We also require an individual with good communication skills and the ability to work under minimal supervision in a team-oriented work environment.

If you have the necessary skills and experience, please submit your application in writing to Debra Witt, Superintendent, White Pine Golf Club, P.O. Box 123, Northville, Ontario N0B 1Y0 or by fax (519) 555-1212.

Only those individuals being considered for an interview will be contacted. We wish to thank all applicants for their interest in employment with our club.

Interviewing

After reviewing all of the applications select a maximum of five or six candidates to interview. Make a list of the behaviours (job stability, attendance, ability to work with minimal supervision, ability to work as a team member, etc) and knowledge (technical skills, etc.) you are looking for in an employee. Develop a list of questions that will assist you in evaluating each of these abilities and characteristics. These questions may ask a candidate to describe situations on previous jobs where they had to exhibit one of the abilities or skills you require, as well as situational questions which ask the candidate what they would do in specific situations. Once you have developed a set of questions, create a simple rating scale that will allow you to rate their response to each question. Avoid any questions that could be considered discriminatory. For jobs that have a significant technical skill requirement (such as mechanics and equipment operators) it may be valuable to include a skills testing segment as a portion of the interview process. Ensure that you have sufficient insurance coverage in case of any mishaps.

If possible, interview candidates with another of your senior staff members asking questions. This will afford each of you the opportunity to observe the candidates while they respond to a question from the other individual. It also provides an additional opinion when the time comes to select the successful candidate for the position.

When selecting the successful candidate, consider the following:

- How well will this person represent your business or operation?
- Will he/she be able to work well with your other staff?
- Has the person shown a great deal of interest and enthusiasm in getting the position?
- Does the person have the necessary physical ability to do the job?

In some cases where you have an exceptional individual who does not have all of the skills or requirements for the job, you may want to adjust the job description to fit the skills and abilities of that person.

Employee orientation

As in most situations, first impressions are often the most important. A new employee usually has four basic concerns or questions:

- Will I like working here?
- What exactly am I expected to do?
- Who else works here and how do I fit in?
- Who do I ask or where do I go when I need information?

You can address these questions through the orientation process. Greeting the new employee with a warm welcome will help to inspire confidence and acceptance. Providing the employee with a detailed job description and describing and/or demonstrating procedures for work, equipment and safety procedures will help to develop their expertise and knowledge of their job. It is important to introduce new employees to other staff and management, and to let them know what other people do and how their job may relate to other jobs in your operation. Introduce them to their immediate supervisor, if it is someone other than yourself, and provide them with access to all orientation documents, employee handbooks, and safety and equipment manuals. The employee should also be made aware of all legal and safety rules, behaviour rules, dress code or any other rules or regulations that govern them while in your employ.

Training, motivating and evaluating

Productivity of employees is influenced by a number of personal and workplace factors. Supervisors have a significant influence on employee productivity by the way in which they manage the workplace to create an atmosphere where employees are motivated and have the skills and tools to do the job. Most successful workplaces have four common features:

- appropriate training and instructions are provided
- employees are motivated to provide their best effort
- employee and employer performance is evaluated
- employee contributions are recognised through a variety of means

Training

Proper training is essential to ensure that employees know how to properly carry out tasks and for them to do tasks in an efficient manner. Supervisors should set training objectives through on-going review of performance and safety information. Training methods will vary depending upon the task but your training methods should be geared to the abilities and needs of the employee. Employees should be shown new tasks or techniques and given the opportunity to do the new task or technique with feedback from the individual doing the training. Follow-up evaluation will ensure that the training has been successful. Other longer term training, such as support for an employee pursuing distance education or evening courses, helps reinforce your efforts to support the personal and professional growth of your employees.

Motivating

Individuals are motivated based on their needs. Our most basic needs are physiological, safety and social needs followed the more personal and internalised needs for self-esteem and self-actualisation. Physiological needs are basic life support, such as food, water and shelter. In terms of the workplace, the physiological needs of individuals are provided by a comfortable working environment along with sufficient pay to provide support for food, clothing and shelter. Safety needs are supported by having a safe work environment, job security and job benefits like health care and disability insurance. Social needs can be met by a supportive team environment and a compatible supervisor. Most jobs provide some support for these three basic needs.

The key to effective motivation of employees is to provide a work environment that not only meets their basic needs, but also addresses their need for self-esteem and self-actualisation. Self-esteem can be supported through the provision of fair wages for work, praise and recognition from supervisors, performance evaluation and pay increases based on performance measures. Self-actualisation needs or the need to be in control of one's own fate can be realised by providing creative and challenging work, giving employees a role in the decision-making process and providing them with flexibility and autonomy.

Performance evaluation

The performance evaluation process fulfils needs of both the employer and the employee. The employer gets information from the employee, evaluates the contribution of the employee to the overall operation, helps build a collaborative approach to solving common problems, provides motivation for the employee to improve their skills and attitude, and allows the opportunity to identify training needs. The employee has an opportunity to evaluate how successful they are at doing their job, provides direction for improvement and, perhaps most importantly, provides an official avenue through which they can share any concerns they have with regards to their job and their future.

Successful performance evaluation can only be conducted in an atmosphere of mutual respect and trust. It is very important not to create a stressful environment where the employee feels that they are subject to criticism and reduced job security. A positive, open and honest exchange of information should be the goal of the performance evaluation process.

The most effective way to conduct a performance evaluation is to develop a list of the skills, attitudes and work habits you want to evaluate. This list should be developed with the input of the employee. In fact, providing the employee with the first opportunity to list the tasks they perform will help you as the employer gain a better understanding of how they view their role within your organisation. Skills include the variety of specific and routine tasks the employee does—mowing, fertilising, irrigating, maintaining equipment, etc. These skills can be evaluated using several criteria including quality of work, safety and productivity. Attitudes include the ability to work within a team, leadership skills and how they interact with supervisors, customers and other employees. Work habits include attendance, punctuality, interest in personal and professional growth, and self initiative. Each of these skills, attitudes and work habits can be given some form of rating system for the purposes of evaluation.

Employers should establish skills and productivity levels expected for each employee's position. When approaching the evaluation interview, it is valuable to first provide the evaluation criteria to the employee and allow them to rate themselves. The interview can then be undertaken as a review of their personal rating as compared to the supervisor's rating to help identify areas requiring improvement. Prior to the interview, the employer should randomly monitor employee performance levels so that they have an accurate and documented reasoning for their performance ratings. During the interview, listen to the employee's comments and concerns—they provide valuable feedback as to where you can work with the employee to improve their skills, attitudes and work habits.

Performance evaluations that are conducted annually and filed are a waste of everyone's time. The performance evaluation process is an ongoing process. Monthly evaluations will assist in tracking progress of employees over time. Clearly stated, realistic and measurable goals should be the outcome of the performance evaluation process. Progress towards these goals will improve the quality of work, productivity and general atmosphere of your workplace. The performance evaluation also helps to identify employees who do not fit well with your operation and can provide the information that leads to retraining, discipline or dismissal.

Employee recognition

The logical extension of successful performance evaluation is the provision of recognition to good employees. The most obvious form of employee recognition is promotion to a more responsible position and/or a monetary raise or performance bonus. It is important to have an equitable system of employee pay and rewards to ensure that there is no opportunity for good employees to feel they are being treated unfairly in comparison to others. Wages are generally based on market conditions in comparison to other competitive organisations. Benefits such as medical and dental coverage and life insurance can help you attract better quality employees to your operation. Incentives may also include non-monetary rewards such as time off and the free use of the facility, such as a golf course, if appropriate. Incentives should be clearly defined and have a purpose in increasing employee productivity or quality of work. They should be applied evenly and fairly to all employees.

One of the most inexpensive yet under used methods of providing employee recognition is the simple thank-you. As mentioned previously, self esteem is an important personal need that can be fulfilled by employment. Recognising employees when they do a good job is a strong motivator for them to perform at a high level. Other non-financial rewards can include a letter of thanks, taking an employee out for a meal or as a group recognising one member of the team who has provided exceptional work.

Team building

To be completely effective, individual employees must work together to optimise performance and productivity. A team attitude where individual members pull together towards a common goal is desirable. To help encourage the development of a team attitude remember to:

- Hire for attitude and train to obtain skills
- Remember if you want staff to change—you need to change first!
- Motivate employees by making each player feel that they belong
- Ask questions then actively listen to the response
- Make your staff feel proud of what they do and the way they are doing it
- Continually thank individually for individual efforts
- Reward the team for team successes
- Focus on solutions
- Keep the atmosphere positive (at almost any cost)
- Implement a program of employee participation and recognition

Supervision

Leadership

Effective leadership is based on a number of factors. Leaders must have clearly defined goals and direction, and the ability to pass their vision of how to accomplish these goals on to their employees. Good leaders are good listeners and communicators. They have to be able to clearly communicate with employees and be able to stop talking and actively listen to employee's suggestions and concerns. Effective leaders know the business they are in and the challenges associated with the business. There is much debate in the turf industry, particularly among golf course superintendents, as to the amount of hands-on work a superintendent should do. There is no finite answer to this question and it will vary significantly from club to club depending upon budget and skill level of available employees. However, there is some merit to the fact that employees will gain respect for a supervisor who, in certain situations, is willing to get his or her hands dirty to get a job done. Employees also respect a leader who is honest and fair in their dealings with all employees.

A leader must also have the power to effectively lead. He or she should have the respect and admiration of employees. Leaders should understand the technical aspects of all jobs and be able to deal with complex situations. Leaders should be in a position where their authority is recognised and acknowledged as owner of the business or by the person or persons who are ultimately responsible for the operation. To be effective, leaders must also have the authority to approve raises, promotions, training and other employee incentives. They should also have the ability to make good on necessary disciplinary actions such as demotions or firings.

A good supervisor has the following characteristics:

- Provides leadership
- Able to organise available staff, equipment and materials to get a job done
- Provides motivation to employees
- Able to make decisions quickly
- Self-motivated

A critical leadership skill is delegation. A supervisor who feels that a job cannot be completed effectively unless he or she is personally involved is not able to do the job as effectively as the supervisor who is able to assign tasks through delegation. There are four distinct steps in effective delegation.

The first step is simply telling employees what they have to do. It is fast, effective and productive in the short term. It is particularly appropriate when the task or employee is new, or the team is falling behind in production and there is a sense of urgency. However, it does not empower staff, prevents employee growth and can mean that good employee ideas for completing the task are lost.

In situations where employees are more experienced, a supervisor can take the second step by having the employee share in the decisions or selling them on the decision. The employee must have some understanding of the technical aspects of the job but still requires some direction and encouragement. There are still clear lines of authority. It is particularly effective with younger and less mature employees who need direction while still being provided with some input into the decision-making process.

The third step is democratic or participating delegation. In this scenario, the employee takes most of the responsibility in making decisions. It provides opportunity for employee growth and is most appropriate when the employee is familiar with the task and has a high level of competence. It can be more time consuming and may not always encourage positive change, as individuals have an innate tendency to do things the way they have always been done rather than look for new and innovative ways to do routine tasks.

The fourth step is complete delegating or empowerment. The authority to make decisions is turned entirely over to an employee or group of employees. It is useful when working with experienced employees who have a interest in taking responsibility and making and implementing decisions. Empowerment assists in the development of mature employees, stimulates creativity, helps build team strength and frees up supervisors to pursue other tasks. It does not work well without the support of employees and must be monitored to prevent employees from abusing the power they are being given.

CASE STUDY: From equipment operator to assistant superintendent

The leading hand at your golf club has left to take an assistant superintendent's job at another course. You decide to promote Bill, an equipment operator who has been with you for several years, to the position of leading hand.

Step 1: Telling
Bill is informed of your decision. He is happy to receive the promotion but at the same time is concerned that he may not be able to handle the added responsibility. You provide him with assurance and let him know that you will provide all of the support he needs. You begin by training him to perform one of the new tasks, keeping time records, and observe his work.

Step 2: Selling
You take Bill aside and discuss with him how you make certain decisions relative to the maintenance of the golf course. He may be concerned about supervising some of his formerly equal co-workers. You provide him with support and reassurance, perhaps by explaining how supervision is more being supportive and positive with employees rather than simply ordering them around.

Step 3: Participating
You share several tasks with Bill and ensure there is good communication between the two of you.

Step 4: Delegating
You give Bill full responsibility. You remain available when he needs your support but stay out of the day-to-day decision making in his area of responsibility. His performance is reviewed regularly. You spend your time on the larger management issues of the golf course and allow Bill to manage the activities of his crew.

Communications, conflict resolution and discipline

Communications

Effective and open communication is the key to the success of your organisation. The development of an employee manual that clearly states working conditions, hours of work, expectations, chain of command, safety and other rules of conduct will help establish clear lines of communication within your organisation.

Good listening skills are critical for effective interpersonal communication. Too many individuals spend the time another person is speaking formulating their next response and waiting for a pause so that they may jump into the conversation. Active listening provides you with the opportunity to really hear what the other person is saying. When listening actively stop talking and concentrate on what the other person is saying. Respond by summarising what the other person has said or requesting clarification of comments that you do not understand. Try to gain a sense of how the person feels as well as what they are saying. Refrain from making any value judgements while listening. Active listening is a critical skill for supervisors as it shows that you respect and appreciate the feelings and concerns of your staff.

The other aspect of communication is speaking. Use language that is clear and concise and be open and honest with your employees. When seeking information, use open-ended questions that get more than a yes or no answer. For example, after explaining a task to an employee, rather than asking 'Do you understand?' ask 'What part of this job would you like me to explain in more detail?'. This type of non-confrontational questioning provides the employee with an opportunity to clarify their understanding of the task.

A key moment for success or failure as a supervisor occurs when you have to criticise an employee. It is an unpleasant task but procrastinating does not resolve the situation and, in most cases, makes it worse. Effective delivery of criticism can quickly alleviate a problem with a minimum impact on relations with your staff. Always carefully consider your response before saying anything. Abrupt and angry comments are rarely effective in altering employee behaviour or improving your relationship with your employee. The purpose of your criticism is to improve the performance of the employee. Exhibit concern and a genuine desire to assist the employee in improving their performance. Rather than denigrate the individual, look for ways in which you can strengthen their self esteem. Instead of saying 'Don't be such an idiot!' try 'I know that you can do better work—can you think of any better ways to do this job?'. Never attack an individual personally but focus on performance. Always deliver criticism in private and personally, not in front of other employees or customers.

Staff meetings afford an opportunity to review and report on the general status of your operation, discuss plans and goals, and solve problems. Staff meetings should be regularly scheduled at a time and place that is convenient to all staff members. Meetings should have a set agenda and time limit. The discussion, decisions reached, and follow-up actions required should be recorded. Staff meetings are opportunities to encourage teamwork by allowing employees to share their views and concerns, and work together to solve specific problems.

Dealing with conflict

Conflict is a reality of dealing with people in a work environment. Conflict resolution must be immediate to prevent serious long-term consequences. Conflict management will depend to a great extent on the nature of the situation and the people involved in the situation. The first step is to get each person's reasons for the conflict so that the problem may be identified. This information should be used to evaluate the problem taking into account all of the circumstances and the personalities of the individuals involved. Recommendations for alternative behaviours should be developed and the best solution selected. Develop an action plan to implement the changes and carry out the solution.

CASE STUDY: Conflict resolution

John, your mechanic, has a problem with Bill, your leading hand, who disrupts John's work by talking about non-work related subjects when he comes into the shop to drop off or pick up a piece of equipment. As a result of Bill stopping by on a regular basis over the past week to talk about the football, John had to work extra time to get all of his routine work for the week completed. John prefers working alone on equipment but Bill likes working with a crew of people. They usually get along but John is angry because Bill's interruptions are forcing him to put in extra hours to get his work done.

1. Listen to John's complaints without comment. Let him fully explain his anger.
2. Indicate that you are concerned and ask questions that get more information but do not make John defensive. Repeat and summarise what John has told you to show that you have listened and understand his concerns.
3. Tell John that you have to be objective in the situation and the problem must be resolved.
4. Ask him if he would mind meeting with Bill to discuss the problem and develop a solution to ensure that John gets his work done on time. Tell him that you plan to speak with Bill about the problem and find out if it is okay

with John to tell Bill that you have already spoken with him. If so, ask John to meet with Bill after you have had a chance to speak to him.

Preparation

1. Review Bill's performance record.
2. Plan for your meeting and formulate responses to address any concerns you may anticipate.

Meeting with Bill

1. Begin on a positive note regarding his work and contributions.
2. Explain the situation clearly without being judgmental.
3. Listen to Bill's response and acknowledge any concerns he may have.
4. Tell him that you would like him to get together with John to develop a solution. Let him know that John has agreed to meet him and you will be checking back with both of them in a few days to see how things went.

After meeting Bill

1. Tell John that you have spoken to Bill; he is expecting John to meet him and you will be checking up afterwards.

After they meet

1. Follow up to see if they were able to successfully resolve the problem.
2. If they have not been able to resolve the problem, arrange a joint meeting to help them find a solution.

Discipline

Perhaps the most difficult and often avoided aspect of management is discipline. The goal of discipline is to prevent problems and create productive performance and behaviour, not to punish employees. Once again, immediate action is more effective than a delayed response that usually results in escalating problems that can endanger the productivity and quality of work, as well as the safety of your employees.

Progressive discipline is a term used to describe the systematic method of administering discipline in relation to the severity and frequency of the offence. Minor offences such as disputes with fellow employees, foul language, inefficiency or incompetency, and late arrival to work may be dealt with initially by informal discussion with the employee. If the behaviour persists, the second offence may incur a verbal warning with each subsequent offence receiving an escalating response, such as written warning, suspension and finally dismissal. Moderately severe offences such as careless or dangerous use of equipment, not showing up for work without notice or failing to comply with written rules may incur a written warning for the first offence with a suspension for the second offence and dismissal for a third offence. Major offences such as drinking or using illegal drugs on the job, fighting, theft or serious insubordination may result in a suspension for a first offence with dismissal for a second offence. The discipline policy should be clearly stated in your employee's manual and communicated to employees in the event of a first offence. All discussion should be recorded in the employee's file and, in the event of written warnings or suspension, the employee should sign and return a copy of the warning or suspension letter to indicate that they have read it.

Dismissal and termination

There are several ways an employee can leave an operation. An employee may leave by mutual agreement in the case of retirement of a long-time employee or layoff of a seasonal employee. In other instances, an employee may leave on their own accord to go to another job, return to school or because they dislike the work. The final way an employee leaves is through dismissal.

Dismissal is the action of last resort in the

event of a serious employee problem. In many jurisdictions, improperly handled dismissals can result in legal action by the employee. The first rule is to have clear, written records as a defence against legal action. The reason for dismissal must be work related. Having written policies for performance and behaviour with a standard procedure for discipline and dismissal, along with written records and performance appraisals, are critical.

In all cases of employees leaving, it is helpful to conduct an exit interview if possible. The exit interview provides an opportunity to evaluate the working conditions, morale and attitude of workers towards both your operation and the way in which it is managed. This information can be used to make your business run more effectively and better compete for quality employees. The interview should be conducted in a confidential fashion to allow the employee to give honest answers to your questions about working conditions, supervisors, training and compensation. Often if employees leave with a positive feeling about your operation, they will return, in the case of seasonal employees, or recommend your operation to their friends as a potential employer.

Employee records

Records must be kept for both payroll and employment purposes. Personnel files for each employee should also include job description, application, interview report, reference check results, letter of offer with terms and conditions of employment, starting salary and any increases, tax exemptions and deductions, performance evaluations, grievances, disciplinary actions, dismissal notice, letter of resignation, exit interview report, record of employment, record of education and training and any other confidential information. Employee records should be kept in a secure file cabinet and access restricted to senior management. In most jurisdictions, the employee also has supervised access to their own file.

Customer service

The bottom line in any business is the customer or client. All decisions should be made with the needs of the customer/client in mind. Some of you may think that you don't have customers or clients to serve, compared to someone working in a garden centre who has very easily identified customers. However, we all have customers or clients to serve.

In general, a customer is someone who purchases a product from you. A client can be defined as anyone who uses your services. An immediate client might be your work supervisor. Obviously you must address the needs of this person in your work. The customer is the person who plays the golf course you maintain, owns the lawn you maintain, or pursues recreational activities on the sports field or in the park you maintain. It is easy to see that we all have customers and clients. Once you have identified the different customers and clients that you serve, you must then identify the needs of these people in order to better serve them.

Customers/clients will have different needs depending upon their perception of the service you offer and their impression of the service offered by your competitors. You must tailor your service to these needs. If people want to control weeds in their lawn, you will not have a successful business if you only fertilise lawns. However, you also have a responsibility to educate the public about your business. Provide weed control, but let people know about the benefits of proper fertilisation as well.

How do you go about identifying customer/client needs? The most obvious solution is communication. Talk to the people you serve and, most importantly, listen to what they say.

Provide them with information that they can understand. Do not use the same technical language or jargon that you would use when speaking with a fellow horticulturist.

Most people have very poor listening skills, yet listening is perhaps the most important aspect of communication. You can improve your ability to communicate with people, both professionally and personally, by enhancing your listening skills. Listening is the simplest form of flattery.

Active listening allows you to communicate effectively with customers and clients. When listening to customers use non-verbal signals to show interest in what they are saying. Nod your head in agreement or acknowledgement, maintain eye contact and assume a relaxed body posture facing them. Verbal signals such as 'uh-huh' and 'yes' show people that you are listening to what they have to say. Use questions to encourage customers to reveal their thoughts and interests. This is particularly effective when people are angry because it allows you to quickly get to the root of the problem. Another effective listening tool is to restate what they have just said. This saves time in identifying their questions or complaints before you answer them and also shows people that you are indeed listening to them.

There are several traps that many people fall into when dealing with customers and clients. Many people don't listen well. They assume they know what the person wants. This can be a particular problem with people who think they know everything about the trade. Most customers or clients are intimidated and turned off by this approach. They prefer dealing with someone who shows interest in what they think and shares their knowledge in an informative way, rather than dictating what they want. You cannot simply assume that you know what a customer or client wants.

Other people may talk too much. This can overwhelm a person with excess information. When dealing with an angry customer or client it may create new questions or concerns that can make the situation worse than it already is. As well, avoid saying anything that could be interpreted as sarcasm.

The following steps illustrate a method of dealing with a customer/client.

1. Listen attentively.
2. Guide the conversation with open-ended questions (Who? What? Where? When? Why? How?).
3. Restate the main points of what they have said in your own words.
4. If they agree with your summary of their question or concern, deal with it in a manner that best satisfies their concerns. If they disagree with your summary, listen again to identify their question or complaint.

Some other factors that help you provide exceptional customer service.

1. Try to remember names as people appreciate being recognised.
2. Treat customers/clients as individuals.
3. Follow up. Be sure that they are satisfied with the service you provided.

The following case studies illustrate the importance of proper communication when speaking with a customer or client. These case studies are offered simply to illustrate some of the points we have discussed. It is important to realise that each encounter with a customer or client will be different, and while there are general guidelines to follow you must exercise judgement about the best way to respond to the question or complaint of an individual customer or client.

Case studies in customer/client service

CASE STUDY 1

Cathy works on a golf course. One day in early autumn she is aerating the greens. A member of the club is out for a round of golf and comes upon Cathy pulling cores out of the turf with the aerating machine. The golfer is upset by the fact that the green is covered with holes, making it difficult for him to putt. He complains to Cathy, asking her why she has to aerate the greens; they looked fine to him when he played yesterday. How should Cathy respond?

Answer A. 'The superintendent told me the greens had to be aerated, so I'm doing it.'

Answer B. 'This green has a lot of *Poa annua* in it, which doesn't grow well in a soil with a low aeration porosity. We use this aerator to relieve the compaction and introduce a medium that will alleviate the problem.'

Answer C. 'I can understand your concern and I'm sorry that we are affecting your game. The soil in this green gets very compacted and it is difficult for air to get to the roots of the grass. If we were to leave it the grass could die or we could get a lot of weedy grass instead. This machine removes a plug of the compacted soil. We then spread sand on the green and work it into the holes to help improve the structure of the soil and relieve the compaction. The sand will help level the green and the grass will cover the holes in a few weeks. In the long run it will improve the quality of this green.

Answer C is obviously the best response. Cathy apologises and explains what she is doing in terms that he can understand. Response A would probably just make the golfer angry at both Cathy and her superintendent. Answer B is full of technical jargon that the golfer may not understand.

CASE STUDY 2

Jim works for a lawn care company. An elderly woman comes up to him and asks what kind of grass she should put in her yard. Which response do you think is most appropriate?

Answer A. 'Well, we have a sale on Kentucky bluegrass this week. Why don't you try it?'

Answer B. 'We have a variety of different grass mixtures that I am sure would look nice in your lawn. But before I can help you choose one, you'll have to tell me a few things about your yard.'

Answer C. 'I personally like Poa pratensis, Baron is a great old cultivar. Would you like me to get some seed for you?'

Answer B is the most appropriate response. Jim indicates that he can help her select a grass seed but needs more information before he can make a decision. Response A may appeal to the customer's sense of value but does not really address her needs. C is a little too technical and does not provide adequate information to the customer.

CASE STUDY 3

Bill, a park employee, is pruning the branches of an eight-metre tree at the edge of a city-owned park. While on the site, Bill is approached by a neighbour adjacent to the park who requests that he remove the tree completely. The leaves fill his swimming pool, excretions from insects on the tree dirty his garden shed and he fears that the roots will soon start to lift his patio. What do you think is the most appropriate response?

Answer A. Bill should agree with the neighbour and remove the tree.

Answer B. Bill should tell the neighbour that the tree is on city property and he has no say in what the city does with it.

Answer C. Bill should sympathise with the problems the neighbour is experiencing but he should also explain the importance of trees in the urban environment and point out that mature trees increase the value of adjacent properties. He may also mention that city trees are protected by specific bylaws, but if the neighbour has a complaint, he may wish to contact the city arborist.

Answer C is the most appropriate response. Bill sympathises with the neighbour and provides him with some information about the value of trees as well as the person to call should he have further questions or complaints.

These case studies are greatly simplified for the purpose of illustrating some basic principles of customer/client service. In most situations there may not be a single correct resolution to the complaint or concern of the customer/client. Common sense and a basic understanding of how to effectively deal with people are your best tools in providing exceptional customer/client service (Witherspoon, 1994).

Dealing with complaints

An important aspect of dealing with customers and clients is addressing their complaints. For every person who voices a complaint there are many others who don't. They just take their business elsewhere. Therefore it is extremely important to listen to the customer or client who does complain to you. In addressing their complaint you may also be correcting a problem in your business that may have been costing you other business. When responding to a complaint, consider the value of having someone who is satisfied with your service and tells their friends and neighbours, as opposed to having someone who will never use your service again and will probably encourage their friends to do the same. Inevitably there will be some people who have an unreasonable complaint or demand that you may not be able to resolve to their satisfaction. However try not to aggravate the situation further by arguing with them.

Here are some strategies for dealing with complaints.

1. Don't return aggression with aggression. Remain calm.
2. Listen without interruption and repeat their complaint in your own words.
3. Empathise with their concern.
4. Apologise when appropriate.
5. Discuss resolution of their complaint:
   —tell them what you are going to do
   —ask them what they think you should do
   —suggest alternative ways of resolving the problem
6. If you must decline their request don't simply say, 'It's our company policy.' Instead:
   —decline with a reason
   —suggest alternatives
   —appeal to their sense of fair play
7. Thank them for their complaint and follow up later to ensure that it was resolved to their satisfaction.

Hopefully this section has provided you with some 'food for thought' that will help you deal more effectively with your customers and clients. Always remember that they are the most important part of any organisation. Without them the organisation would not exist and we would not have a job.

There are many excellent publications that deal in detail with financial management, human resource management, and customer/client service. Training companies offer a wide range of seminars and other training aids covering these important subjects. Often professional trade associations include articles about these often overlooked, yet critical, aspects of turf management in their journals or newsletters and also have experts speak at their conferences. Keeping pace with developments in these areas is as important as staying abreast of new developments in the science of turf management.

References

Daniel, W.H. and R.P. Freeborg. 1979. *Turf managers' handbook*, Harvest Publishing Company, New York.

Graham, D.D. 1997. 'When the summer staff goes away', *Golf Course Management* 65(9): 38–44.

Hannebaum, L.G. 1995. 'Labor analysis and pricing', *Grounds Maintenance* 30(10): 12–16.

McEwan, K. and L. Owen. 1995. *Employers' handbook for owners, managers and supervisors in agriculture and horticulture: Books 1–5*, Ontario Agricultural Human Resource Council.

Ostmeyer, T. 1994. 'Planning for purchases', *Golf Course Management*, 62(1): 210–15.

Vander Kooi, C. 1997. *The complete estimating book with production times for the green industry*, Vander Kooi and Associates Inc., Littleton, Colorado, USA.

Witherspoon, W.R. 1994. 'Professional development and customer service', Chap. 6 in *The Horticulturist I*, Independent Study Course, University of Guelph, Ontario.

CHAPTER 12

Contract establishment

V.A. HAINING, Parks and Recreation, The City of Melbourne, Melbourne, Victoria, Australia

Introduction

Today globalisation and attendant specialisation for comparative advantage brings with it massive opportunities for corporations to substantially improve both their effectiveness and their efficiency. As companies actively pursue these aims, traditional contractual mechanisms often appear to be increasingly deficient. Today new paradigms are needed which comprehend the subtlety and sophistication of the global marketplace.

Purpose of a contract

A contract is a legal agreement between two or more legal parties giving rise to mutual obligations and rights that are enforceable in a court of law. A contract inherently involves agreement between these parties. The parties must have reached or be deemed to have reached agreement. The process will necessarily involve an offer being made by one party and accepted unconditionally by the other. The contract need not be in writing to be enforceable. Many enforceable verbal contracts can be made. The intention of a contract is to set out the distribution of business risk, the description of work, scope, and requirements or outputs, as well as the consideration or payment to be made. What is important is that all parties have a clear and unequivocal understanding of all the delivery elements of their work, the associated payments and the quality of outputs needed, as well as the timelines involved.

Types of contracts

Contracts can take many forms and historically, different contract types have carried favour in different business communities depending on the circumstances which prevail. The simple item priced contract for the supply of goods is one form of contact that is used often and with great success. Its key attributes are a clear description of the item, a clear description of the price, delivery date, any guarantees, warranties, and any defect liability periods that apply.

Another common contract applied is a schedule of rates contract. Here the parties agree to an arrangement whereby labour is supplied at an agreed hourly rate with services provided at a set rate. Here the purchaser generally has the flexibility to vary the amount of labour sought, however the supplier or deliverer is limited to the scheduled price which he has tendered. Schedule of Rates Contracts are appropriate where the scope of work, the complexity of it, and the exact nature of the requirement are unknown or uncertain. In these circumstances it is not reasonable to expect a purchaser or bidder to price a particular parcel or package of work where the level of effort and expertise required to deliver that work is not readily or reasonable quantifiable. Traditionally, where there are high levels of uncertainty or unknowns this schedule of rates approach removes that element of risk from the bidders and delivers to the purchaser a tender price which is consistent with the works required with the exclusion of that uncertainty.

A cost plus contract similarly is an appropriate mechanism to use in circumstances where revenue streams are unknown, the cost structures are uncertain or in dispute, and a desirable business way forward is for both parties to establish cost and revenue streams over time and then make commercial decisions based on that information. The cost plus contract, therefore, allows both parties to enter into an agreement where the risk of significant revenue or cost volatility is not initially distributed to either party. Often the bidder is offering management value-added services, where the expertise is the skill of the seller or the project manager to achieve economies for the business of the buyer.

A further method of contracting is the lump sum approach. A lump sum contract has distinct advantages to both parties because each knows exactly what consideration or revenue will be available to the provider on a regular basis, subject to performance. Under this scenario, the contractor receives a set return irrespective of how much labour, plant, equipment and resources is applied to the particular contract objectives. The lump sum approach is particularly useful where all contributors want certainty for budgeting purposes into the future, and where there may be uncertainties and the purchaser has made a conscious commercial decision to have those uncertainties passed, up front and carried into the contract environment by the other contracting party. In addition to lump sum, contracts often contain provisional sums. Provisional sums are incorporated where further works may be incorporated at a later point depending on the exact requirements that emerge.

Business decisions to outsource

Initially a business decision must be formulated with due regard to your company's overall objectives, mission and vision, together with a good knowledge of the capability of the potential market in which you wish to operate, and the capacity of the industry to deliver the types and scope of services and/or products which you are seeking. Obviously that initial business decision must recognise the strategic worth of the product and/or service being considered for contract. Issues of control of intellectual property, product branding, customer management, asset monitoring, output needs, community service obligations and so on are fundamental and must be considered up front. Having considered those fundamental issues a risk analysis distribution and management process should be put in place to execute your contracting strategy.

Contract formation: risk management process

Risk is the chance of something happening that will have an adverse impact upon our business or operational objectives. It is measured in terms of consequences and likelihood. Figure 12.1 illustrates a generic model dealing with the process of risk management.

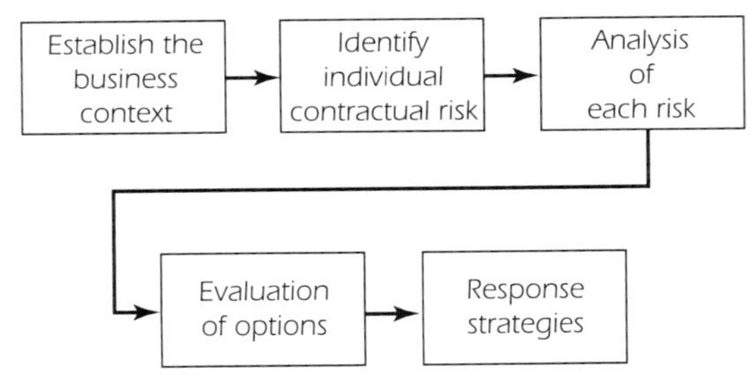

Figure 12.1 Key steps in the process of risk management

Establishing the context

In establishing the context and relationship of the business, the environment must be clearly understood. Is the industry undergoing massive reform or re-engineering? Are technology changes occurring so quickly as to challenge the viability of the core products and services? Obviously the importance of establishing the context cannot be understated.

In the case of The City of Melbourne's activities the driving context was the Corporation's Competitive Business Strategy and the State Government's Compulsory Competitive Tendering (CCT) legislation.

Risk identification

The categories of risk are fourfold:

1. Risks to council (buyer risk)
2. Risks to contractors (bidder risk)
3. Risks to customer (consumer risk)
4. Risks to assets (capital or owner risk)

A risk and response list for each of the above categories should be developed. A suggested format is set out in Table 12.1.

Table 12.1 Example of a risk and response list.

Element	*Council*	*(Buyer)*	
Risk	Description	Potential Responses	Comments
Contractor	Bankruptcy/scheme of arrangement	Terminate/curtail Exercise Guarantees Sue Renegotiate up Redraft and amend Scope of works	

Risk analysis

Risk is analysed by combining estimates of likelihood and consequences in the context of existing and/or potential control measures. Here the objective is to rank risks on a graduated basis so that critical risks can be recognised and managed with greater emphasis.

The risk analysis process will take available historical data, information on the business environment, bid and/or expression of interest data and, where necessary, quantified judgement to determine the 'merit order' of risk. A table of 'Likelihood and consequences' will provide the defining framework for this to be determined (Table 12.2).

Table 12.2 The scale and depth of likelihood and consequences.

	Scale	*Definition*
Likelihood	Rare	Highly unlikely in life of contract
	Unlikely	Possible but not in the first twelve (12) months
	Likely	Will occur in the life of the contract
	Highly likely	Likely in the first twelve (12) months of the contract
	Almost certain	Likely in the first three (3) months and repeatable over contract
Consequences	Catastrophic	Significant loss of assets, reputation of agency seriously damaged
	Very high	Critical event, major cost implications, reputation tarnished
	Moderate	Significant event, some cost, manageable within contract
	Low	Impact minor, capable of being handled as a routine issue within contract
	Negligible	Impact may be safely ignored

Risk evaluation

Risks can be classified with a range of graduations, but a useful approach is to devise a classification which is relatively simple and therefore capable of being worked with relative ease. A range might be as expressed in Figure 12.2.

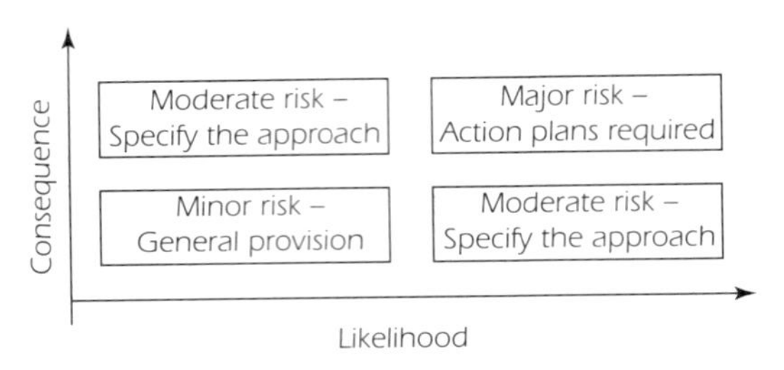

Figure 12.2 Risk range

Risk treatment

The aim of risk treatment is to identify the potential ways of treating risk, with a view to reduce the potential loss using a map process (Figure 12.3).

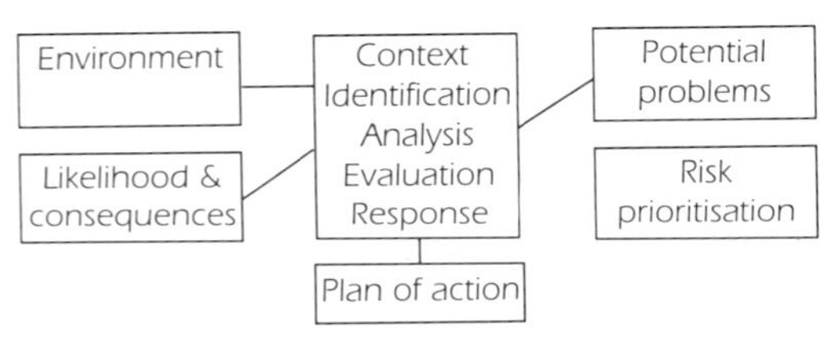

Figure 12.3 Risk process/outputs map

Risk reduction and/or avoidance

In the first instance, prevention and/or avoidance options should be examined. It needs to be recognised however, that high confidence levels of risk prevention often bring with them attendant high costs. Purchasers need to be careful in specifying the standards required so that inordinately large cost imposts do not accrue for relatively minor increments of risk reduction or service increment. This (actuarial) trade-off must be dealt with at both the conceptual (bid structure stage) and practical (bid evaluation phase) stages within the purchasing process.

A failure to correctly and adequately address risk and its attendant potential distribution at the time one considers the business context, may ultimately result in bid pricing which is totally inconsistent with budget provision or which carries with it such draconian imposts for pricing, including rate setting, as makes it unworkable.

Risk transfer

For a contract to be successful, the distribution of risk must not only be clear and precise in the tender documentation, it must also be commercially realistic. It is not sensible, for instance, for a purchaser to seek to shift a risk to the seller where it is not possible to price the risk or obtain requisite insurance.

For example, where a contractor is providing tree maintenance services in a high risk fire danger area, a requirement for the contractor to carry $200 million plus insurance to cover his potential liability, should his negligent actions give rise to a bushfire, may be unrealistic and unobtainable. A more feasible solution would be for the purchaser to set up safety and control mechanisms which demonstrate 'leading edge' performance in this area, and then seek insurance coverage themselves. Obviously, where the consequential and direct liabilities are reasonable and quantifiable, then a business decision to have the tenderer price these may be very reasonable.

An issue of cause or fault also regularly arises with maintenance contracts. Contractors often argue that gaps between perceived performance and actual asset condition reflect 'natural' declines in the assets' utility, rather than a specific maintenance failure. Specification drafting must carefully address what constitutes capital and maintenance works, so that the risk of graduated asset decline does not lead to a shift of cost back to the owner or a loss of utility to the consumers.

For The City of Melbourne's Parks and Gardens, liability for remedial works, where the maintenance contractors through no negligence of their own are required to make good, is currently a risk with associated costs borne by those contractors. This particular risk distribution feature is a fundamental attribute of 'total asset' management contracts (Table 12.3).

Table 12.3 Key risks that may apply to historical contracts.

Risk	*Likelihood*	*Consequences*
Catastrophic asset failure	Rare	Catastrophic, significant loss of assets
Occupational health and safety	Likely	Moderate, significant event, some costs but manageable within contract
Contractor insolvency	Rare	Very high, critical event, major cost implications
Degradation in service quality	Highly likely	Low
Poor customer focus	Likely	Low (short term)

Comments on contractual risk categories

Any capital asset or system relies for its well-being on the underlying engineering asset framework. Primarily, for Parks and Gardens, this framework often comprises the systems of irrigation and drainage. The key historical asset is the tree stock or turf areas. Losses in these asset areas can be catastrophic.

An important horticultural asset category where failure has a catastrophic potential impact is irrigation. Here one should not only be concerned with the condition of the asset on a day-to-day basis but also with ensuring that the assets, their type and location status are adequately maintained. The probability of major failure is increased where poor preventative maintenance systems exist and condition monitoring is non-systematic.

Good contract placement and subsequent management systems seek to systematically reduce the probability of failures. Some practical actions are recommended in response to the above possibilities:

1. Inform tenderers as to the limitations of your systems. Instruct them to appoint an irrigation 'champion' who is appropriately skilled and dedicated to, and accountable for, systems and operations management of the high risk asset category.
2. Require a specific risk management plan to be submitted as part of the tender, identifying response strategies to cover the risk of system breakdowns or (local) asset failures.

For turf, the risk of disease and pest is ever present. To shift the risk to the contract may provide the principal and/or superintendent with a secure, safe feeling, but will not necessarily assist if an outbreak of disease occurs.

If the risk is significant and the asset strategic, you would do well to consider retaining the risk management within your business.

Occupational health and safety

Council is committed to ensuring that a safe environment is provided by contractors for their staff and subcontractors. Risks which arise within this category for horticultural works include:

- Untrained, unlicenced, uncertified employees (e.g. chemical applications, traffic management, machinery operators)
- Plant and equipment adequacy and conformance
- Inadequate supervision and/or reporting

Risks to the public at large include:

- Playground equipment which is not to the Australian Standard
- 'Trip' hazards on pathways and roads
- Injury arising from poor work methods (e.g. poor mowing practices or branch trimming)
- Inappropriate chemical usage

A response strategy to these risks might incorporate the following elements:

- Undertake a comprehensive analysis of the potential contractors OH&S (Occupational Health and Safety) systems, performance and capability, as part of a tender prequalification process

- Require current compliance with industry published standards (e.g. in Victoria, Australia, 'Safety Map') or substantial compliance with a stringent plan to achieve certification within three (3) months
- Embed a schedule of works into the tender specification to cover all necessary upgrading for playgrounds, as required by current codes and standards
- Include playground assessments in your own or your contractors' quality assurance monitoring systems. Undertake OH&S code compliance auditing through contract management processes.

Contractor insolvency/underpricing

The likelihood of a key contractor becoming insolvent is low if you have undertaken an adequate selection process. However, if this was to occur the consequences would probably be significant for you and your customers. A more likely event is for a tenderer to underprice the works, then trim back services to deliver to a price and not perform to the nominated standards.

The strategy recommended for dealing with these risks is:

- Undertake a full comprehensive financial evaluation of each prospective tenderer
- Require works insurance and public liability insurance ($10 million minimum) a condition of contract
- Require an unconditional bank guarantee for timely and effective performance of the works

In respect to underpricing giving rise to potential service delivery shortfalls, inconsistencies and under-delivery of standards, the following strategy is worth considering:

- Use of an incentive payments scheme
- Consider a short-term contract with options built in to extend, subject to satisfactory performance

Degradation in service quality

Degradation in service quality can eventuate as a result of:

- Lack of information on assets, their current conditions and rates of attrition
- Lack of contractor familiarity with the sites and usage patterns for assets
- Inadequate specification of the maintenance definition and capital works requirements, with an attendant risk of graduated asset quality reduction.

A strategy to manage these risks is to:

- Ensure potential contractors have appropriate asset management knowledge, systems, procedures, task allocation mechanisms and conformance recording processes.
- Specify the standards governing the recording of assets, associated data exchange needs and the obligations of potential contractors for the provision and collation of requisite data.
- Require continuity of involvement by key personnel from successful contributor bid teams through the contract implementation stage for at least first six months.
- Define maintenance standards and highlight areas within the specification requiring major maintenance works.
- Ensure that the risk of unforeseen events, which give rise to additional works, is clearly assigned to the contractor and that they have the capability to 'flex' up to deal with such contingencies.

Poor customer service

Loss of customer loyalty or loss of consumer confidence in your product or service can quickly cripple a company financially and take years to remedy.

Often contractors are skilled at promising slick service systems, however it is imperative that you satisfy yourself that the potential contractor's systems are capable of providing a quality response at all times.

Risks to be considered in servicing the contract are:

- Lack of contractor capability to deal effectively with customer enquiries. This may be technical, such as a lack of systems support, or may be personnel related, caused by inadequately trained staff.
- Lack of an agreed process for dealing with customer enquiries.
- Absence of records on customer complaints/issues.
- Poor or inadequate remedial mechanisms to quickly address and 'close out' customer issues.

Customers' perceptions are important. If they believe that your contracting activity introduces a relatively disinterested party into the process, the likelihood is that they will feel dissatisfied in complaining to that party.

The strategy for dealing with these risks should embrace the following components:

- Limiting customer interface by contractors to routine non-contentious, localised issues.
- Ensuring that bidders present a clear and documented process for managing and recording customer enquiries/complaints. Find out where they provide similar services. Visit those sites and seek the views of the customer. Talk to other purchasers of their services and be prepared to form your own views.
- Requiring customer service training to be undertaken by the successful contractor within the first three months of the contract commencement and to provide training updates to staff as an output of their quality systems.
- Requiring the 'customer service standards' for the business to be embedded into the contractors' customer management practices.
- Ensuring that an escalation mechanism is made available to the public to enable recourse to you if the customer is dissatisfied with information supplied or methods used by the contractor.
- Ensuring that the contractual reporting processes require data provision which identifies resolution times, trends and critical issues.

Customer service accountability should remain with you, the principal. However, day-to-day enquiries of a routine nature should be handled by the contractors to the standard specified by the contracts manager.

Contract formation: requirements determination

The 'requirements determination' phase of the contract establishment process is the most fundamental. Here purchasers must clearly be able to document what service or product is required to be delivered by the marketplace. Requirements determination often involves a re-survey of customers or user groups to confirm current services or, where appropriate, to delete these and to specify new services where customers are clearly demonstrating a desire that these be provided. The requirements that are identified must be expressed in clear and simple unambiguous terms. A critical task of any specification writer is often to de-jargonise the work or the requirement so that the market at large can interpret, understand, price and bid without fear that they have misconstrued or misunderstood.

In engineering environments, the difficulty arises because of the frequent tendency of the users to specify processes and details, and adopt a level of specificity which necessarily limits competitive bids. It should be pointed out that, if such specificity is essential, the contracts manager is probably in the realm of dealing with proprietary services and the tendering process is not an appropriate mechanism. Indeed, it is an inappropriate mechanism to apply to those situations. Specifications therefore, which focus on inputs to the detriment of overall output requirements, are generally only preferable where specific engineering and/or technical reasons exist for their maintenance. Otherwise performance-based specifications enable a far more challenging and enlightened bidding process to occur where new concepts, new methods of service delivery and service provision are directly generated as a consequence of innovative bids. Requirements determination process then can often identify opportunities for re-engineering within a business or re-engineering of a process across a business or series of businesses. Those opportunities should be captured, if possible, through this process.

'Packaging' of work is also of critical importance. Work can be packaged in individual parcels of set tasks such as mowing of the turf, grass cutting, turf grass renovations, mechanical and equipment maintenance, spraying programs, garbage removal, building painting, cleaning, etc, or can be expanded to the full responsibility for the provision of the total asset maintenance management process. This may include all of the specific tasks, as well as the provision of a whole range of information and work command and control aspects. Packaging becomes critical as it directly impacts on the size of the bidding market and the potential of bidders to bring a re-engineering strategy to the business at hand.

Having recognised that there are some fundamental differences between performance output-based specifications and detailed input specifications, there are some distinct disadvantages of performance-based specifications which the reader must be conscious of. Essentially once you have entered into a contract based on output requirements, you have given up certain rights over the detailed technical configuration of the work. If circumstances arise where the technical configuration being delivered, or the way in which services are being executed, is entirely contrary to your expectation and entirely contrary to the expectation of your customers, you may have a problem. Hence if a contractor introduces, for example, a customer service management system that brings with it a range of unforseen disadvantages, you may not be able to direct that contractor to amend, vary or to indeed delete that system unless you are prepared to entertain a variation for the work. Such variations will invariably bring with them a cost.

Specification characteristics

A specification must necessarily be detailed, accurate, current and achievable. A specification which does not contain these elements will not enable a potential contractor to adequately understand, price and deliver. Consequently a contract will invariably bring with it significant problems if you do not take the time up front to get the details right. Accuracy and detail are critical. Achievability is equally important in that the specification must not require the contractors to deliver to standards which are technically unattainable. There is often an expectation to build specifications to standards and levels of performance which are at the leading edge. The risk is that those standards may be in excess of customers reasonable expectations. Often the incremental costs associated with delivering extra additional increments of service standards are disproportionate to the level of utility delivered to the end consumer, that is large costs are often associated with low increments of service as you enter the exponential component of the cost curve.

Conditions of tendering

Conditions of tendering govern the processes under which the bid will be issued, returned, evaluated and awarded. It is critical that probity is not only conducted throughout the process but is demonstrably evident throughout. Probity necessarily brings with it the need for a transparent rigorous ethical and defensible methodology which any external agency, party or individual can come and examine at their will to satisfy themselves that the award of the contract has been made on a fair and equitable basis. The conditions of tendering provide administrative control within the process such as issuing date, closing date, lodgement details and whether or not facsimiles are acceptable.

A failure to follow the procedures, as outlined in the conditions of tendering in the awarding of the contract, may give rise to an administrative breech of law with potential to liability. It is essential that the published procedures are followed without exception. In this regard a particularly important element relates to the acceptance or non-acceptance of late tenders. In the author's view, late tenders should never be accepted.

Use of expressions of interest, registrations of interests, and pre-qualification processes

It is often difficult to develop a specification, especially where the actual outputs to be delivered, or the product to be acquired, is relatively difficult to specify or is unknown. An expression of interest is often a simple mechanism to assist the contracts manager. This is where the marketplace is approached with the idea of obtaining potential solutions to a specified problem. Hence a problem may be the technical constraints that apply to a particular horticultural problem that has remained unsolved for many years. The expression of interest would seek methodologies, ideas and concepts for addressing those problems. An issue for consideration, however, is that of intellectual property. Should a registrant offer a new innovative solution, then it would be entirely inappropriate for the principal to re-offer that intellectual or bidder knowledge to the marketplace for all to bid on.

Expressions of interest are also used for determining who is in the marketplace and their broad capabilities. This model is applied where the registrant company capability is assessed up front in a two-stage procurement process. This is then applied as a basis of short listing tenderers. It is necessary to draw reader's attention to the difference between a pre-qualification and a registration of interest. A pre-qualification process is that a registrant is either acceptable or unacceptable and is therefore a clear yes/no scenario, whereas an expression of interest involves capability analysis and determination and generally a short listing of prospective tenderers, where the capability information is taken forward into a further process of evaluation and generally applied in conjunction with price data to give a net value analysis.

Use of evaluation criteria

It is particularly important that the evaluation criteria reflect the objectives of the purchaser. If the purchaser seeks a cost reduction, then they should state this clearly and unambiguously, and the selection criteria and evaluation criteria must reflect this. If, however, the purchaser wishes the best value for money, then the evaluation criteria similarly must reflect this objective. Unfortunately it is often considered a good business decision to simply reduce the cost of the service without a business undertaking an analysis of the implied or explicit service consequences of such actions.

Value, on the other hand, brings in the concept of quality of service, after sales service, life cycle costing, reliability of service, consistency of delivery and responsiveness to problems. A contract which does not have these elements often will not deliver a quality result for consumers. A contractor bidding for such works cannot, and should not, be expected to second-guess the value propositions which purchasers may wish to infer in their documentation. If you wish to contract on a value basis, you must establish in the specification and the associated commercial documentation the exact method by which you will evaluate value. This is particularly difficult when dealing with such issues as consistency, reliability, responsiveness and quality of service, because you are invariably dealing with issues that are difficult to put into specifics and yet are critical if you want a quality product and/or service delivered at the end of the day.

If one fails to adequately and properly specify the quality and value dimensions of the work and/or service, then it is unfair and unrealistic to consider that the contractor is underperforming when indeed he is performing precisely to the specifications.

Having established the value matrix which you wish to apply, it is also necessary that you establish a rigorous, impartial, sustainable and defendable methodology of evaluation. This is often best done through the establishment of an evaluation team. Generally this is a cross-functional team built from within your business or indeed from a diverse range of businesses and/or customers. In this way you are able to bring a multi-skilled and cross-functional focus to complex offers and bids.

Evaluation process

The evaluation panel will need to read and critically analyse the registrations or bids put forward. These are some issues which warrant comment.

Any obvious error should be questioned with the tenderer. If pricing is substantially below the average for one tenderer, this can often indicate that the tenderer has not understood the nature and/or scope of the works. In the first instance, you should confirm that the tenderer has understood the nature and scope of the works. You should ask them to explain how they intend to undertake the works, with what resources and to what level of quality. This process, however, should not be taken as an opportunity to rebid. Additionally, new offers should not be accepted. The option that should be made available to a tenderer, in those circumstances where clearly they have made an error in their pricing for instance, is to withdraw the tender. That may present you as the purchaser with some significant problems if, for instance, it is the in-house team bidding for the work or it is your existing satisfactory contractor. Under those circumstances you must look at your conditions of tendering to see if you have included a provision whereby you can, at any time up until the formation of contracts, not proceed, and restart the process. Obviously this is a last resort, but if you have put up front that provision, then tenderers have accepted this condition.

Post tender negotiations should also be used to clarify issues around quality, value, responsiveness, consistency, reliability, questions of systems and questions of capability. While we may have specified predominantly the outputs to be achieved, nevertheless the contract manager should satisfy himself that the potential contractor has systems capable of delivering, with some degree of rigour and reliability, the products and services required.

Having established that process, there is an obvious need to maintain documentation throughout. The documentary trail must be auditable, complete and true. Good probity requires that all decisions are recorded, documented and supported, especially during panel decisions. Where the panel is not able to agree, this should also be recorded.

Having determined through your panel the recommended tenderer, the next stage in forming a contract is to prepare the documentation. The recommendation must be expressed in clear, cogent and simple english. It must stand-alone. It must be capable of being read by a board, council or committee member, who must understand what it is you are recommending and why. It must be commercially sound and demonstrate how the risks have been distributed with the contract, and between the parties, and it must convince the reader that the decision recommended is in the best interest of the consumer and the business.

Contract formation

The process of forming a contract is relatively straightforward but, nevertheless, has a strong history of problems. The elements in a contract are: offer, acceptance, consideration, intention and capability. The offer must have definite terms. It must be communicated. It may be communicated by conduct from time to time, although this is not common and it may be withdrawn prior to acceptance at any time. This is of critical importance to you if you are bidding and it becomes evident to you that any obvious error has occurred. You are quite able to issue a withdrawal of that offer. Although there is some commercial embarrassment to you, if you choose to take that path, it may be preferable to endure that short-term commercial embarrassment than to sustain a long-term financial ruin. An offer also lapses with the issue of a counter offer, hence if the other party were to come back with another bid or a different variation from what you have offered, your offer lapses.

The parties must actually intend to create legal relations. In a commercial environment, this is invariably the case. It needs to be said that social agreements are not enforceable in that sense. Acceptance must be unqualified, it must be communicated and it must be clear. It is necessary that acceptance be unqualified, and often problems arise because, in the simple commercial world, offers and counter offers are exchanged in different companies. Documents flow between the parties and it may be unclear, in many instances, whose contract conditions are actually being accepted and what the key terms of a contract actually are. You cannot assume, simply because you lodge your conditions with your bid, that the document that is ultimately returned to you will reflect your terms and conditions.

Consideration concerns itself with the issue that each of the parties to a contract must convey a benefit to each other or suffer detriment in return to promises made. If there is no consideration, there is no contract. A method of achieving the execution of a promise or set of promises, without consideration, is to enter into a deed of agreement. A deed does not require consideration.

When does a contract actually end? There are three methods by which a contract is finalised. First, by agreement, you can agree to end the contract. A mutual agreement is adequate and sufficient. Secondly, by specific performance, when the contract works are completed, practical completion has been executed, defects liability is ended, all retentions have been released and no security deposits are held. Thirdly, the contract is ended by frustration. It is not possible to continue to perform the contract.

A final word

Throughout this process of forming a contract, great care must be taken to avoid innocent misstatements. Here it must be said that officers, at whatever level, who choose to make off-handed comments which a tenderer or contractor relies on and executes certain actions pursuant to, is placing the company in a liability position, because that officer will generally be held to be ostensibly empowered to issue those directives. You must therefore be aware of the limitations and delegations within your business and/or company. You must be aware that no commitments should be made until contracts are agreed and signed, and that work should not commence until those contracts are properly executed.

CHAPTER 13

Contract management

V.A. HAINING, Parks and Recreation, The City of Melbourne, Melbourne, Victoria, Australia

Introduction

Contracts will never manage themselves. A dynamic and positive agreement requires considerable effort to manage. Failure to plan your contract strategy is a clear plan to fail. In contract management it is important to consider what you want from the contract, recognise the three built components of the operational platform, programming and control, and strategic development, and have an understanding of the overall role that each group plays in the management of the contract.

Decide what you want from the contract

Deciding what you want invariably involves critical assessment of how much emphasis is placed on direct up-front dollar savings as opposed to longer term 'quality of services' value:

- Demonstrated commitment to excellence
- Responsiveness—can do attitude
- Commitment to occupational health and safety
- Strong on the ground performance
- Strong customer focus
- Consistent delivery—on time every time—no surprises
- Technical competence

Recognise the three build components required

A contract comprises three key components: its operational platform, its programming and control component, and its strategic and development aspects. At each level in the process of contract administration, the superintendent must be mindful of how the contract has been formed and its inherent limitations. Figure 13.1 describes a methodology for contract management, with particular reference to local government and the services sector.

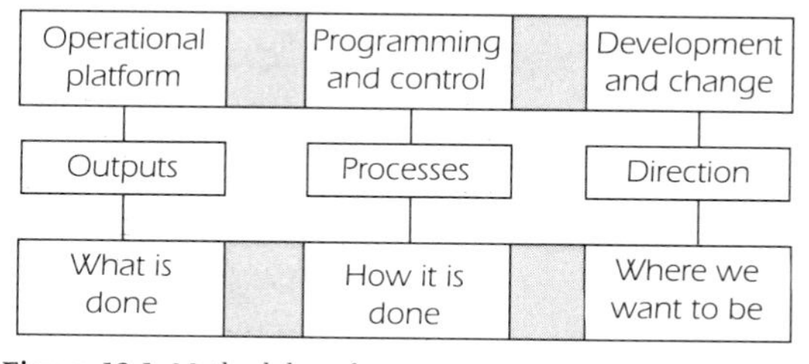

Figure 13.1 Methodology for contract management

Complex models are not required to understand basic contracting methods. It is important, however, for you together with your contractor to have a very clear picture of what the landscape or environment will look like in three or five years. Your shared vision is paramount and should be developed early in the contract process.

Secondly, it is important that the achievement of goals is not accidental but rather driven by controlled processes. This aspect of business rigour ensures that changes are driven systematically rather than by reliance on individuals.

Thirdly, the contract must deliver what has been agreed by the parties. It need not deliver more than this.

Understand the overall roles that each group plays

A good contract creates the environment where the contractor can achieve both value for the client and a satisfactory rate of return. A contract often provides the opportunity for change processes to be embedded and new skills to be developed by employees. Figure 13.2 illustrates the key drivers and enablers that each group plays in contract management.

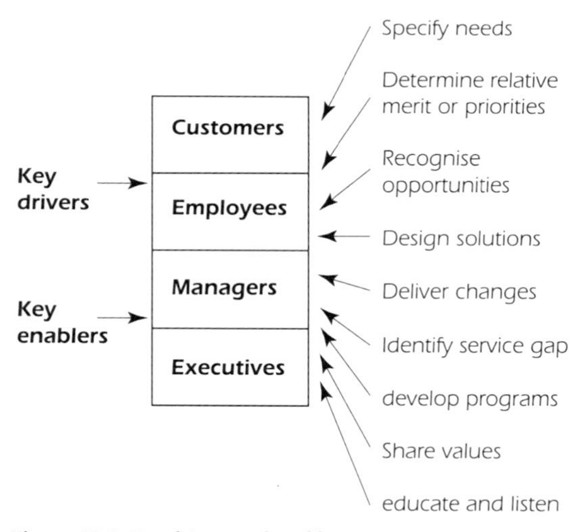

Figure 13.2 Key drivers and enablers in contract management

The day-to-day management of the contract must recognise and exploit the possibilities for achievement of excellence. An effective service delivery process must therefore link the 'enabling' and 'facilitating' functions of management and executive (in both the purchaser and provider sides) with the operations and market response ('drivers') side of the business. Quality service delivery requires a resilience and responsiveness that draws on the strengths of both systems and people.

A dynamic contractual arrangement enables employees at and across all levels to make a relevant contribution while optimising systems and processes.

A good contract positively fosters the development of communication throughout the 'stakeholder' groups so that excellence can consistently be achieved.

Specific issues

The contract superintendent

The superintendent is the focal point of the contract. He reports to the principal on the performance of the contract and is ostensibly accountable to the principal for delivery under the contract. He or she must be part negotiator, part motivator, part lawyer and part beggar. Contract administration is often complex, involving a whole range of alternative solutions and opportunities where every opportunity or solution brings with it potential for a range of dysfunctions which need to be recognised and managed. Above all else, the superintendent has a responsibility under the contract to act fairly and impartially and to allow the parties to perform the contract.

The superintendent should be aware of their role as superintendent, rather than their role as an employee of the principal. From time to time, these roles may be in conflict. The superintendent, for instance, is required to evaluate fairly claims for extension of time, even though these may not be in the principal's interest. Should the superintendent not act fairly or within the words of the contract, then the contractor is entitled to make a claim and enter into the dispute process. If the superintendent has not followed the processes, procedures and words of the contract, the processes stipulated in the contract must be followed.

When establishing a contract, the first thing the superintendent must do is to read the contract. While this sounds self evident, there are many examples where disputes arise simply because neither party has adequately read the text. Often contracts are a conglomerate of different documents cobbled together after a complex series of negotiations and exchanges of correspondence. Those exchanges of correspondence and negotiations often contain alternate meanings to important issues. It is essential that the parties develop a joint understanding of the contract details early in the administration of the contract.

In this regard, it is worth noting an emerging trend in service contracting to 'contract dilution' on the part of many contractors, who seek to withdraw the extent and range of liability and service by offering the logic that the bid never really envisaged the extent of service actually needed. In other words, that the bid inherently contained a degree of 'puffery' or padding which any reasonable person would have known was an amplification or overstatement of what was to be provided. The contract superintendent must respond to this in the negative and not accept a dilution from the specification. The superintendent must deliver the specification irrespective.

This problem is exacerbated when companies undertake bids through bid teams that have no relationship to the operational and implementation teams that follow on from a successful bid. Implementation teams often have a somewhat different view from their bid team colleagues.

It is also important, as a first step, that the superintendent immediately on being appointed moves to meet and greet the contractor and their personnel. A non-adversarial common framework is essential.

The superintendent will need to manage the contract on a day-to-day basis in a manner which is consistent with the type of contract formed and its essential attributes.

Contract types

Contracts will vary in their degree of sophistication and complexity. At the higher end of the complexity chain is the 'Total Quality Management' (TQM) contract. High degrees of inter-company dependency exist, planning and process control is seamless, and the services delivered are strategically critical to both parties. The 'strategic alliance' model is often put forward as representing the TQM contract. Unfortunately this is seldom the case. More often than not contracts that are written with a view to being an alliance TQM type contract fail. This is because such contracts are complex entities based on considerable trust between the parties. The emergence of the requisite trust and understanding by key managers evolves through time as the contract management processes occur. They rely on a series of behavioural attributes rather than a written set of prescriptive interactions.

Whereas the TQM contract is quite rare, high levels of effective teamwork giving rise to significant mutual benefits can be achieved through the adoption of 'value' based contracting. The essential difference for this model is that it presumes the purchaser/provider split, whereas the TQM model requires the adoption of the partner/partner framework.

Both these 'types' are significantly advanced on the other options—output and traditional.

The 'output' based contract focuses on deliverables. Process management is with the contractor as are most of the commercial and operational risks.

Of course the 'traditional' contract has a proven track record. In many instances it is entirely appropriate to apply the 'type'. Nevertheless for complex service delivery needs it has many deficiencies. Table 13.1 summarises the alternative attributes of different types of contracts.

These different 'types' will require the application of differing management techniques within each of the contract 'build' components. Further, the attributes exhibited through the life of the contract will be rather different. The focus of each of these contract types will also vary significantly.

Contract administration: focus matrix

To 'walk the talk' is a critical process in delivering a successful contract. This does not mean however that the contract superintendent should undertake an instructional role with the contractor. Indeed it is imperative that contract staff, particularly those who have moved from semi-government or local government environments to the private sector, understand that it is no longer the council or the principal who will provide the detailed instructions for the work. Also critical for the superintendent is that they do not innocently undertake instructional activity in lieu of the contractor's management staff, as this will lead to the contractor avoiding potential liability on the basis that he has complied with the principal's instructions to the letter, rather than having achieved specified outcomes to the service delivery standard as contained in the contract (Table 13.2).

The next critical aspect is the establishment of a regular communication process. This will involve regular meetings where clear and open communication is established on all key issues. It is in this role also that the superintendent must recognise and give positive feedback to the contractor. Contract administration is not about finding out what goes wrong in the contract alone. It is axiomatic to the successful management of the contract that the superintendent makes onerous efforts to discover whether the contractor and their staff have achieved excellence and to recognise that whenever possible. This positive reinforcement, particularly in an environment of dynamic change, provides a very powerful sustaining mechanism and helps the contractor continue to embed and improve changes which are often a critical component in service quality improvement and efficiency gain.

Successful contracting also demands that

Table 13.1 Alternative attributes of different types of contracts.

Traditional	*Output specification*	*Value propositions*	*TQM model**
No innovation	Low innovation	Local re-engineering	Service chain re-engineered
Risk with client	Risk with contractor	Risks shared	Risks analysed and placed with party best able to manage
Systems independent and secret	Systems understood but separate	Systems aligned	Systems integrated
Residents/stakeholders as extras	Limited customer feedback	Customer focus	Customer relationship management
Penalty based systems	Disincentives	Joint value analysis	Total process cost minimisation
$/hour delivery	Lump sum	Incentive based	Value gains shared
Operational control	Client control	Joint control	Customer driven

* Total Quality Management

Table 13.2 Specific outputs, processess, directions and attitudes in the schedule of rates, output specifications, value propositions and total quality management.

	Schedule of rates	*Output specification*	*Value propositions*	*TQM Model*
Specific Outputs	Contract allocation (may be absent)	Clear and unambiguous point of delivery Measured QA Measurement supports	Cross functional Measurement Systematic key performance indicators	Driven by process management
Processes	Unspecified or 'as is today' Man-hour allocation Unconnected activities	Asset management focus Task management Work scheduling Focus managed activities	Contractor re-engineers Preventative maintenance systems Interlinked activities	Mutual re-reengineering Maintenance/capital policies developed. Interlinked and aligned activities
Directions	Planning unnecessary Static and set direction Instructional/following work ethic	Plan for today Static set direction Introspective and incrementalist	Plan around customers/ stakeholders Develop value Involve employees	Best of breed strategies Dynamic relationship Learning culture
Attitudes	Master/servant Reactive management Interventionist	Provider/client Command and control Observational	Collaborative Impact oriented Co-operative	Integrated Growth and strategic oriented

there must be measurement and monitoring systems and processes in place. In The City of Melbourne, for instance, in its Parks, Gardens and Recreation facilities, extensive contract management systems are applied. You will need to tailor a particular monitoring and measurement system that meets your needs for your particular circumstances. The critical thing to bear in mind is that without installing a measurement and monitoring process, which you and your contractor agree upon, there can be no 'evidential' basis of determining the achievement or non-achievement of goals or critical success factors.

In addition to installing a measurement and monitoring process, you, as a contract superintendent, must satisfy yourself that the way in which the data is being assembled, provided and presented is, in the first instance, reasonable and truly reflective of the condition of the service being delivered. Hence, for contracts which concern themselves with services associated with major assets, the contractor should provide information on the condition of assets in a manner and form which enables you, as a representative of the principal, to record and ascertain critical strategic information which is then available to apply in policy determination. Contractors who seek to preserve such information for themselves do so with a clear strategic intention of creating a knowledge monopoly where they are the only people capable of bidding reasonably and informatively for your downstream business. While many contractors offer extensive services in this area, the practitioner should avoid collaborating with this process unless there is an absolute understanding about the ownership of the data, access to it and its future application.

The need to communicate openly and frequently has been mentioned. It needs to be said that any contract where the relationship between the parties is not open, honest and regular will invariably have problems and more than likely fail. The contract superintendent must have strong people skills. He or she must be able to deal with awkward, difficult, taxing situations on a regular basis and do so in a way that is non-threatening and fair. This does not mean that the contract superintendent has to be reasonable. It does not mean that the contract superintendent has to agree to dilute the standards which may seem hard or difficult to achieve. It simply means that the superintendent must show, at times, a degree of empathy with their counterpart in recognising the difficulties that often exist in the provider side in achieving complex, challenging targets.

It is also important that for a successful contract to be ongoing, there is a process of joint review of the relationship between the parties. This extends beyond the immediate contract and brings in elements such as potential business opportunities that exist in the longer term, and opportunities to reflect on shared values, shared missions and visions. Your contractor is your partner in this process. You should be open, honest and listen carefully to submissions that they make. A formal process of joint review of the relationship undertaken at least annually is my recommendation to keep the relationship dynamic.

Often you will be involved in minor dispute resolution on a day-to-day and ongoing basis. This is a normal, natural process and should not be seen as a contract failure. The contract will fail when the relationship fails. The relationship will fail when you, as the superintendent, do not provide the process control and review opportunities that are an essential ingredient. An important part of this is the need to resolve disputes in a quick, effective and non-confrontational way. Disputes should be resolved as close to the workplace as possible and your employees or agents should be empowered to resolve them. This may mean from time to time that you do not achieve the result, in minor instances, that you would personally believe to be the right one under the contract. However, you must also recognise that a contract which continually escalates minor issues to major disputes, through a formal facilitated review process, may develop into a highly bureaucratic relationship lacking real commitment. Many such disputes arise because of fundamental differences in interpretations in specifications and, again, the failure of the words and specifications often to be an adequate basis for the practitioners in the field.

One should also assume that your contractor is not a 'carpet bagger' coming along to steal from you in the night. You should assume that your contractor is intending to deliver a quality service. In managing and maintaining a relationship, trust is fundamental. Notwithstanding the foregoing, the contract superintendent must know the legal basis on which the contract is formed, the consequences of breach, and must ensure that the work is undertaken in accordance with the specification. Where this does not occur, the contract superintendent must take forward the issues through the dispute resolution processes that apply within the contract.

The superintendent must have at his or her disposal accurate records at all times. While we all have an inclination not to maintain records, and we tend to keep only sketchy reference data, it is imperative that if you are involved in a dispute of any magnitude quite clear records are available. Many contractors have avoided their contractual obligations because the principal and/or the superintendent were unable to establish the evidential basis of their breach claim. Records, while often a trouble and a trial, should nevertheless be fully maintained at all times.

Summary

The key steps to achieve a good contract are:

- To know what you want from the contract and where you wish to be in three or five years
- To agree with your contractor on what the contract means and to actively commit to working together to achieve this
- To manage by the use of objective data and facts
- To rely on trust and recognise that minor differences do not represent failures
- To be able to listen carefully to your contractor and customers, and plan to achieve their needs as well as yours.

CHAPTER 14

The playing quality of turfgrass sports surfaces

S.W. BAKER, Sports Turf Research Institute, Bingley, West Yorkshire, England

Introduction

Management of turfgrass sports areas should be directed primarily at producing a playing surface that encourages skilful play, maximises a player's enjoyment of the game in which he or she is participating and minimises the risk of injury (Baker and Canaway, 1993). As such, it is essential that we are able to assess the performance of sports surfaces in terms of quantitative measurements of factors such as ball bounce, ball roll, grip and surface hardness and have criteria by which measurements of playing quality can be interpreted. Testing playing quality should be an important component of research on sports turf areas as it is useful to be able to assess how different construction methods or management practices affect the sports player (Baker *et al.*, 1988a; Lodge and Baker, 1991). In addition, performance specifications are now being developed for the construction and particularly the maintenance of sports turf: again it is important that the playing quality and surface safety are included in such specifications.

The objective of this chapter is to consider which aspects of playing quality are relevant to a variety of sports, methods of quantifying playing quality, and how playing quality measurements can be interpreted. Measurement techniques are restricted to those used for natural turf surfaces, although it is acknowledged that in the future certain methods developed for artificial turf surfaces may also be incorporated in testing programs for natural turf areas.

Components of playing quality relevant to different sports

It is inevitable that the requirements for the playing surface vary for different sports. Table 14.1 lists aspects of playing quality that are relevant to the main turf-based sports played in the United Kingdom. The components of playing quality can be divided into two main parts, firstly ball–surface interaction and secondly player–surface interaction. Consideration of the physical processes involved and appropriate definitions are given in greater detail by Bell *et al.* (1985) and Baker and Canaway (1993).

Ball rebound

Ball rebound properties form an integral part of many sports. Rebound height, for example, influences the shots that a player can make in cricket and tennis, while for soccer both very low bounce on a muddy, waterlogged pitch or high bounce on a dry, compacted, poorly grassed pitch or some of the early artificial turf surfaces (Winterbottom, 1985) can be unacceptable to players. Vertical

Table 14.1 Principal components of playing quality for various sports.

	Soccer	*Rugby*	*Hockey*	*Tennis*	*Cricket*		*Bowls*	*Golf*	
					Pitch	*Outfield*		*Greens*	*Fairways*
Ball–surface interaction									
Ball rebound	++	+	–	++	++	+	–	+	+
Ball spin	–	–	–	++	++	–	–	++	+
Ball roll	++	–	++	–	–	++	++	++	+
Player–surface interaction									
Traction (grip)	++	++	+	++	+[1]	+	–	–	+[2]
Hardness–firmness	++	++	+	+	+	+	–	–	–
Surface evenness	+	–	++	++	++	++	++	++	+

++ Major importance + Important – Minor or of no importance

Notes: 1 Important for batsmen running between wickets but stability of foot holes is particularly important for faster bowlers.
2 Important particularly on slopes and banks especially in wet conditions.

rebound properties are often expressed in terms of ball rebound resilience which is the ratio of the height bounced to height dropped. Thus a ball which bounces to a height of 4 metres when falling from a height of 10 metres would have a rebound resilience of 0.4 or 40 per cent.

Angled ball behaviour

In practice, true vertical bounce is rare in virtually all sports so in most situations angled ball behaviour needs to be considered. Angled ball behaviour is influenced by the mechanical properties of the surface that govern rebound resilience but also by frictional characteristics that influence how the ball reacts to any spin that has been imparted onto it. For example, with tennis a shot hit with topspin will bounce to a lower height than a shot with no spin, and one hit with backspin will bounce higher than normal. In both cases the variation from normal bounce height will depend on the amount of spin imparted, the frictional properties between the ball and the surface, and the contact time of the ball and the surface.

Angled ball behaviour may also be influenced by any deformation of the ball on contact with the surface or by deformation of the surface. A good example occurs on wet, soft cricket pitches where the ball can depress the surface causing scars up to about 4 to 5 millimetres deep in poor conditions. Loss of energy in deforming the surface will reduce the velocity of movement in both horizontal and vertical directions giving a generally 'slow' batting surface, but sometimes the ball can rise unexpectedly if a steep plane is created at the end of the pitch mark.

The combined effects of incoming and outgoing ball velocity, including both horizontal and vertical components, and the changes in the angle of incidence and the angle of departure influence what the player perceives as the 'pace' of the surface in sports such as tennis, cricket and, to a lesser extent, soccer.

For golf, consideration of angled ball behaviour is particularly important as it will influence how far a ball will travel on a fairway and how quickly it will stop on a green. For any given shot with, for example, a wood or low iron the distance travelled in the initial bouncing phase of movement will depend on the mechanical properties of the surface as influenced by, for example, moisture content or grass length, the general slope of the fairway and local undulations. This means energy loss is greater if the ball pitches on an up slope section rather than an adjacent downslope area.

On golf greens, the interaction between the ball

and the surface will influence how quickly a chipped shot will stop. For any combination of incoming angle, velocity and amount of backspin, the distance a ball travels will be controlled by those surface properties that influence deformation on impact, contact time between the ball and the surface, and friction between the ball and the surface (Haake, 1991b; Lodge, 1992; Baker, 1994).

Lateral spin

The interaction of topspin and backspin with angled ball behaviour has already been discussed. Lateral movement in response to imparted spin is particularly important in cricket. The amount of turn that occurs is influenced by the rate of ball rotation and its direction relative to the direction of ball movement, and the frictional properties and contact time between the ball and the surface. These properties may be determined by soil type, moisture content, grass cover, soil cracking and the rigidity of the surface below the point of ball impact. The effects of spin are also influenced by other components of playing quality, for example it is more difficult to play a turning ball if the bounce is high than a pitch taking spin but with low bounce.

Ball roll

The speed of ball movement across the playing surface is relevant to sports such as soccer and hockey and on cricket outfields, but critical on golf and bowling greens. In addition, smoothness of the surface is important in ensuring that balls do not deviate unfairly, although this is discussed in more detail in the section on surface uniformity and evenness.

For golf and bowls, one of the main factors that distinguishes a good player from an average one is the ability to assess how quickly the ball or bowl will stop when projected at a given initial velocity. This will be influenced by many factors such as cutting height, grass species or moisture content. As the surface becomes 'faster' (i.e. the rate of ball deceleration falls) it is more difficult to control the finishing position of the ball and this becomes crucial for downhill hit putts in golf and in bowls where the curvature of the arc of travel of a biased bowl increases on faster surfaces. Therefore, for both golf and bowls a reasonably fast surface is desirable, especially for the better players as it increases the level of skill that is needed. A desire for very 'fast' playing surfaces may have important effects on long-term turfgrass quality, particularly if the grass becomes stressed by prolonged periods with a very low cutting height. One of the major skills of greenkeeping, and indeed groundsmanship (as similar problems occur in terms of the pace of hockey and soccer pitches), is maintaining a surface of acceptable pace for the players while still retaining good grass growth, rooting and turf density.

Traction and grip

Where sport involves running and turning, adequate grip between the player's footwear and the playing surface is essential. The term 'friction' is usually used in situations where a player has smooth-soled footwear, but where the amount of grip is enhanced by the use of studded, cleated or spiked footwear the term 'traction' is generally used (Bell *et al.*, 1985). Different aspects of friction or traction can be recognised, for example whether motion is primarily linear or rotational. In addition, shoe–surface interaction can be considered in terms of the force required to initiate movement (i.e. static friction or traction) or the forces required to maintain motion once movement has started (dynamic friction or traction).

As well as influencing the running and turning movements that are a fundamental part of many sports, traction properties are also important in terms of safety. A surface with inadequate grip may lead to a greater number of falls or inadvertent collisions between players which increases the risk of injury. Alternatively, excessive grip may increase the incidence of knee and ankle injuries, which has been a concern with some artificial turf surfaces (Winterbottom, 1985) and some forms of reinforcement systems for natural turf surfaces (Baker *et al.,* 1988b).

In some sports the capacity to slide on the surface is an integral part of the game. This is particularly evident in tennis and in the sliding tackle in soccer.

Hardness–firmness

The term 'hardness' can be used to cover the conditions of player–surface impact that relate to running, falling and injury potential (Baker and Canaway, 1993). Alternatively, these parameters can be considered in terms of the mechanical properties that govern the 'firmness' of the surface, and these include the stiffness or compliance of a surface (i.e. how much it deforms under a given load) and the resilience of the surface (i.e. how much energy is returned to the player) (Bell *et al.,* 1985). Subjective assessments of surfaces by players reflect a combination of both stiffness and resilience. A surface which is very stiff causes jarring of limbs, muscle soreness and a greater risk of injury in the event of a fall. A surface which is excessively compliant or lacking in resilience is likely to cause fatigue.

From the player's point of view, the hardness of the surface is particularly important in any sport that involves running or falling on the surface, for example soccer or rugby. There is clearly also a strong link between hardness–firmness properties involved in player–surface interaction and properties affecting ball rebound resilience.

Surface uniformity and evenness

With virtually all sports, it is important that playing quality is uniform across the playing surface or at worst the mechanical properties only change gradually, for example in response to wear gradients on a pitch. Uniformity is critical in terms of ball–surface impacts, for example alternating high and low bounce from adjacent areas on a cricket pitch or tennis court detract considerably from the quality of a game. Similarly, marked variations in surface stability can have safety implications as an unexpected reduction in traction may cause a player to fall.

Surface evenness is important on most turfgrass sports areas as unevenness can affect ball rebound and roll, and unexpected changes in levels can cause a player to stumble or fall. Some gradient, as long as it is within prescribed limits, is acceptable in most sports and indeed may be desirable because of the implications for surface drainage. In sports such as crown green bowls and on golf greens an ability to 'read' the levels and adjust shots accordingly is one of the main skills of the game. What is not acceptable is sudden changes in surface evenness as this may cause a ball to bobble or deviate in an unpredictable manner.

Agronomic factors influencing playing quality

Some of the factors influencing playing quality such as soil type, grass species, cutting heights, and moisture content have already been referred to in the preceding sections. Inevitably changes in rootzone composition and management regime will have a profound influence on playing quality and many aspects of this are now covered in the sports turf literature. There is not, however, sufficient space to include this work in any detail in this chapter but extensive reviews of agronomic factors influencing playing quality are given by Bell *et al.* (1985) covering a range of sports, Canaway and Baker (1993) for soccer pitches and Baker (1994) for golf greens.

Measurement of playing quality

A variety of test methods have been developed to measure the playing quality of turfgrass sports surfaces. Where possible, the apparatus should be simple because, for example, complex electronic apparatus suitable for measuring player–surface impact properties on artificial turf (e.g. the Berlin Artificial Athlete or Impact Severity equipment (Baker, 1990)) are less reliable on wet, muddy, natural turf surfaces. At the same time, test methods should replicate the physical processes and forces that occur in play as closely as possible.

Vertical ball behaviour

Ball rebound properties are frequently assessed using a vertical ball bounce test. The main requirements are a method of releasing the ball without impulse or spin, and a measurement scale graduated as percentage rebound. Thus, for example, for soccer where a release height of 3 metres is generally used, the units on the scale are marked every 30 millimetres (i.e. 1 per cent of 3 metres) with the zero position coinciding with the top of the ball (Fig. 14.1). The drop height will vary for different sports but the recommended release heights for natural turf surfaces in the United Kingdom are as follows: soccer 3 metres; rugby 3 metres; hockey 3 metres (1.5 metres for artificial turf); and cricket 3 metres or 5 metres.

Where the ball is relatively large and the rebound height normally reasonably high (e.g. soccer), it is usually possible to record rebound height by eye. However for hockey and cricket balls it is usually necessary to record ball impacts by video then use slow motion or freeze framing to assess the maximum rebound height.

In most cases the ball used for testing should conform to regulations imposed by the governing body of sport in question, although it is often necessary to standardise the ball further in terms of inflation pressure or rebound properties on concrete. The oval shape of a rugby ball means that a soccer ball is recommended for rebound testing for this sport (McClements and Baker, 1994b).

Figure 14.1 Ball rebound measurement for soccer

Angled ball behaviour

Assessment of angled ball behaviour is inevitably more complex as a method to project the ball at a realistic velocity and sometimes imparting spin is needed, along with a means of recording the impact in two or even three dimensions.

A number of approaches have been adopted for different sports. For example for golf, Haake (1991a) developed a method of projecting the ball with a range of velocities, impact angles and levels of backspin using two rotating wheels from a modified baseball/cricket bowling machine. Ball impact was recorded using a stroboscope to produce a succession of images on each photographic print that was made. Analysis of the images allowed changes in angle, velocity and spin before and after impact to be calculated.

For routine measurements of trials on putting green turf and to assess the performance of actual golf greens, this technique has been simplified (Figure 14.2). Balls are still projected using the modified bowling machine with different velocities, angles and backspin but impact is recorded in terms of stopping distance, i.e. the distance from the initial pitch mark to the final resting place of the ball (Lodge, 1992; Baker and Richards, 1995; Baker *et al.*, 1996). This simulates the situation in real play as the golfer is concerned with how quickly the ball will stop after being pitched onto a green.

Alternative methods of recording angled ball behaviour have been developed for tennis and cricket. For tennis, compressed air has been used to project the ball (International Tennis Federation, 1997) and for cricket a modified crossbow has been used (McAuliffe and Gibbs, 1997). In both cases, incoming and outgoing velocities and angles have been recorded using banks of light sensitive receivers which detect the position of the ball.

Figure 14.2 Measurement of stopping distance for golf. The ball is fired with specified values of velocity, angle and backspin to simulate landing conditions for different shots, and stopping distance is measured as the distance between the pitch mark and final resting point of the ball.

A more recent modification for cricket, used on current trials at the Sports Turf Research Institute, uses a bowling machine for ball projection but records the position of the ball using stroboscopic photography with two digital cameras, one mounted at the side in the direction of travel and one above the path of the ball (Carré *et al.*, 1998). Images can be downloaded directly from the digital cameras to a computer, greatly increasing the number of images that could be analysed compared with manual techniques used in the earlier studies for golf. The use of two cameras allows lateral deflection to be recorded as well as changes in velocity, angle and backspin.

Ball roll

A variety of methods have been used to assess ball roll properties, including distance rolled after release of the ball down a standard ramp (Figure 14.3), deceleration of the ball using electronic timing gates and, in the case of bowls, measurement of the trajectory of the bowl.

For golf, the most widely used method has been based on the Stimpmeter (Radko, 1980). This consists of an aluminium bar 0.91 metres long with a 145° v-shaped groove extending

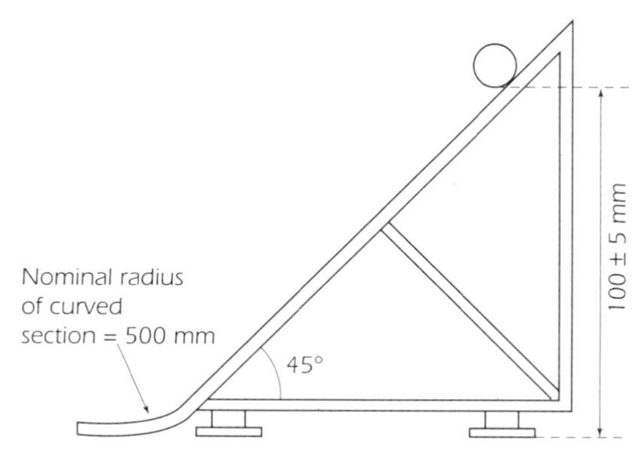

Figure 14.3 Assessing ball roll properties including distance rolled after release of the ball down a standard ramp

its entire length. It has a ball-release notch 0.76 metres from the tapered end which rests on the ground. In use, the ball is placed in this notch and the angle of the bar gradually increased until the ball starts to roll, which should occur when the angle of the Stimpmeter is 20° to the horizontal. The distance rolled in opposing directions is then measured.

Similar principles have been used for other sports such as soccer, hockey and bowls but the dimensions of the ball-release ramp have varied (Table 14.2).

Table 14.2 Details of ramps used for ball roll measurement for various sports.

Sport	*Ball release height (m)*	*Spacing of bars (mm)*	*Angle of ramp (°)*	*Reference*
Soccer	1.0	105	45	Canaway *et al.* (1990)
Hockey	1.0	50	45	McClements & Baker (1994a)
Bowls	1.0	45	30*	Bell & Holmes (1988)

* Rubber strips placed on bars to prevent ball slippage.

On sloping sites, calculation of the arithmetic mean from readings in opposing directions can create an over-estimate of distance rolled had the site been flat. Brede (1991) has proposed the following correction:

$$\text{Distance rolled} = \frac{2 \times S(\text{uphill}) \times S(\text{downhill})}{S(\text{uphill}) + S(\text{downhill})} \quad \ldots (1)$$

where S(uphill) = distance of ball roll uphill and S(downhill) = distance of ball roll downhill.

It is suggested that this correction is used when downhill values are more than 25 per cent greater than corresponding up slope measurements. Where a relatively light ball, which can be subjected to windblow, is used (e.g. soccer) or the measurement area is restricted (e.g. experimental plots), it may be more appropriate to measure ball deceleration or the change in velocity over a specified distance, for example by using infra-red beams. For example, Baker and Canaway (1991) calculated deceleration of a ball released down a standard ramp. Initial velocity (u) was calculated from two beams 0.3 metres apart and final velocity (v) was measured at a distance (s) of 2 metres again using timing gates 0.3 metres apart. Deceleration (in metres per second per second) was calculated using the following equation:

$$\text{Deceleration} = \frac{u^2 - v^2}{2s} \quad \ldots (2)$$

For lawn bowls, a traditional method of measurement of green speed has been to record the time for the bowl to travel from the bowler's hand to a jack placed 27.4 metres (30 yards) from the front edge of the bowling mat. Because on faster surfaces the bowl can be projected with less force and travels in a wider arc, the time taken to reach the jack is higher on faster surfaces. However, to avoid the requirement of having to stop the bowl within 0.15 metres of the jack, use of a ball roll ramp is simpler for non-bowlers and green speed results can be calculated from the following equation given by Holmes and Bell (1986). Green speed is in seconds and distance rolled in metres.

$$\text{Green speed} = 6.01 + 0.36 \text{ Distance rolled} \quad \ldots (3)$$

An additional assessment of relevance for bowls is the consistency of draw, that is the curvature of movement caused by the bias of the bowl (McAuliffe and Gibbs, 1993). A variety of methods such as the use of chalk marks or video recording can be used to chart the motion of a bowl, and analysis of the trajectory can be used to check whether there is consistent draw on the left-hand and right-hand sides of the rink.

Traction

A variety of methods have been used to measure traction-related properties of natural turf areas, including shear vanes and the loss of energy of a pendulum with a studded plate which made contact with the turf. However the two main techniques used in the United Kingdom have been based on the rotational force required to initiate movement of a studded disc (Canaway and Bell, 1986; Canaway *et al.*, 1990); and sliding distance using a trolley with a studded foot released down a ramp (Baker *et al.*, 1988b).

Details of the traction apparatus are given in Figure 14.4. The equipment is dropped from a height of 50 to 75 millimetres to ensure that the studs penetrate the turf surface, and the torque wrench is used to measure the rotational force needed to induce movement. In the sliding distance test a 45 kilogram trolley is released down a ramp allowing the test foot to make contact with the turf. Sliding distance is measured as the

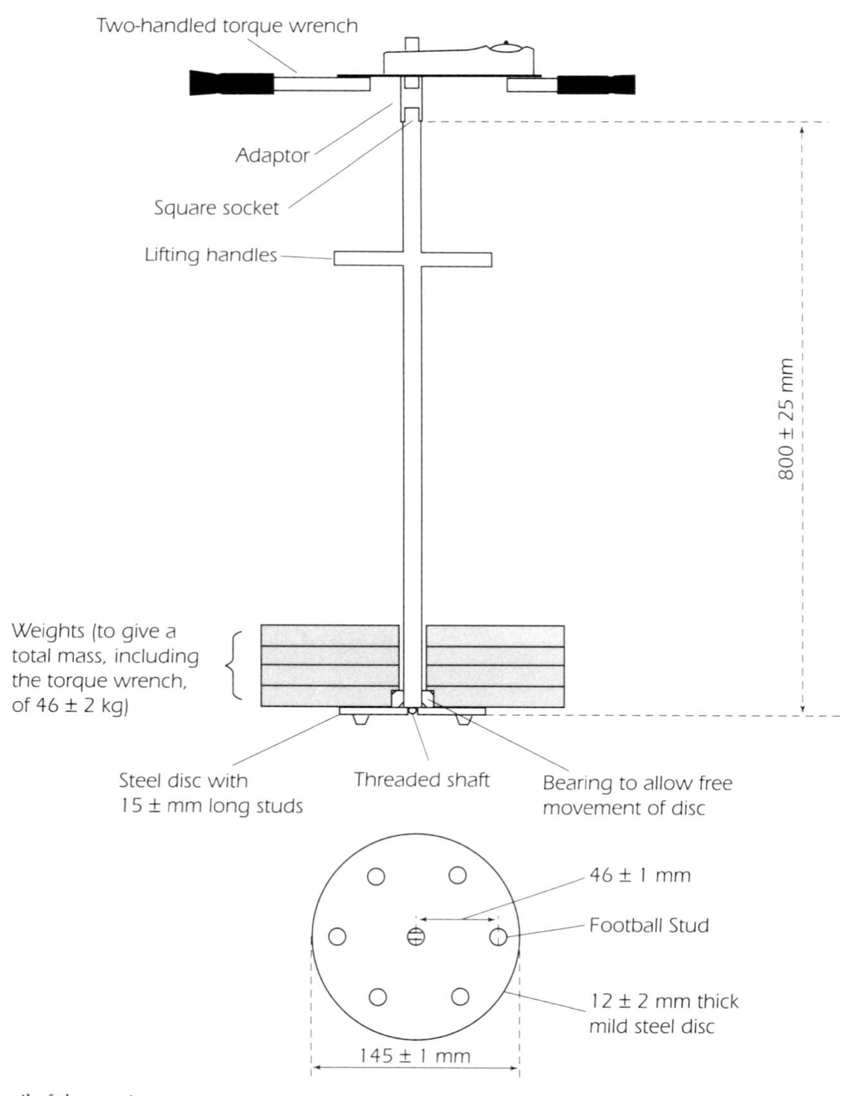

Figure 14.4 Detail of the traction apparatus

distance travelled by the test foot while in contact with the turf. Although the sliding distance test can provide useful information on the sliding properties of the turf, the size and weight of the equipment and the slow speed of operation mean that routine measurements of grip for the player are usually made using the traction apparatus with the studded disc.

Hardness

Impact properties for players running or falling on the surface have been measured in several ways including subjective assessment by kicking the heel into the turf surface, studies using a penetrometer and measurement of surface deflection or deceleration of a falling weight using apparatus initially developed for artificial turf surfaces, for example the Berlin Artificial Athlete, the Stuttgart Artificial Athlete or Impact Severity tests (Winterbottom, 1985; Baker, 1990).

Because of its relative simplicity, the most widely used method for natural turf has been based on the Clegg Impact Tester (Clegg, 1976). In the United Kingdom the method is based on a 0.5 kilogram test mass which is released down a guide tube with an accelerometer housed at the top of the test hammer used to record peak deceleration during impact with the turf. The drop heights now in use are 0.3 metres for golf and cricket and 0.55 metres for winter games pitches (soccer, rugby and hockey).

In the United States the use of 0.5, 2.25 and 4.5 kilogram impact hammers has been studied using a vibration analyser to record deceleration time curves (Rogers and Waddington, 1990). This allowed a number of parameters of the impact response to be recorded, such as maximum deceleration, time to maximum deceleration, the total duration of the impact and Severity Impact (an integral of deceleration over time). Rogers and Waddington (1990) suggest that although in principle one factor may be sufficient to characterise impact properties, in situations where there are both grassed and bare areas it may be sensible to measure both peak deceleration and the duration of impact.

Surface evenness

Although not necessarily a direct measurement of playing quality, surface evenness is so important in many sports that some reference to its measurement is needed. Two main methods are available. First, the deviation of the turf surface under a straight edge can be measured. In the United Kingdom a 3 metres straight edge has commonly been used, but a 1 metre straight edge has also been employed to record more localised changes in surface levels (McClements and Baker, 1994a). An alternative method has been a profile gauge based on a series of independently moving rods in a frame which are displaced by surface undulations. Spacing of the rods has ranged from 200 millimetre for soccer (Canaway *et al.*, 1990), rugby (McClements and Baker, 1994b) and hockey (McClements and Baker, 1994a) to 50 millimetres for golf (Baker *et al.*, 1996).

Performance requirements for different sports

There is relatively little advantage of having a suite of tests to measure the playing quality of a sports surface if there is no means of interpreting the results in terms of players' perceptions of the quality of the surface. A number of studies have therefore been carried out to relate the various mechanical and electronic tests to players' assessments of the surface. The normal format has been to make the various physical measurements shortly before a game is due to take place and then afterwards carry out a questionnaire survey to determine players' opinions about the surface. For example, results from a ball rebound test could be related to questions about the height of bounce, such as was it too high, high but acceptable, ideal, low but acceptable, or too low. Clearly a range of surfaces need to be examined for each sport and, given the vagaries of human response, a large number of sports participants need to be questioned. Details of the principal studies for each sport are given in Table 14.3.

Table 14.3 Studies relating to players' perceptions of playing quality to physical measurement of sports surface characteristics.

Sport	*Number of sites investigated*	*Number of questionnaire returns*	*Study details*
Soccer	20 pitches	444	Canaway *et al.* (1990)
Rugby	43 pitches	412	McClements & Baker (1994a)
Hockey	31 pitches	249	McClements & Baker (l994b)
Golf (Stimpmeter)	> 1500 greens	–	Radko (1980)
Golf	147 greens	787	Baker *et al.* (1996)
Bowls	74 greens	774	Bell & Holmes (1988)

One specific problem with this form of analysis is that expectations of pitch quality may vary between different standards of players. For example, a soccer pitch that may be acceptable to players in an amateur league may be heavily criticised by professional players. For many properties, requirements for high grade pitches are in the central part of the range considered acceptable for lower standards of play, in other words the more extreme values are eliminated. For example, for soccer most professional pitches in the United Kingdom have ball bounce values in the range 30 to 45 per cent (Baker, 1990) but a preferred range of 20 to 50 per cent and an acceptable range of 15 to 55 per cent have been proposed for a reasonable standard of amateur play. There are, however, several exceptions to the principle that sports surfaces for professional or high grade use require performance values in the central part of the range of potential values. It has already been mentioned that faster greens for golf and lawn bowls provide a greater challenge, and thus faster surfaces are usually expected for competitions involving the top players. A similar principle applies in cricket, with hardness and ball bounce on pitches used for the professional game generally being greater than that on pitches used by amateur players. Most of the performance requirements developed in the studies listed in Table 14.3 relate to a high standard of amateur play. However, in the case of Stimpmeter readings for golf greens (Radko, 1980), values are interpreted in the two categories of regular membership play and tournament play.

Winter games pitches

Recommended performance requirements for winter games pitches are given in Tables 14.4, 14.5 and 14.6 and are based on the studies of Canaway et al. (1990) for soccer, McClements and Baker (1994b) for rugby and McClements and Baker (1994a) for hockey.

Table 14.4 Preferred and acceptable ranges for playing quality measurements for soccer (Canaway *et al.*, 1990).

Parameter	*Brief details of test method*	*Preferred range*	*Acceptable range*
Ball rebound (%)	3 m vertical drop test using soccer ball	20–50	15–55
Ball roll (m)	Distance rolled by ball released from a height of 1 m down a 45° ramp	3–12	2–14
Traction (N m)	Torque to cause tearing using studded disc	≥ 25	≥ 20
Hardness (gravities)	Clegg Impact Soil Tester, 0.5 kg hammer released from 0.55 m	55–140*	35–200*
Evenness (mm)	Standard deviation of profile gauge measurements	≤ 8	≤ 10

* Hardness values recalculated for 0.55 metre drop height compared with 0.3 metre drop height in original publication.

Table 14.5 Preferred and acceptable ranges for playing quality measurements for Rugby (McClements and Baker, 1994b).

Parameter	*Brief details of test method*	*Preferred range*	*Acceptable range*
Ball rebound (%)	3 m vertical drop test using Rugby ball	20–50	15–55
Traction (N m)	Torque to cause tearing using studded disc	≥ 35	≥ 25
Hardness (gravities)	Clegg Impact Soil Tester, 0.5 kg hammer released from 0.55 m	50–100	30–180
Evenness (mm)	Standard deviation of profile gauge measurements	≤ 8	≤ 10

Table 14.6 Preferred and acceptable ranges for playing quality measurements for hockey (McClements and Baker, 1994a).

Parameter	*Brief details of test method*	*Preferred range*	*Acceptable range*
Ball roll (m)	Distance rolled by hockey ball released from a height of 1 m down a 45° ramp	≥ 5.5	≥ 3.5
Traction (N m)	Torque to cause tearing using studded disc	≥ 35	≥ 25
Hardness (gravities)	Clegg Impact Soil Tester, 0.5 kg hammer released from 0.55 m	90–140	65–200
Evenness (mm)	Standard deviation of profile gauge measurements	≤ 4	≤ 5*

* ≤ 6 mm in goalmouths.

Golf

Values for interpreting Stimpmeter readings given by Radko (1980) are given in Table 14.7, the values having been converted from imperial units to the nearest metric equivalent.

Table 14.7 Values for interpreting Stimpmeter readings taken from Radko (1980) but converted to metric equivalents.

	Average distance rolled (m)	*Green speed*
For regular membership play	2.55	Fast
	2.25	Medium Fast
	1.95	Medium
	1.65	Medium Slow
	1.35	Slow
For tournament play	3.15	Fast
	2.85	Medium Fast
	2.55	Medium
	2.25	Medium Slow
	1.95	Slow

Proposed limits for interpreting the playing quality of golf greens under British conditions are given in Table 14.8. It should be noted that no attempt was made to define the range of values that are likely to occur at different times of the year and the limits have been set to allow for the seasonal variation that occurs. The broad range in the limits that are given also avoids the problem of over-prescribing the requirements for golf greens, as one of the joys and challenges of golf is playing a wide variety of courses under a range of weather conditions. It should also be noted that an upper limit has been given for green speed values as it is unlikely that green speed above the given range could be sustained without detrimental long-term effects on sward quality under British conditions.

Flat green bowls

Performance requirements for flat green bowls have been given by Bell and Holmes (1988) and these are presented in Table 14.9. It should be noted, however, that these requirements apply specifically to British conditions where green speed values are often considerably lower than, for example, Bermudagrass surfaces in Australia or Cotula surfaces in New Zealand, or indeed synthetic turf.

Other sports

Comprehensive studies of playing performance for other sports played on turfgrass surfaces and listed

Table 14.8 Proposed limits for interpreting the playing quality of golf greens under Bntish conditions (Baker *et al.*, 1996).

Parameter	*Brief details of test method*	*Preferred range*	*Acceptable range*
Green speed (m)	Stimpmeter	1.6–2 8	1.5–3.0
Hardness(gravities)	Clegg lmpact Soil Tester, 0.5 kg. mass dropped from 0.3 m	70–100	55–120
Stopping distance* (m)	'Five iron' simulation (angle 53°, velocity 22.7 m s^{-1}, backspin 750 rad s^{-1})	0.5–5.0	–0.5–8.0
Stopping distance* (m)	'Nine iron' simulation (angle 53°, velocity 18.8 m s^{-1}, backspin 880 rad s^{-1})	0.0–2.0	–1.0–3.5
Surface evenness (mm)	Standard deviation of proflle gauge measurements	≤ 1.0	≤ 1.25

* For routine measurement on greens it is recommended that measurements of stopping distance are omitted and assessments confined to those of green speed, hardness and surface evenness .

Table 14.9 Preferred and acceptable ranges for playing quality measurements for bowls (Bell and Holmes, 1988).

Parameter	*Brief details of test method*	*Preferred range*	*Acceptable range*
Green speed (s)	Time taken for bowl to travel 27.4 m (30 yards)*	212	210
Evenness (mm)**	Standard deviation of profile gauge measurements	≤ 1.5	≤ 2.0

* For testing purposes this is best measured by releasing a bowl down a 30° ramp from a height of 1 metre and recording the distance travelled. This distance can be converted to green speed using Equation 3 given earlier. An alternative of measuring the travel time after release of the bowl by a player has also been discussed earlier.
** A number of additional requirements for height differences within the green are also given.

in Table 14.1 have yet to be carried out. Some information is available and, for example, Stewart and Adams (1970) have proposed a scale for interpreting ball rebound against the pace of cricket pitches. For example very fast pitches would have rebound values above 15.6 per cent, an easy paced pitch would have values between 7.8 and 10.4 per cent, and below 7.8 per cent the pitch would be considered slow. Similar performance requirements have been proposed for artificial turf cricket pitches, e.g. Sports Council (1984).

Less work has been undertaken on the playing performance of natural turf tennis courts, although requirements have been proposed for artificial turf surfaces (International Tennis Federation 1997).

Conclusions

Our understanding of the playing quality of natural turf surfaces has increased considerably in the last fifteen years. This has come about through a series of research studies focused on the needs to set practical objectives for research evaluation, as well as for application in performance and safety requirements for construction and maintenance. Some of the test procedures are inevitably relatively simplistic to keep testing costs to reasonable levels and to ensure that the equipment can be used in all weather conditions in which play is likely to take place. However there have been considerable advances in the last five years, for example, in studies of angled ball behaviour and technological improvements in monitoring techniques should further increase our knowledge of the interaction between the sports surface, the ball and the player.

References

Baker, S.W. (1990). 'Performance standards for professional soccer on artificial turf surfaces', *J. Sports Turf Res. Inst.*, 66, 42–69.

Baker, S.W. (1994). 'The playing qualify of golf greens', in *Science and Golf. Proceedings of the World Scientific Congress of Golf*, Eds A.J. Cochran and M.R. Farrally, E. and F.N. Spon, London, pp. 409–18.

Baker, S.W. and Canaway, P.M. (1991). 'The cost-effectiveness of different construction methods for Association Football pitches. II. Ground cover, playing quality and cost implications', *J. Sports Turf Res. Inst.*, 67, 53–65.

Baker, S.W. and Canaway, P.M. (1993). 'Concepts of playing quality: criteria and measurement', *Int. Turfgrass Soc. Res. J.*, 7, 172–81.

Baker, S.W., Cole, A.R. and Thornton, S.L. (1988a). 'Performance standards and the interpretation of playing quality for soccer in relation to rootzone composition', *J. Sports Turf Res. Inst.*, 64, 120–32.

Baker, S.W., Hind, P.D., Lodge, T.A., Hunt, J.A. and Binns, D.J. (1996). 'A survey of golf greens in Great Britain. IV. Playing quality', *J. Sports Turf Res. Inst.*, 71, 9–24.

Baker, S.W., Isaac, S.P. and Isaac, B.J. (1988b). 'An assessment of five reinforcement materials for sports turf II. Playing quality', *Z. fur Vegetationstechnik*, 11, 12–15.

Baker, S.W. and Richards, C.W. (1995). 'The effect of rootzone composition on the playing quality of Festuca/Agrostis/*Poa annua* golf greens', *J. Turfgrass Management*, 1 (3), 53–68.

Bell, M.J., Baker, S.W. and Canaway, P.M. (1985). 'Playing quality of sports surfaces: a review', *J. Sports Turf Res. Inst.*, 61, 26–45.

Bell, M.J. and Holmes, G. (1988). 'Playing quality standards for level bowling greens', *J. Sports Turf Res. Inst.*, 64, 48–62.

Brede, A.D. (1991). 'Correction for slope in green speed measurement of golf course putting greens', *Agron. J.*, 83, 425–6.

Canaway, P.M. and Baker, S.W. (1993). 'Soil and turf properties governing playing quality', *Int. Turfgrass Soc. Res. J.*, 7, 192–200.

Canaway, P.M. and Bell, M.J. (1986) . 'Technical note: An apparatus for measuring traction and friction on natural and artificial playing surfaces', *J. Sports Turf Res. Inst.*, 62, 211–14.

Canaway, P.M., Bell, M.J., Holmes, G. and Baker, S.W. (1990). 'Standards for the playing quality of natural turf for Association Football', in *Natural and Artifical Playing Fields: Characteristics and Safety Features. ASTM STP 1073*, (eds R.C. Schmidt, E.F. Hoerner, E.M. Milner and C.A. Morehouse), American Society for Testing and Materials, Philadelpia, USA, pp. 29–47.

Carré, M.J., Haake, S.J., Baker, S.W. and Newell, A. (1998). 'The analysis of cricket ball impacts using digital stroboscopic photography', in *The Engineering of Sports* (ed. S.J. Haake), Blackwells Science, pp. 379–86.

Clegg, B. (1976). 'An impact testing device for *in situ* base course evaluation', *Australian Road Res. Bur. Proc.*, 8, 1–6.

Haake, S.J. (199la). 'The impact of golf balls on natural turf. I. Apparatus and test methods', *J. Sports Turf Res. Inst.*, 67, 120–7.

Haake, S.J. (1991b). 'The impact of golf balls on natural turf. II. Results and conclusions', *J. Sports Turf Res. Inst.*, 67. 128–34.

Holmes, G. and Bell, M.J. (1986). 'The playing quality of level bowling greens: A survey' *J. Sports Turf Res. Inst.*, 62, 50–66.

International Tennis Federation (1997). *An Initial ITF Study on Performance Standards for Tennis Court Surfaces*, International Tennis Federation, 41 pp.

Lodge, T.A. (1992). 'A study of the effects of golf green construction and differential irrigation and fertiliser nutrition rates on golf ball behaviour', *J. Sports Turf Res. Inst.*, 66, 95–103.

Lodge, T.A. and Baker, S.W. (1991). 'The construction, irrigation and fertiliser nutrition of golf greens. II. Playing quality assessments after establishment and during the first year of differential irrigation and nutrition treatments', *J. Sports Turf Res. Inst.*, 67, 44–52.

McAuliffe, K.W. and Gibbs, R.J. (1993). 'A national approach to the performance testing of cricket grounds and lawn bowling greens', *Int. Turfgrass Soc. Res. J.*, 7, 222–30.

McAuliffe, K.W. and Gibbs, R.J. (1997). 'An investigation of the pace and bounce of cricket pitches in New Zealand', *Int. Turfgrass Soc. Res. J.*, 8(1), 109–19.

McClements, I. and Baker, S.W. (1994a). 'The playing quality of natural turf hockey pitches', *J. Sports Turf Res. Inst.*, 70, 13–28.

McClements, I. and Baker, S.W. (1994b). 'The playing quality of natural turf Rugby pitches', *J. Sports Turf Res. Inst.*, 70, 29–43,

Radko, A.M. (1980). 'The USGA stimpmeter for measuring the speed of putting greens', in *Proc. 3rd Int.*

Turfgrass Res. Conf, ed. J.B. Beard, Am. Soc. of Agronomy, pp. 473–6.

Rogers, J.N. III and Waddington, D.V. (1990). 'Portable apparatus for assessing impact characteristics of athletic field surfaces', in *Natural and Artificial Playing Fields: Characteristics and Safety Features. ASTM STP 1073*, eds. R.C. Schmidt, E.F. Hoerner, E.M. Milner and C.A. Morehouse, American Society for Testing and Materials, Philadelpia, USA, pp. 96–110.

Sports Council (1984). *Specification for Artificial Sports Surfaces. Part 3: Surfaces for Individual Sports, Section I: Cricket*, Sports Council, London, 19 pp.

Stewart, V.I. and Adams, W.A. (1970). 'Soil factors affecting the control of pace on cricket pitches', in *Proc. 1st Inter.Turfgrass Res. Conf., Harrogate, England*, July 1969, ed. Sports Turf Res. Inst., pp. 533–44, Sports Turf Res. Inst., Bingley.

Winterbottom, W. (1985). *Artficial Grass Surfaces for Association Football*, Sports Council, London, 127 pp.

CHAPTER 15

Prescription surface development: Golf course management

P. RYAN, Pacific Coast Design, Sandringham, Victoria, Australia

Introduction

The preparation of surfaces for the playing of golf varies greatly from country to country with local climate, site conditions and budget being the primary factors causing this variation. However within like climatic zones, many surface preparation practices will be similar as appropriate management information and suitable turf machinery are available in most parts of the world.

Site conditions even within alike climatic zones may differ greatly depending on the structure of the natural ground and the standard of golf feature construction (greens, tees, fairways and bunkers). Many of the world's best golf courses are build in areas where natural sand deposits give advantages in drainage and subsoil aeration. On such sites management practices differ from nearby golf courses built on heavier clay or silt deposits. International turf management wisdom at present dictates that most newer golf greens are built to USGA or USGA modified standards which incorporate a predominantly sand profile with some organic matter. This allows for a more free draining profile but needs a greater awareness of good turf management practices to maintain good turfgrass surfaces.

Over the years many golf greens have been built on local top soils pushed up and planted. These more traditional style greens will have many different management needs depending on the nature of the soil and historical management practices. Across the world budgets and staff for golf course maintenance vary so much that it is impossible to specify a 'standard' in looking at preparing surfaces. New international standard golf resorts in India, such as Eagleton Golf Village, Bangalore, have a staff of 60 with an annual budget of $US380 000 while courses established forty years ago in the same country may have only four staff and an annual budget of $US12 500. In the USA, Europe, Africa and Asia similar comparisons can be made with annual budgets varying from $US10 000 to over $US1 300 000 for an 18 hole golf course. This range obviously has a great effect on the ability of turf managers to prepare surfaces for the playing of golf. Given that there are so many different surface preparation models that could be utilised, this chapter will be largely dealing with two types of surface preparation, first for tournament play (International Standard), and secondly for a 'typical' members golf club in normal circumstances.

Local adaptions of the methods described here will occur that compensate for either the non-availability of modern turf equipment, restrictions in budget or even lack of basic turf requirements (water, nutrients and drainage).

The main issue to keep in mind is not what cannot be done to prepare a turf surface but what can be done to produce a consistent playing surface. Nothing is more frustrating to a golfer than to be playing on inconsistent surfaces on a golf course. While teeing grounds are usually not critical as golfers can tee the ball up almost irrespective of the turf quality, the same cannot be said of fairways. A golfer who drives the ball well but has a poor playing surface for his next stroke will not be at all pleased if he sees his playing partner has an excellent lie just off the fairway on good grass.

Golf greens however cause the greatest frustration and the most critical comments. Every golfer playing the golf course must putt on the greens and when a short putt counts for as many strokes as a 240 metre drive, tempers can get frayed. Inconsistent putting surfaces and/or poorly prepared putting surfaces have on many occasions led to the dismissal of a turf manager. Most golfers love the game and understand the many variations already mentioned which may affect their golf, but when they get variations in prepared surfaces across the same golf course on the same day then understandably they wonder why.

Standards and expectations in preparation of turf golf surfaces

For 'normal' club play

Most golf clubs conduct a minimum of one competition per week with at least four competitions (Saturday, Sunday, mid-week mens and mid-week womens) per week being commonplace. Standards for the preparation of the golf course turf surfaces may vary during different times of the year due to changes in regional climatic conditions (such as summer versus winter). General preparation standards for 'normal' club play are outlined in Table 15.1.

The USGA Greens Section has developed a comparative chart to guide golf clubs as to what Stimpmeter readings constitute a fast or slow green (Table 15.2).

For golf tournament play

For many clubs the highlight of the calendar year is a major tournament. This event is often planned for months in advance or in the case of a National 'Open' tournament, years in advance.

Table 15.1 General standards in the preparation of the turf golf surface for 'normal' club play.

Zone	*Item*	*Climate/Grasses*	
		Warm-season	*Cool-season*
Tees	Turf height	8 to 10 mm	10 to 12 mm
	Surface	Both : Firm and level	
Fairways	Turf height	10 to 12 mm	16 to 18 mm
	Surface	Both : Good grass cover, firm, tight and consistent	
	Widths	Both : varies from 30 to 55 m	
Rough	Turf height	40 to 60 mm	60 to 80 mm
	Surface	Both : As consistent as possible	
	Rough should be maintained to penalise golfers about a quarter to half a stroke.		
Greens	Turf Height	3.5 to 4.5 mm	3.8 to 4.8 mm
	Turf Height also varies due to rolling or severity of greens contouring		
	Surface	Both : Firm, smooth and consistent	
	Speed	Both : *Stimpmeter 198 cm (6' 6") to 228 cm (7' 6")	

* Stimpmeter - used as a measure of the 'speed' of a golf ball as estimated by the distance a ball rolls across the surface when propelled at a set speed. The further the ball rolls, the 'faster' is the green. Instructions for use of the stimpmeter can be obtained from the USGA or supplier.

Table 15.2 Stimpmeter readings guide to golf clubs.

	Regular Membership Play	*Tournament Play*
Fast	259 cm (8' 6")	320 cm (10' 6")
Medium fast	228 cm (7' 6")	289 cm (9' 6")
Medium	198 cm (6' 6")	259 cm (8' 6")
Medium slow	168 cm (5' 6")	228 cm (7' 6")
Slow	137 cm (4' 6")	198 cm (6' 6")

Most National Golf Associations have either a set of guidelines for the presentation of a major tournament or a team of qualified persons to advise clubs on the preparation for such an event. Over the past fifteen years most National Golf Associations have developed these guidelines to create a more consistent event so as to allow the golfing skills and abilities to determine who should win an event rather than variation in prepared turf surfaces.

Long acknowledged for the finances, time and effort put into developing such guidelines is the United States Golf Association, which through the USGA Greens Section, and in association with Golf Course Superintendents, have developed methods of construction, maintenance and presentation which are frequently used throughout the world. The USGA have also developed a 'How to conduct a Competition' manual, which covers administration, rules, set up and preparation for such an event (USGA, 1996).

It is rare that a major tournament is played on a golf course other than at a time of the year when the golf course can be best presented. Nature does sometimes provide turf managers preparing for such events with subtle reminders as to just who is really in charge and there is nothing that can counter this other than commonsense and good humour. Table 15.3 provides a guide on the preparation of golf course tournament play (Beard, 1982; USGA, 1996; Zontek, 1997; and Pacific Coast Design, 1997a, b).

Preparation of turfgrass for the playing of golf

We have looked at the standards and expectations desirable for differing forms of golf play, but

Table 15.3 Guidelines on the preparation of golf course tournament play for both cool- and warm-season golf courses.

Zone	*Item*	*Climate/Grasses*	
		Warm-season	*Cool-season*
Tees	Turf height	6 to 8 mm	8 to 10 mm
	Surface	Both : Firm and level	
Fairways	Turf height	8 to 10 mm	14 to 16 mm
	Surface	Both : Good grass cover, firm, tight and consistent	
	Widths	Both : varies from 25 to 35 m	
Semi-rough	Turf height	24 to 26 mm	38 to 40 mm
	Surface	Both : Good grass cover, firm and consistent	
	Widths	Both : About 2.5 m in width from fairway to rough	
Rough	Turf height	50 to 60 mm	70 to 80 mm
	Surface	Both : As consistent as possible	
	Rough should be maintained to penalise golfers about a half a stroke.		
Greens collar	Turf height	6 to 8 mm	8 to 10 mm
	Surface	Both : Firm, smooth and consistent	
Greens	Turf height	2.5 to 3.5 mm	3 to 4.5 mm
	Turf height also varies due to rolling or severity of greens contouring		
	Surface	Both : Fast, firm, smooth and consistent	
	Speed	Both : Stimpmeter 10' 6" (320 cm)	

they are an end product of the preparation process. With any golf course the administration, management and golf course superintendent should be following a predetermined plan with respect to maintenance and preparation of the course. A general five-year plan is normally developed with specific detailed annual plans which are reviewed as required.

The plan or program for management and maintenance of the golf course should take into account all aspects of maintenance as they are all interdependent. An integrated management program caters for and utilises the interdependency of maintenance practices. This approach maximises the investment return from maintenance and sets a clear path with substainable goals. If the golf club has scheduled a tournament then often the annual maintenance program will work back from that specific event so as to present the golf course in the best possible condition at that specific time. Without such maintenance programs, budget and planning it is impossible to present a good consistent golfing surface. Once again the local climate, site conditions and budget will dictate many of the preparation tasks at any given time of the year. Golf course superintendents often have the assistance of an industry association, Golf Association sponsored research, or specialist consultants to assist in the preparation and monitoring of an integrated management program.

Basic agronomic conditions for turf

It is very difficult to prepare good turfgrass playing conditions for golf without first addressing the basic principles for good quality turf growth. As stated previously, the structure in which turfgrass grows ('the growing medium') may vary greatly but the fundamentals for good turf growth generally remain constant across all climates. The fundamentals for good turf growth in growing mediums is found in Table 15.4 (Jarrett, 1985; Pacific Coast Design, 1997a, b).

Table 15.4 Fundamentals of growing turf for golf courses.

Growing Medium	*Desirable*	*Comment*
pH	between 6 to 7	Required for optimum availability of nutrients
Drainage		
– Greens	10 to 15 cms/h	Saturated infiltration rate to maintain firm putting surface required to maintain good playing surface (water should not sit on the general turfgrass surface but flow away into non-core golf areas)
– General	good drainage	
Moisture retention capacity		
– Greens	desired range is between 12 to 18% (by weight)	
– General	maintain moisture (through irrigation) between wilting point and field capacity	
Nutrient levels		
– Greens	growing consistently on the leaner side for firm, smooth surfaces general fertiliser advice guided by test results is 'a little and often'	
– General	maintain nutrient levels for maximum turf consistency.	
Aeration/Porosity		
– Greens	total pore space of between 35 to 50% ideal mix is 25% capillary, 25% noncapillary	
– General	root hairs require oxygen for healthy turf growth	
Bulk density (compaction)		
– Greens	desired range is between 1.5 to 1.7 g/cm^3	
– General	while a firm golf turf surface is mostly desirable, a compacted growing medium will not produce quality turfgrass	

In addition to the above it is desirable to control competing factors which may prevent optimum development and growth of the desired turf within the growing medium. That may be weed infestation (inclusive of undesirable grasses), root invasion, shading from larger species (trees, shrubs) or insect and/or fungal infestation.

Setting the standards: 'Keepers of the Greens'

In seeking to control competing factors that may prevent optimum development and growth of the desired turf within the growing medium, some administrators, managers and golf course superintendents may be tempted to overuse available technology. Golf has come a long way since its inception as a game. However let us not forget that the game evolved from humble beginnings being played over grass pasture areas generally unsuitable for crops.

Those involved in the administration of golf and the maintenance of golf courses have a very real responsibility to the wider community as 'Keepers of the Greens'. The large tracts of land required for golf are also key contributors to the environment of our planet and greatly benefit both cities and the people living in them. Setting the standards for a golf course means accepting that some of the competing factors will affect the golf turfgrass surface at varying times. Every time a weed is sighted or minor insect damage reported should not be an automatic signal for application of chemicals. Getting the basic agronomic conditions right for the turf and maintaining a healthy plant is the best way to control most competing factors. Preventative measures for these basic conditions should be included when planning the maintenance programs rather than adopting curative measures only, as they will achieve control for only a certain period before reoccurrence. It is important for golf administrators, managers and golf course superintendents to plan together and acknowledge an agreed 'Standard' for the maintenance of the golf course. It is also important that this is communicated to the golfers and wider community so as to fully promote the golf course as an integral part of our environment.

Primary maintenance practices

In most golf courses across the world, these practices will be undertaken to varying degrees in the preparation of turf surfaces for golf. These practices should not be viewed in isolation but as key contributors to the integrated management system of the golf course. Components of integrated management systems include golf course design, construction, correct turfgrass selection, growing medium management, surface maintenance, mowing, fertilising, irrigation and pest management (Balogh and Walker, 1992). Alterations to one of the maintenance practices may have an effect on other primary maintenance practices, and in turn the quality of the prepared golf surfaces.

Surface maintenance

Topdressing of areas to achieve a uniform surface for growth of turf, not necessarily a flat surface, is required so that when mowing a uniform turf height and density is achieved. The frequency of topdressing will be dependent on many factors. Topdressing should be with the same growing medium as the area to be topdressed. This is to avoid layering of growing medium materials which may lead to other problems. The depth of topdressing is dependent on the level of irregularity of the surface, however as a general rule no more than one-third of the turfgrass plant being topdressed should be covered. If the need for surface uniformity is greater than this, then an approach of lifting of the turf area and redeveloping the surface may need to be adopted. If the lifting, renovation or rebuilding of an established turf area with heavy soil type is to be undertaken then incorporating into the growing medium a better, more friable sand may also be carried out.

The benefits of this incorporation process, however, need to be evaluated as this improved growing medium may have differing nutrition and irrigation needs to surrounding areas leading to possible inconsistent surfaces. Topdressing should be dry when applied and be worked into the grass to achieve desired uniformity. Other benefits of topdressing is the minor improvement in infiltration rates with improved medium texture and the assistance in biological control of thatch (Handreck and Black, 1984). A very light topdressing termed 'dusting' is often undertaken on the greens to prepare for tournaments. As a guide, any final dusting should be completed 12 to 14 days prior to the commencement of the tournament to allow the grass to fully recover (Zontek, 1997).

Mowing

The golf course will need to be mown on a regular basis to achieve suitable playing surfaces. Mowing patterns should be varied to promote upright growth. Greens, tees and fairways are commonly mown with cylinder mowers. Roughs may be mown with rotary mowers. Mowing of grass is a physical action directly separating parts of the grass plant. It is important for the health of the grass plant that this action is undertaken with well maintained and sharp mowers set at the desired height of cut. A massive opening of the grass plant by a blunt mower creates a site for infection from fungal and viral pathogens, loss of water through transpiration across a greater area, and a general loss of vigour for the plant. A sharp mower will create a clean cut assisting in both a uniform surface as well as a healthier turf plant (Beard, 1973).

The frequency of mowing is dependent on the growth of the turfgrass plant, climatic conditions and the required standard of surface presentation. Table 15.5 provides a mowing guide for golf course operations.

Fertilising

It should be the intended aim of the fertiliser program to achieve consistent growth of turfgrass in line with the required standards for the golf course. Fertiliser programs should be linked to regular monitoring of the growing medium. Monitoring of pH is available worldwide through inexpensive testing kits. More advanced monitoring of nutrient, salt and cation exchange levels (in greens) is also desirable if availability and budget allow. Such monitoring will ensure that fertiliser programs are providing required turf growing conditions without

Table 15.5 Mowing guide for golf greens, tees, fairways, semi-rough and rough.

Zone	*Minimum*	*Normal*	*Tournament*
Greens	3 times per week	6 times per week	double cut daily
	A walk behind cylinder mower is normally used for tournament preparation with daily double cutting commencing the week prior to the tournament.		
Tees	once per week	2 times per week	daily
Fairways	once per week	1–2 times per week	daily if possible
Semi-rough	not required for normal play		once per week (8 weeks prior)
Rough	dependent on growth	2 times per month	once per week (8 weeks prior)

Figure 15.1 Mowing manually or by powered machine consistently is still important in the maintenance of the golf green

wastage. Programs to maintain pH or improve growing medium structure (i.e. gypsum) should be undertaken in association with the fertiliser program.

Over-fertilising of turf produces a soft fluffy turf, while under-fertilised turf produces a weak open plant giving inconsistent surfaces for a golf ball (Rist and Gaussoin, 1997). The aim is to provide the correct nutrient levels and the environment in which these nutrients are available to the turfgrass plant. Correctly fertilised turf produces desirable surfaces which allow golfers to direct and 'work' the ball (fairways and green surrounds) and putt smoother consistently (on greens), greatly adding to enjoyment of the game. Irrigation water applied to turfgrass also needs be regularly monitored as this water may often have excessive nutrient or salt content (i.e. ground water or effluent water) (Beard, 1982).

Over recent years, a range of slow release type fertilisers has been developed which assists in nutrient retention and is used in many high rainfall golf course situations throughout the tropics. Opinions vary however as to their overall benefits, especially in milder climate zones, primarily due to their cost (Handreck and Black, 1991). However, this technology is developing rapidly and will likely play a role in turf nutrition in years to come. Organic fertilisers are used in many forms around the world. Care should be taken as these may affect the structure and performance of the growing medium (Beard, 1982). When incorporated throughout the growing medium during construction or renovation, organic fertilisers can greatly assist turf establishment and growth (Pacific Coast Design, 1997a, b). A global trend in turf management seeking improved environmental conditions will see more emphasis placed on the use of organic fertilisers, and more research is being undertaken in this regard. General commercial fertilisers are still the most common form of nutrient supplement used in golf course situations. These are usually applied as a liquid (mixed with water), powdered form dissolved in water and applied as a solution, or applied in a granular form through mechanical spreader. It is advisable to wash applied fertiliser off the plant leaf to avoid foliar burn during high temperatures and to ensure the solution reaches the growing medium where a higher degree of nutrient take-up will occur.

Most elements (nutrients) are taken up in solution by the turf plant root system, however above ground parts of the plant can also absorb some nutrients. Macronutrients carbon (C) and oxygen (O) are mainly gained from the atmosphere, while hydrogen (H) enters the plant via water, once again primarily through the root system. These macronutrients make up a large portion of organic dry matter in turf plants, which in turn makes up about 20 per cent of the plant. The balance of the plant is composed primarily of water (Beard, 1973). Other macronutrients include nitrogen (N), phosphorus (P) and potassium (K) which form the major component of most commercial fertilisers. Turf also requires the macronutrients calcium (Ca), magnesium (Mg) and sulphur (S) in large amounts relative to micronutrients (Beard,

1973). Although required in small amounts, micronutrients are normally absorbed in solution through the growing medium and are necessary for good turf growth, as indicated in Table 15.6 (Handreck and Black, 1991).

Table 15.6 Macronutrients and micronutrients of benefit to turf (after Beard, 1973; Pacific Coast Design, 1997a, b).

Nutrient	*Symbol*	*Importance*
Nitrogen	N	Main nutrient required by turf All growth is dependent upon an adequate supply of this element
Phosphorous	P	Important for root development
Potassium	K	Important for disease resistance
Magnesium	Mg	Essential part of chlorophyll
Sulphur	S	Needed for protein formation and chlorophyll production
Calcium	Ca	A major constituent of plant cell walls
Iron	Fe	Connected with chlorophyll synthesis
Boron	B	Acts as a catalyst
Manganese	Mn	Connected with chlorophyll formation
Zinc	Zn	Catalyst and regulator
Copper	Cu	Catalyst and regulator
Molybdenum	Mo	Connected with nitrogen changes in plant tissues
Chlorine	Cl	Assists in maintaining cation balance in plant cells

Irrigation

The correct selection of grass for a golf course is a key factor in minimising the need for supplementary water (irrigation). Cool-season grasses selected for a hot climate more suited to warm-season grasses place greater strain on the management and resources needed to maintain a suitable turf surface. Research is continuously being undertaken in many countries to select turfgrasses with greater drought tolerance while still providing a good golf playing surface. For optimum turf growth, water is required by the plant as water is the main carrier of nutrients (in solution) and directly makes up about 80 per cent of the turf plant structure. Water in the growing medium is required for microorganisms which decompose dead root matter and other plant residues. Sufficient water must be maintained in the growing medium to prevent wilting of the turf plant.

While good drainage is necessary for optimum plant growth, both air and water are also required. A balance between the competing requirements is needed for the production of a good turf surface for golf. Irrigation should be applied to golf surfaces/growing mediums to maintain moisture levels between wilting point and field capacity (Jarret, 1985). The degree of watering to achieve this will depend on site conditions, evapotranspiration rate, rooting depth and turf species (Beard, 1973; Pacific Coast Design, 1997a, b). A weather station linked to an automatic irrigation system is helpful in determining the degree of watering required, as are some soil moisture sensors linked to the irrigation system. However at the majority of the world's golf courses irrigation decisions are still made by turf managers.

Irrigation water is best applied in the early morning (pre-dawn period is best) as the water is taken up and used by the plant during the day (Jarret, 1985). The application of water to a golf course during the day may disrupt golf activities, however if it cannot be avoided then turf maintenance should take precedence over the golf. If an automatic irrigation system is in place then a 'syringe' cycle or quick cooling of the turf surface will be able to be undertaken during periods of heat stress.

Application of supplementry water through irrigation is today mainly undertaken with in-ground irrigation systems. These systems, while expensive, supply water to the growing medium in a far more efficient manner with more uniform application. With greater demands on global water supplies, turf managers need to have an irrigation application system that does not waste this valuable resource. Fixed type sprinklers relying on physical human movement or mechanically creeping sprinklers are less effective methods of irrigation, and can often lead to overwatering problems. Hand-held water application through hose and suitable nozzle can be an efficient applicator of water when applied by a motivated turf manager.

Overwatering is not only a waste of a valuable resource, but actively contributes to poor playing conditions through creating an environment which promotes compaction, weed infestation, disease invasion and soggy turf (Beard, 1982). Many sand profile golf greens today are constructed in a manner to achieve a perched water table. A perched water table develops when a relatively impermeable layer is placed above a more permeable layer. In the case of golf greens, this is when a sand growing medium is placed on top of a highly permeable drainage layer. Because water adheres to itself, creating a reasonably strong surface tension, it does not drain completely from within the sand growing medium but will leave a residual, or perched saturated water table, immediately above the drainage layer. This water is then available for the use of the grasses which are growing on the surface (Pacific Coast Design, 1997a, b).

Additional water which is applied to the surface will pass quickly through the top zone of the growing medium but remain in the lower zone, thus the root system is encouraged to grow downwards. Many plant nutrients will also be held within this saturated zone and will further encourage deep rooting of grasses. Deep rooted greens turf is more able to tolerate wear and to resist short-term water deficiencies. This allows for the greens to be dried out to some degree in order to provide a faster and harder putting surface (Pacific Coast Design, 1997a, b).

With water an increasingly valuable resource and quality water being restricted for turf use, many managers are utilising treated effluent water sources for supplementary irrigation. While monitoring the composition of the effluent water is essential, most treated effluent water has many nutrients beneficial to turf. Currently, biological products designed to increase water clarity while reducing solids, odour and nutrient levels in treated effluent, are being promoted through industry sources and trials are being conducted to assess their overall performance. Wetting agents designed to assist moisture retention in sands and relieve hydrophobic conditions are commonly available to the turf manager, and should form a part of the irrigation program in these situations.

Physical conditioning

The term physical conditioning covers physical actions taken on turfgrass or the growing medium, and is a primary maintenance practice which has many implementation variations across the world.

The terms used to describe aspects of physical conditioning may also vary with geography.

Aeration

Aeration is a coverall term used to describe the act of physically interfering with the growing medium to assist air circulation into the growing medium below the surface. Carbon, oxygen and hydrogen are obtained by the turfgrass plant from the air (and water) in the growing medium. The act of aeration also assists in the decompaction of the growing medium, improves drainage, water penetration into the growing medium, and allows the incorporation of materials such as lime, gypsum, organic fertiliser or wetting agent (Beard, 1982; Handreck and Black, 1991). Many machines are available for the aeration of different zones of the golf course. Some of these are described in Table 15.7.

Scarification and grooming

Scarification or vertical cutting is a physical practice primarily used to remove excess thatch. Thatch is an organic layer of living, decaying and

Table 15.7 Aeration equipment and machinery used in golf course management.

Zone	*Machine*	*Comment*
Greens	Hollow tyning machine	cores removed from green of varying depth and size
	Drilling machine	holes drilled into green of varying depth and size
	Hydrojet (water injection)	water under pressure injected into the green creating small holes – Normally smaller self propelled machines over greens
	Hand held tyner	for small area treatment of dry patch and aeration
Tees	As per greens	– Sometimes tractor mounted tyning machines cover tee areas and around greens
Fairways	Tractor mounted tyning machines inclusive of hollow tyning and solid tyne machines to a greater depth than other areas	

In addition, all zones can utilise various types of spiking implements for a lesser degree of aeration and assistance in water penetration. Aeration devices are commonly used to alleviate compaction with coring being the most common treatment utilised.

dead roots, stems and shoots between the growing medium and vegetation zones. Generally, warm-season grasses tend to develop greater amounts of thatch than cool-season grasses (Beard, 1982). A certain thickness of thatch or mat is desirable within a golf course turf system, as it creates a more receptive surface for golf balls allowing spin and reducing run. 'Mat' is a more descriptive term used to describe the zone of thatch and topdressing medium on a golf green. The desired thickness of mat or thatch depends of the grass species and general climatic conditions, however as a guide mat or thatch on a green should not exceed 8 millimetres (Beard, 1982). Scarification on greens is usually undertaken by a powered walk-behind scarifier. Surface disruption will occur and after the thatch removed is cleaned off the green, a light dusting (topdressing) may be undertaken to restore the surface. Manual raking or mechanical sweepers are used in the removal of the thatch material from the putting green surface. The depth of scarification will depend on the depth of thatch as the blades of the scarifier should just reach the growing medium beneath the thatch zone. Scarification of green surrounds and tees may be undertaken with either a powered walk-behind scarifier or a tractor mounted (power take-off driven or hydraulic driven) scarifier. Care must be taken to avoid inground sprinklers, drainage pits or other services. Scarification of fairway areas is normally undertaken by tractor mounted scarifiers.

Powered grooming reels mounted on cylinder mowers to be used in association with the mowers are now being used in the maintenance of putting greens. These 'grooming reels' are like mini-scarifiers and when regularly used assist in maintaining a desired mat depth without significant disruption to putting green surface. However if the putting green already has excess thatch, the use of grooming reels should be replaced by the scarifier.

Weed, disease and insect control

The last but not least of the primary maintenance practices is weed, disease and insect control. As stated previously, if good conditions for turf can be maintained within the growing medium and a dense, healthy cover of turfgrass maintained over the various golf areas, then minimal opportunities for weed, disease and insect invasion of turf will occur. At times however, conditions may occur which will allow one or more of the above problems to get out of control on the golf course. The 'standards' mentioned previously in respect to an acceptable golf surface must be set at a realistic level taking into account environmental, budgetary and site condition constraints.

It is very important to note that weed, disease and insect control on the golf course cannot be viewed in isolation. It must be treated as part of the management approach taken for the golf course and linked with basic agronomic conditions and other primary maintenance practices. The term IPM or Integrated Pest Management has been used to best describe this approach of treating pest management as part of an overall integrated management program scrutiny (Figure 15.2) (Balogh and Walker, 1992).

Figure 15.2 Manual removal of weed species as part of the natural intergrated pest management program

In some countries poor management practices have led to concerns over potential problems in relation to the use of pesticides. While general agricultural practices may involve greater use of pesticides, golf courses are generally situated in areas of direct contact with large population centres and as such are under far greater scrutiny. Potentially adverse problems include contamination of water resources with residual pesticides, effects on human and wildlife, and development of insect and disease populations resistant to currently used pesticides (Balogh and Walker, 1992). IPM should an integral part of developing any golf club's 'standards' if they are to maintain the support of the wider community. If pesticides are used at the golf course then both management and staff must have a full understanding of all aspects of the issues. According to Balogh and Walker (1992), Integrated Pest Management in general terms means:

- all of those involved in the process of managing the golf course (regulators, community, management and staff);
- the development of clear 'standards' which are acceptable to all. As previously mentioned, these standards will be the benchmark by which the golf course is not only maintained but also judged by the wider community;
- whenever possible, the creation of good basic agronomic conditions to avert or minimise potential problems. Prevention is far less expensive in the long term and usually produces better surfaces for golf;
- the creation of action thresholds based on the club standards (specific levels of weed, pest or disease damage);
- regular monitoring of the golf course (climate, growing medium condition, pest populations or disease damage) for early warning and possible early solutions to potential problems. An early solution to a problem will be less expensive and less disruptive to the golf course;
- the use of specific club historical references in evaluation of current monitoring. It is important to learn from the successes and mistakes of the past;
- the use of regional data in association with current monitoring. Other golf courses or turf institutions may give an early warning or supplementry information on potential problems. Communication between golf courses in a region is very important, not only for the assistance given but also in communicating as a body to the wider community;
- the correct identification of potential problems is critical, that is physical condition, weed, pest or disease. Turf managers should utilise research institutes, golf authorities or other

experienced turf managers to ensure the identification of problems is correct. This avoids expense, potential environmental problems and poor playing surfaces;
- the selection of appropriate control measures for problems such as physical conditioning, management (e.g. irrigation), biological control, chemical control or a combination of one or more of the above;
- if chemical control is involved (by itself or in combination) it is important that the correct or best chemical at the correct rate is chosen. It then must be applied accurately by trained personnel using correctly calibrated and maintained equipment to the intended target. Timing is important (dependent on the target) and the correct climatic conditions should be in place with other management practices being tailored to the situation. Golfers using the golf course during this period should be adequately advised and given specific instructions if appropriate;
- maintain written or digital (computer) records of any chemical applications inclusive of comments on all of the above actions. Copies of such records should be regularly stored in a place other than the original location to ensure that these vital club records are not destroyed by accident (e.g. fire, flood);
- monitor the results of any applications or combination treatments and evaluate the selected management response and result against intended or desired results;
- add the evaluated information to the historical information for the golf course to assist in future management decisions; and
- communicate evaluation results to management and access response against standards.

The availability of chemicals for the treatment of weeds, pests and diseases varies from country to country. Chemicals are usually registered by an individual country for use in particular turf situations, and this registration is normally granted after sufficient information is supplied to the regulatory authority from independent research on the effects of using the chemical on its intended target.

Many countries do not have such a regulatory process for chemicals relating to turf so managers often have to rely on chemicals supplied to them by the agricultural industry. The Chemical and Pharmaceutical Press (CPP) in association with the Golf Course Superintendents Association of America (GCSAA) have produced a reference covering the majority of chemicals used in the golf course turf industry. This reference is updated on a regular basis and includes cross references, active ingredients and material safety data sheets (Anon, 1991). While primarily concerned with chemicals available in the US, this information would be an invaluable guide to turf managers in countries without a registration process.

Research on chemicals, evaluation of chemicals, and rules governing their use are changing the situation relating to their use every day. It is imperative that golf clubs support their management team in keeping abreast of changes. In the long term it will have an important bearing on the wider community perceptions of golf. Using historical information and turf management expertise, an integrated pest management program can be budgeted for within the overall management program for the golf course. While there may be a budget allowance for use of chemicals on the golf course, this should not be taken as an automatic decision that they will be used.

An integrated pest management approach will best serve the golf club and form the nucleus of the club's overall integrated management approach.

Secondary maintenance practices

This term covers maintenance practices that are not yet globally accepted as being primary practices in the preparation of golf turfgrass playing surfaces. In some countries or at some golf courses however, some of these practices may be a primary maintenance tool. Some additional surface maintenance practices are discussed.

Rolling of greens

Rolling of turf surfaces to produce a smoother surface has been around as long as green keeping itself and is practised in many countries. With the increased play on many golf courses during the 1950s and with greens being primarily formed with natural soils, the practice of rolling was criticised for adversely compacting the greens. Over the past twenty to thirty years however, with a strong movement towards sand-based greens, the practice has made a return to golf courses. Many of the original rollers were manually pulled across greens, towed behind mowers or, if self propelled, cumbersome to operate. Today the rollers designed for golf greens are more mobile, self-propelled machines. The use of the golf greens rollers on the newer sand-based greens increases the distance a golf ball will roll. This allows turf managers to maintain a desired speed of greens without the need for excessively close mowing heights (Beard, 1997). The benefits or otherwise of rolling will vary depending on the structure of the growing medium, amount of thatch, turf species and other cultural practices. Research into rolling, conducted since 1992 has been summarised by Beard (1997), see Table 15.8.

Much of the research on rolling has been carried out on USGA style sand-medium greens. Where research has involved heavier growing mediums, the negatives resulting from rolling have been far greater (Beard, 1997). The rolling of greens does not replace regular mowing. This

Table 15.8 Effects of rolling on green quality (after Beard, 1997).

Item	*Effect*
Ball roll distance	10 to 12% increase generally duration and distance affected by other cultural practices
Duration of effect	from 24 to 76 hours after rolling
Single versus multiple distance	consecutive single rollings in one event produced greater ball roll
Rolling interval distance	dependent on growing medium, regular rolling increases ball roll
Roller operational effects	direction of roller, speed of roller or mow/roll priority had no effect on ball roll distance
Turf effects – sand greens	dependent on growing medium no visual effects, no additional thatch, no adverse effects on turf quality, some wear stress on turfgrasses
Disease and moss responses	studies not yet concluded, to date trends indicate: decrease in severity of dollar spot increase in pink snow mould (*Microdochium nivale*) substantial reduction in extent of moss invasion decrease in localised dry spots
Root zone effects – sand root zone	dependent on growing medium no effect on soil bulk density in top 25 mm, no effect on saturated infiltration rate
Surface smoothness	improved surface smoothness

means that rolling, while a definite aid to green surface condition, is an added maintenance expense that may improve a golf surface but does not affect the basic agronomic condition of the turf. Golf managers will need to evaluate the benefits of rolling for each individual situation, however in the preparation for a major tournament the use of a modern golf greens roller will certainly give benefits (Zontek, 1997).

Additional physical conditioning devices

Several machines to assist in decompaction, aeration and physical conditioning have been developed for turf. It is interesting to note that smaller countries, notably New Zealand and South Africa, with more isolated markets and traditionally smaller budgets have developed very innovative machinery for these purposes. One such machine is the tractor mounted mini-vibrating mole plough. Normally PTO driven and fixed to the tractor via a three point linkage, this machine moves four to six, depending on machine, solid bullet-like tynes mounted parallel to the surface at the end of a vertical slicing blade, at a depth varying from 10 to 30 centimetres. The blades, through an offset system, vibrate as they move through the ground.

This action both decompacts and assists in aeration, but most importantly is valuable in establishing temporary drainage lines within the growing medium. When combined with an elevation drop across a fairway, this machine exiting into the 'rough' areas creates these temporary drains and assists the movement of excess water from the main golf playing surface (fairway) to a site of secondary importance (the rough). Another such machine is the tractor mounted deep tyning machine (vertidrain) utilising a solid tyne which has a deeper action than traditional coring style machines. The mechanical action of this deep tyning machine is being used to assist decompaction of fairway areas which in the past have remained untreated. Other machines like those above are available in different countries for general physical conditioning of the golf course. Turf managers have also often used machines developed for agricultural purposes to solve turf problems where specific turf maintenance machinery is not available, and this remains a good option.

Golf feature preparation and maintenance

Golf course management primarily deals with growing medium and turfgrass surfaces. However the surface development on a golf course while being of prime importance is often affected by the preparation and maintenance of non-turf areas of the golf course.

Bunkers

Bunkers are designed as hazards for the golfer and in some cases to assist in protecting non-golf zones. They are usually well drained and composed of sand to a depth of 30 centimetres (Pacific Coast Design, 1997a, b). This sand profile puts great strain on turfgrasses immediately adjacent to the bunker and special care should be taken in dealing with these grass areas. Due to the nature of bunkers and their role in golf, a great deal of sand is regularly moved from the bunker onto the forward surrounding turfgrass areas around the bunker. This regular topdressing with free draining sand may lead to moisture retention problems in the surrounding growing medium and stress on turf within these zones. Compaction of areas where golfers enter the bunkers may also occur, with the regular traffic adding additional stress. Turf managers should be aware of these localised problems and within their maintenance programs allow for decompaction, adequate nutrition and moisture within the growing medium.

Trees, shrubs, plants and ground covers

Trees, shrubs, plants and ground covers all play an important role in defining the unique qualities of a golf course, and their impact on a golfer may often be greater than that of the turf surface. As assets of the golf course they should be managed in harmony with the turf surfaces and decisions relating to management should take into account their aesthetic, historical and presentation value. The primary disruption to turf surfaces from trees is likely to be root invasion of growing medium and competition for nutrients and moisture. A tree root pruning program may need to be undertaken after careful assessment of the effect on the tree. Trees should be monitored for shading effect on turf surfaces as well as adequate movement of air throughout the golf areas.

Shrubs and plants may also compete with turfgrasses for sunlight, nutrients and moisture so their effect will need to be monitored. Trees are normally relatively long living when compared to shrubs and plants, so physical location decisions need to be made carefully. Ground cover issues relating to turf surface preparation mostly relate to invasion of turf areas, and these issues should be dealt with through normal maintenance practices.

Guidelines for construction and selection for golf courses

Key factors in basic agronomic condition of the growing medium and best practice integrated management systems for the golf course are the construction of the golf course, selection of materials and selection of turfgrasses. Most decisions have already been made with golf courses around the world selecting methods, materials and grasses through advice, budget and availability. In many cases these decisions, whether good or bad, will need to be lived with and the situations managed to produce a consistent golf surface of the best possible nature.

Guidelines given with respect to construction and selection can be utilised when undertaking reconstruction, renovation, upgradation or new construction. Great care must be taken by golf club management when deciding on construction methods and selection for reconstruction works at the golf course. If there is an agreed reconstruction program in place to reconstruct the majority of the golf course, then new methodology, materials and turfgrasses can be utilised. Any newly constructed golf zone utilising new techniques and grasses will most likely be different, from a golfing and turf management point of view, to that of a more established or 'old' zone. A new zone may present a better golf ball lie and provide more spin (fairway). A newly constructed green may putt faster and truer than the more established greens.

From the turf management point of view, newly constructed areas utilising new methodology and grasses may need different management, mowing or physical conditioning, and have different nutritional and irrigation needs to that of older areas. The effect of these decisions needs to be understood by management when deciding to reconstruct or renovate areas of the golf course and communicated to the entire golfing membership.

Base construction of golf turf features

Most decisions relating to base construction of features for the golf course relate to greens, tees and bunkers.

Greens

Throughout the USA where new golf course construction easily outnumbers that of any other region, the United States Golf Association, through its Green Section, has developed guidelines (Hummel, 1993) for the construction of a golf green. The use of these guidelines produce what is commonly termed a 'USGA green'. These guidelines were developed to assist golf course management in constructing greens that would provide a good growing medium for turf and consequently good playing surfaces for golf: when followed they do just that. A great deal of literature is available on the USGA green and this can be obtained from the USGA. However the success of the system is reliant on availability of materials and regular testing of materials used in the construction by a competent laboratory (Anon, 1993) of which there are many throughout the US. In countries where a laboratory with suitable experience in what is required may not be available, the regular testing of materials may not be an economic option. In such circumstances or where specific materials are not available, golf managers should not despair but look to the general principles of turf management and select the best available materials.

As discussed earlier, golf first started on sites used generally for pasture. Many of the worlds best golf courses are built on local sands 'pushed up' to form greens, tees and fairways. Greens not constructed to USGA greens recommendations are normally still constructed from a selected sand. Sand provides a growing medium that drains freely, has good aeration and is less prone to compaction. These qualities normally outweigh the disadvantage of poor moisture retention and poor cation exchange (Beard, 1982). In general terms, qualities required by a sand selected for greens construction can be found in Table 15.9 (Pacific Coast Design, 1997a, b).

The general procedure for the construction of such greens is as follows (Pacific Coast Design, 1997a, b); more detailed specifications should be obtained from the golf course architect/designer or local golf authority which may be more site specific. A typical USGA 'modified' or sand green will contain three zones: compacted subgrade, gravel bed, and a growing medium sand zone. Create a firm subgrade from material (clay preferred) which can be compacted and will not settle. Subgrade is to be formed to mirror surface finished contours as designed. Into the subgrade, subsurface drainage should be

Table 15.9 Characteristics of sands selected for greens construction.

Item	*Desired qualities*	
pH	between 5.2 and 5.7	
Salt levels	less than 800 ppm	
Particle size analysis	gravel (2 mm)	0%
	very coarse sand (1 mm)	< 3%
	coarse sand (0.5 mm)	0–20%
	medium fine sand (0.25 mm)	60–90%
	sand (0.10 mm)	0–20%
	very fine sand, silt and clay (0.05 mm, 0.002 mm and pan)	preferably < 3%
Infiltration rate	saturated infiltration rate desired range of 10 to 15 cm h^{-1}	
Porosity	total pore space of between 35–50%. Ideal mix is 25% capillary, 25% non-capillary	
Bulk density	desired range is between 1.5 to 1.7 g cm^{-3}	
Moisture retention capacity	desired range between 12 to 18% (by weight)	

installed as per design drawings so that the green does not hold water in any areas. Excavated material from drainage trenches should be taken out of the green zone and disposed of elsewhere. Drainage material (gravel) around the drains is to be 6 to 10 millimetre clean (washed) gravel (free from silt and clay) with a minimum of 75 per cent of the gravel being of the 10 millimetre range. The gravel bed layer is provided to allow for the rapid discharge of water from the sand when excess water has been received. Water will move easily within this layer down into the drainage system which has been installed within the impermeable subgrade layer. The uniformity is important as this layer, combined with the growing medium layer, can form a perched water table which is desirable. Testing of the gravel bed layer and selected sand material is needed to ensure a perched water table is achieved. The gravel bed layer should be a uniform thickness of 100 millimetres across the subgrade and be 7 millimetres minus clean gravel (washed with 60 to 70 per cent of gravel in the 7 millimetre range).

The growing medium layer of sand has been described in detail above. This must be placed over the gravel bed layer in a uniform manner ensuring that the design profile of the green is maintained. The finished surface must be protected from erosion (wind or water) until such time as it is prepared for the planting of grass and actually grassed. The depth of selected sand will need to be confirmed by laboratory if a perched water table is to be achieved. If no laboratory is available then a depth of between 250 to 300 millimetres can be used as a guide (Pacific Coast Design, 1997a, b) however every effort should be made at least once to confirm the preferred profile by sending materials via courier to a certified laboratory which can receive materials from overseas (quarantine clearance). Failure to do so may result in management problems for the finished green.

Tees

A similar material selection as used for greens can be utilised for tees. Similarly, a compacted impermeable subgrade base is constructed with drainage incorporated. Based on laboratory testing, a 250 to 280 millimetre depth of sand growing medium is placed directly over the subgrade base.

Fairways

Fairways in most parts of the world are constructed using the available natural topsoil on the site selected. However in circumstances where budget is not a constraint, 'sand capping' or placement of approximately 250 millimetres of selected sand (Pacific Coast Design, 1997a, b) over the golf fairway areas can be carried out. This importation of material gives full control over the nature of the growing medium creating a profile similar to that of a tee or green. Whenever possible the structure of the fairway growing medium material should be improved if so directed by laboratory testing (with gypsum, lime, etc). If budget allows, underground drainage should be incorporated into the growing medium to collect excess water and direct it into non-core golf areas of the golf course.

Selection of grasses

As previously stated, the correct selection of turfgrasses for the differing zones is an essential part of the integrated management of a golf course. Turfgrass selection plays a major role in the preparation of surfaces and their final presentation for golf. Selections from naturally occurring stands of grass either by seed or stolons was the traditional method of establishing golf course greens, tees, fairways and roughs. Between the early 1900s and 1940s both private companies and research institutions recognised the benefits and commercial potential of selected breeding of turfgrasses with many of their efforts being supported by golf associations. The major cultivar breeding programs occur in America and Europe where evaluation standards have been established for some time (Beard,1973; van Wijk, 1993).

Selection of naturally occurring grasses continues in many countries, but through travel over time many grasses thought to be naturally occurring today in fact may have been introduced to a region many years ago. There are two generally recognised types of turfgrass, warm- and cool-season grasses. Within these two general types there are many different grass species and cultivars, some of which may have a wide tolerance of both warm and cool climatic conditions (Beard, 1973; Taliaferro and McMaugh, 1993). Many new cultivars are being bred in response to environmental concerns with drought tolerance, lower nutritional requirements and disease resistance being primary characteristics sought (Neylan, 1997). However traditional qualities of selected turfgrasses for differing zones on the golf course are still important considerations. These qualities have been summarised by Beard (1982) in Table 15.10.

Table 15.10 Desirable turf characteristics for golf course operations (after Beard, 1982).

Zone	*Desirable turf quality*
Green	uniformity in growth habit, appearance and smooth putting tolerance to close mowing with a low creeping growth habit fine leaf texture with resistance to excessive grain or thatch disease, weed and pest resistance high shoot density, resiliency and drought tolerance
Tee & Green surround	uniformity in growth habit, appearance and smooth surface density, resiliency and drought tolerance disease, weed and pest resistance
Fairway	uniformity in growth habit, appearance and smooth surface density, resiliency and drought tolerance disease, weed and pest resistance
Rough	uniform in appearance and penalty to golfer minimal maintenance requirements

Many new cultivars are being developed and released each year with 40 seeded varieties being released during 1996 alone (Pacific Coast Design, 1997a, b). From 1987 to 1993, thirteen new *Cynodon* cultivars were released with eight of these being propagated by seed (Taliaferro and McMaugh, 1993).

Warm-season grasses

Generally found in tropical, subtropical and temperate regions these grasses have their origins in Africa, Asia and South America. As a group, in comparison to cool-season grasses they are more drought, heat and stress tolerant with lower growth habits. They are usually propagated by vegetative means although some cultivars are established by seed (Beard, 1973; Anon, 1996; Taliaferro and McMaugh, 1993). Warm-season species used extensively on golf courses include bermudagrasses (*Cynodon dactylon, C. transvaalensis, C. dactylon x C. transvaalensis*), zoysiagrass (*Zoysia japonica, Zoysia materella, Zoysia tenuilolia*) (Beard, 1973, 1982; Taliaferro and McMaugh, 1993; Pacific Coast Design, 1997b). Other warm-season grasses utilised on golf courses include many native grasses selected for their tolerance to close mowing, drought resistance and low maintenance requirements. Other species such as seashore paspalum (*Paspalum vaginatum*) are providing superior turf surfaces (in high salt conditions) for fairways at the Thai Country Club, Bangkok, Thailand, and can be maintained in tournament condition.

Transitional zone grasses

In climatic zones where temperatures during summer months are high and either a warm-season or cool-season grass may be appropriate, most consultants are specifying warm-season grasses with either oversowing or an allowed invasion of cool-season grasses for fairways, and a cool-season bentgrasses for greens. The most common warm-season grass selected in this zone is bermudagrass with oversowing of either Rough bluegrass (*Poa trivialis*), especially in the case of greens where bentgrass is not used, or newer types of creeping fescue grass (Pacific Coast Design, 1997a).

Cool-season grasses

Predominant in Europe and North America, cool-season grasses provide a wide range of choice for golf course turf surfaces. Some of the most commonly used grasses include the bent grasses (*Agrostis stolonifera, Agrostis capillaris, Agrostis tenuis*), the bluegrasses (*Poa pratensis, Poa trivialis*), the fescues (*Festuca ruba L., Festuca rubra var. commutata* Gand., *Festuca ovina* L., *Festuca ovina var. duriuscula* L. Koch., *Festuca arundinacea* Schreb.), and the ryegrasses (*Lolium perenne* L., *Lolium multiflorum* Lam.)(Beard, 1973, 1982; Taliaferro and McMaugh, 1993; Pacific Coast Design, 1997a).

References

Anon. 1991. *Turf and Ornamental Chemicals Reference*, Chemical and Pharmaceutical Press, c/o John Wiley & Sons, New York.

Anon. 1996. '1996 Seed Update', *Golf Course Management*, May, vol. 65, no. 5.

Anon. 1993. USGA *Greens Section Record*, March/April, vol. 31, no. 2.

Anon. (undated). *Specification for Course Preparation*, Australian Golf Union, Australian Open Championship.

Beard, J.B. 1973. *Turfgrass: science and culture*, Prentice Hall, Englewood Cliffs, N.J., 658 pp.

Beard, J.B. 1982. *Turf management for golf courses*, McMillian, N.Y., 642 pp.

Beard, J.B. 1997. 'Greens rollers help golf balls roll further', *J. Golf Course Management*, Jan., Golf Course Superintendents Association of America, vol. 65, no. 1.

Balogh, J.C. and W.J. Walker. 1992. *Golf Course Management and Construction: Environmental Issues*, Lewis Publishing, Chelsea, MI, 951 pp.

Handreck, K. and N. Black. 1984. *Growing Media for Ornamental Plants and Turf*, New South Wales University Press, Sydney.

Hummel, N.W. 1993. 'Rationale for the revisions of the USGA green construction specifications', *USGA Green Section Record*, March/April, pp 7–33.

Jarrett, A.R. 1985. *Golf Course & Grounds, Irrigation and Drainage*, Prentice-Hall Inc, N.J., 246 pp.

Neylan, J. 1997. '2010 and Beyond', *Golf and Sports Turf Australia*, vol. 5, no. 2, April.

Pacific Coast Design. 1997a. *Specifications for Golf Course Construction*, Revised edn, Pacific Coast Design, Sandringham, Australia, 200 pp.

Pacific Coast Design. 1997b. *Specifications for Golf Course Construction at the Tianma Golf and Country Club, Shanghai, China*, Pacific Coast Design, Sandringham, Australia.

Rist, A.M and R.E. Gaussoin. 1997. 'Mowing isn't sole factor affecting ball-roll distance', *Golf Course Management*, June, vol. 65, no. 6.

Taliaferro, C.M. and P. McMaugh. 1993. 'Developments in Warm Season Turfgrass Breeding/Genetics', *International Turfgrass Society Research J.*, vol. 7, ch 3.

United States Golf Association. 1996. *How to Conduct a Competition*, USGA, Far Hills, N.J., 61 pp.

van Wijk, A.J.P. 1993. 'Turfgrasses in Europe: Cultivar Evaluation and Advances in Breeding', *International Turfgrass Society Research J.*, vol. 7, ch 4.

Vermeulen, P. 1995. 'S.P.E.E.D - Consider what's right for your Course', *USGA Greens Section Record*, December, vol. 33, no. 6.

Zontek, S. 1997. 'Preparing Your Greens for that All-important Tournament', *USGA Greens Section Record*, July/August, vol. 35, no. 4.

Acknowledgements

Thank you to Ian Chivers, Racing Solutions Pty. Ltd., for his advice and editorial assistance. Thank you to those many turf managers across the world who have shared their ideas and thoughts with me over the past twenty years and who have allowed me access to their golf courses.

CHAPTER 16

Prescription surface development: Lawn bowling greens, croquet lawns and tennis courts

G.W. BEEHAG, *Globe Australia Pty Ltd, Sydney, NSW, Australia*

Introduction

Lawn bowls, croquet and tennis evolved in Europe to become important games in those countries where they are now played. Lawn bowls is a major sport in Britain, Australia, Canada, Ireland, New Zealand and South Africa. In Australia there are 2 218 bowling clubs with in excess of 354 000 registered bowlers. Tennis is an old game and popular in Britain, Europe, New Zealand, South America, South Africa, the United States and Australia. Croquet is enjoyed world-wide despite not having the public profile of lawn bowls or tennis.

Regulation playing surface dimensions

Flat lawn bowls for international events is played on a green which may measure a minimum of 36.58 metres by 36.58 metres up to 40.23 metres by 40.23 metres. The overall dimensions of a lawn tennis court are 23.77 metres by 10.97 metres (Adams and Gibbs, 1994). A croquet lawn measures 32.0 metres by 25.60 metres.

Construction of bowling greens, croquet lawns and tennis courts

Unlike golf, there are no internationally recognised construction specifications for bowling greens, croquet lawns or tennis courts. The construction philosophy of golf greens has generally been adopted for bowling greens and croquet lawns. Early construction techniques utilised natural soils and designs considered within the turfgrass industry to be appropriate at the time (Beale, 1924; Dawson, 1939). Later information was published in New Zealand (Arnold, 1957) and in Australia (Pierce and Rigney, 1946). As knowledge accumulated, more scientifically based information was published by industry funded research organisations (McMaugh, 1968; NZTCI, 1971). The most recent information about the construction of bowling greens, croquet lawns and tennis courts is contained in work by Adams and Gibbs (1994), Baker (1990) and Evans (1992).

Key construction issues

High quality bowling greens, croquet lawns and lawn tennis courts represent a monoculture grassland constructed on artificial profiles using materials which satisfy a performance specification. The most extreme example of an artificial grassland is that of the mesh-element, reinforced sand concept developed in England and recently adopted for a bowling green in Victoria (Neylan and Robinson, 1993). The underlying construction criteria of all three surfaces relates to uniformity, durability and playability. The relative ranking of each of these criteria varies between each playing surface based on the specific management objectives.

The profile of bowling greens and croquet lawns could reasonably be argued to be identical because of similar playing requirements. The depth of specified growing media varies between 100 and 200 millimetres. The combined sand content in the range 1.0 to 0.25 millimetres diameter for bowling green soils in Australia varies between 50 to 60 per cent. The soil typically used in Australia for lawn tennis courts possesses a relatively high clay content (20 to 50 per cent). In California, lawn bowling greens utilise sand which meets the University of California Sand Specification (Davis, 1977). In Britain, the STRI specification for bowling greens calls for a soil depth of 110 to 125 millimetres (turfed surface) and 150 millimetres (seeded surface) (Evans, 1992).

The STRI specification for lawn bowling greens nominates a 150 millimetre depth of hard aggregate with a particle diameter of 8 to 12 millimetres (Evans, 1992). The composition of drainage aggregate materials includes crushed and screened basalt or granite rock, screened and washed river gravel and industrial slag. In Britain, the inclusion of a coarse sand layer as included in the revised United States Golf Association specifications is utilised. The inclusion of a coarse sand layer is uncommon in Australia. A drainage pipe system, which utilises either polyvinyl chloride or polyethylene plastic, is enveloped in drainage aggregate and placed into trenches excavated into a compacted soil base. In Australia, the omission of a pipe drainage system has been adopted for bowling greens in coastal sites which are comprised of deep and free-draining sands. Some bowling greens in Australia are constructed on suspended concrete to allow for car parking underneath (Figure 16.1).

Figure 16.1 Bowling green constructed on suspended concrete to allow for car parking underneath, Gold Coast Bowls Resort, Queensland, Australia

Lawn bowling green ditch construction

Bowls is based on a rigid set of ditch dimensions governed by the International Bowls Board (I.B.B.). The composition of ditch materials is not specified under the rules of the game. The regulations specify the height and width of the ditch and height of sand in the ditch. In Australia, the regulations are contained in 'The Constitution', 'By-Laws' and 'Laws of the Game of Bowls' published by Bowls Australia (1998). The ditch must have a minimum width of 200 millimetres and a maximum width of 280 millimetres. Within the ditch, sand or any suitable loose material must be no greater than 25 millimetres below the level of the green. The face of the bank, as measured from the outer ditch wall, may be either vertical or slope toward the green so that at a point 255 millimetres above the surface the overhang does not exceed 50 millimetres. A variety of ditch designs and materials are used throughout the world (Evans, 1992; Liffman, 1984). In Australia, pre-cast ditches may be of concrete, fibre cement or fibreglass, all with drainage holes. Some designs of pre-cast ditches may have an in-built bowls rest and a protective bowls face of rubber or synthetic grass. Aluminum bowls rests are common. Ditches constructed of in-situ concrete are normally built on a reinforced concrete foundation with the provision of vertical drainage holes. The bank may be of brick, concrete or an outdoor tile. The plinth in Australia is normally of compressed fibre cement. In Britain, the bowling green may have a sloping bank of grass (Adams and Gibbs, 1994).

Figure 16.2 *Cynodon* spp. bowling green at Club Banora, Tweed Heads, NSW, Australia

Utilisation of species and cultivars

Throughout the world, with the exception of lawn bowls in New Zealand and a few bowling greens in Britain, gramineous species are maintained on lawn bowling greens, croquet lawns and lawn tennis courts. The management requirements of low mowing height and high mowing frequency has demanded the selection of cultivars capable of producing a fine leaf texture and high shoot density. High wear tolerance is an additional selection criteria of cultivars for lawn tennis courts. In the northern hemisphere in cooler regions, natural ecotypes or improved cultivars of *Agrostis* spp. (bentgrass) and to a lesser extent *Festuca rubra* (creeping red fescue) are used. In Britain up until the 1930s sea marsh was used on bowling greens in Scotland and northern England. Sea marsh was a blend of *Agrostis stolonifera* var. *compacta* and *Festuca rubra* spp. *rubra*. In Britain today a blend 80:20 by weight composed of *Festuca rubra* and *Agrostis* spp. cultivars is widely used (Evans, 1992). Bentgrass monostands and bentgrass/creeping red fescue polystands are used throughout Europe and in Canada (Anon, 1986). Within the warmer regions of the northern hemisphere interspecific *Cynodon* spp. (bermudagrass) cultivars are maintained on lawn bowling greens and croquet courts. In southern California and Florida, the predominant *Cynodon* spp. cultivar used is Tifgreen (Haley, 1979).

In the southern hemisphere, the predominant species maintained on lawn bowling greens, croquet courts and lawn tennis courts are primarily of *Agrostis* spp. and *Cynodon* spp. The most common ecotypes and cultivars are those of *Cynodon* spp. which are often maintained on bowling greens well outside their climatic range of adaptation. In the southern Australian states which have a cool climate, *Agrostis* spp. is the predominant species maintained on lawn bowling greens and croquet courts. *Cynodon* spp. are used on lawn tennis courts in many locations of southern Australia. Throughout the warmer regions of Australia in Queensland, New South Wales, northern Victoria, South Australia and Western Australia, *Cynodon* spp. selections remain dominant. *Cynodon* spp. selections used include numerous naturalised *C. dactylon* ecotypes, *C. tranvsvaalensis* and the interspecific *C. dactylon x C. transvaalensis* hybrids Tifdwarf, Tifgreen and Santa Ana (Figure 16.2). *Digitaria didactyla* (Queensland blue couch) and *Paspalum vaginatum* (seashore paspalum) are less common in the warmer, frost-free regions of South Australia, Western Australia and Queensland. Up until the early 1970s in Australia the dominant warm-season turfgrass species on bowling greens were the many *C. dactylon* ecotypes, *C. transvaalensis*, Queensland blue couch and seashore paspalum (Beehag and Surrey, 1992). Following the introduction of Tifdwarf into Australia during the 1960s and the development in New South Wales of Greenlees Park Couch, a naturalised *C. dactylon* ecotype, these two couchgrasses are now regarded as the standard on which other cultivars for bowling greens are judged. Morphological variation within Tifdwarf is relatively common in lawn bowling greens in northern New South Wales and Queensland. Two recent *Cynodon* spp. selections used on lawn bowling greens and tennis courts are CT2 from California and Riley's Super Sports, an Australian naturalised ecotype.

In South Africa, numerous naturalised ecotypes of *C. dactylon*, particularly Royal Cape, *C. transvaalensis* (South African bermudagrass), and *C. x magennisii* are used on lawn bowling greens (Brockett, 1982; Louw, 1996). Tifdwarf bermudagrass has recently been used in Durban and *Paspalum vaginatum* from Australia has been used successfully in the Western Province of South Africa.

Non-gramineous species are used in New Zealand on lawn bowling greens. The most common plants are two species of *Leptinella* spp., formerly included in the genus *Cotula*. *L. dioica* and *L. maniatoda* are native species and have been used in New Zealand since the 1920s (NZTCI, 1971; Evans, 1992). The principal reason for the use of *Leptinella* spp. in New Zealand on lawn bowling greens is their ability to remain relatively drier and to produce a greater green speed during inclement weather conditions in comparison to bentgrass greens (Evans, 1992). Other plant species used in New Zealand include *Plantago triandra* (Starweed), *Pratia angulata*, *Hydrocotyle* spp., *Crassula* spp. (Tillaea) and *Colobanthus* spp. (Ormsby, 1993). *Leptinella* spp. plants have also been sent to Britain from New Zealand and were included in field trials by STRI (Evans, 1992). Tifdwarf bermudagrass has been used to a limited degree in Auckland, New Zealand.

Establishment of the playing surface

The establishment phase of lawn bowling greens and grassed tennis courts demands attention to detail in order to achieve the stringent surface requirements. Procedures involve grassing, fertiliser and pesticide application, irrigation, mowing and rolling. In Australia, the establishment of *Cynodon* spp. lawn bowling greens and tennis courts is almost exclusively by vegetative means using stolons from a previously renovated green. The stolons are harvested by conventional scarification techniques and spread by manual means. Pressing of the stolons into the soil surface to achieve an intimate contact is carried out by light hand rollers and the use of an old bowling green mower cylinder. In recent years attempts have been made to establish *Agrostis* spp. bowling greens in Victoria by using washed sod from turf farms. However, the conventional method of establishing *Agrostis* spp. lawn bowling greens in southern Australia is seeding. The seed may be pre-germinated for up to 24 to 36 hours in a fungicide solution for *Pythium* prevention. Application of the seed is normally via hose proportioner devices. It is becoming commonplace in Australia to use a temporary mesh covering to stabilise applied seed or stolons during establishment. In Britain, seed and turf are used to establish lawn bowling greens (Evans, 1992). In New Zealand, cotula bowling greens are initially established by the scarification and collection of vegetative material (NZTCI, 1971).

Pre-establishment fertilisers based on high phosphorus and processed organic products are normally used in conjunction with controlled-release nitrogen formulations. The amount of phosphorus applied may be determined on the results of a soil analysis. Post-establishment fertiliser applications are based on soluble nitrogen and potassium materials. Pre and post-establishment pesticide applications are based on the species being established together with the likely incidence of insect pest and fungal disease occurrence.

Irrigation application at establishment is typically one of short duration and high frequency. During the post-establishment phase, the irrigation frequency is reduced and the duration increased. Mowing of lawn bowling greens and tennis courts commences when the height of the turfgrass cultivar has reached approximately 2 to 3 centimetres. The mowing height is reduced with an increase in mowing frequency until the standard of surface has been achieved. Complementary rolling of the surface begins at a time when the mowing height approximates a centimetre. Topdressing of lawn bowling greens during the grow-in phase is critical in the achievement of a true surface. In Australia, the use of finer-textured soils having a clay content of 10 to15 per cent is common for topdressing purposes. Several applications of topdressing soil are applied after establishment.

Playing surface standards and performance testing

The components of turfgrass quality as defined by Beard (1973) are subjective and do not provide objectivity when defining playing surface quality. The playing surface requirements for lawn bowls, croquet lawns and tennis courts differ. The playability of each surface is governed by the physical condition of the underlying profile construction media. The adoption of synthetic surfaces in some countries which have traditionally played on natural grass, combined with professional play requirements, has highlighted the need for greater objectivity in surface evaluations.

Playing surface standards

The playing surface requirements for lawn bowling greens have been systematically studied in Britain and New Zealand (Adams and Gibbs, 1994). In addition, the relative playing quality between synthetic and natural grass bowling greens have been compared in Australia (Neylan and Robinson, 1993) and New Zealand (Gibbs, 1997). The generally recognised surface evaluation criteria for flat lawn bowls includes a uniform coverage, level playing surface, a fast and uniform green speed, uniform draw on both hands, adequate infiltration for playability, and the grass surface must be wear-tolerant. There are no published playing surface standards for lawn bowling green surfaces outside Britain (Bell and Holmes, 1988) and New Zealand (McAuliffe and Gibbs, 1993).

While a lawn bowling green must be 'level' this is an impossible criteria to realistically achieve. No survey results about the degree of level of lawn bowling greens have been published in Australia. However the degree of tolerance across the entire playing surface of most bowling greens would typically fall within less than 10 millimetres. It is the opinion among some bowling greenkeepers in Australia that having a slightly higher centre compared to the edges facilitates surface water runoff. The recommended standard of level in Britain is that surface levels be within +/– 6 millimetres from the mean. Green speed of bowling greens remains a strong point of contention. There is no standard of green speed adopted by all state bowling associations in Australia. The recommended green speed in Australia is between 14 to 15 seconds. However, faster green speeds in Australia are not uncommon on dry Tifdwarf bowling greens during winter dormancy. The standard for green speed in Britain is 10 to 12 seconds (Bell and Holmes, 1988). In New Zealand, green speeds vary from12 to 14 seconds for grass and between 15 to 19 seconds for cotula and starweed greens (Ormsby, 1993). The recommended green speed in South Africa is 12.7 to 14.2 seconds (Louw, 1996).

Numerous surfaces including synthetic and clay are adopted for tennis courts throughout the world. The demands placed on tennis court presentation are particularly high because of colour television and media coverage during professional events. A grassed tennis court remains the preferred playing surface by club players and for some professional tournaments such as at Wimbledon. The playing surface requirements for lawn tennis courts include a uniform coverage, a relatively level playing surface, a fast court speed, consistent height and trueness of ball bounce, and an extremely wear-tolerant grass surface. There are no worldwide playing surface standards for lawn tennis and very few surveys have been conducted. Surface resilience is less critical on lawn tennis courts compared to a cricket wicket (Adams and Gibbs, 1994). Friction between the surface and the tennis ball determines bounce and pace as well as influencing the sliding action by players. The demands on a croquet lawn surface are less in comparison to those on lawn bowls despite similarities between both games. The playing surface requirements for croquet lawns are uniform coverage, level playing surface, consistent speed, and the surface must have adequate infiltration.

Measurement of playing surface performance

The adoption of alternative surfaces and differing construction profiles for bowling greens and tennis courts has resulted in questions being raised about their relative surface properties and playing performances (Gibbs, 1997). The use of synthetic playing surfaces with various underlying construction materials beneath bowling greens in particular, has produced playing surface standards in Britain and New Zealand using various measurable and objective criteria (Bell and Holmes, 1988; McAuliffe and Gibbs, 1993; Gibbs, 1997). Performance testing of lawn tennis courts has been relatively limited with few published papers (Thorpe and Canaway, 1986; Holmes and Bell, 1987).

Assessment criteria for lawn bowling greens are not universally adopted throughout the world. Many criteria, for example green speed and the amount of draw, are still assessed by subjective means. Green speed measurement by bowlers using the distance travelled by a bowl remains the predominant method in Australia. In Victoria, an inclined plane to measure green speed has recently been devised and marketed. The performance testing on lawn bowling green surfaces using objective measurements is relatively recent and has included green speed, amount of draw, uniformity of level, surface hardness and infiltration rate (Gibbs, 1997; McAuliffe and Gibbs, 1993; Neylan and Robinson, 1993). The measurement of the degree of level of a bowling green utilises conventional surveying equipment, while surface hardness is measured using a Clegg Hammer developed in Australia. Surface infiltration is measured using various designs of disc permeameters. The testing of lawn tennis courts uses various devices, to measure ball bounce, ball–surface friction and spin. The physical characteristics of the court surface have been found to have a lesser impact on rebound resilience compared to the actual tennis ball (Thorpe and Canaway, 1986). There has been no published data for the surface performance measurements of croquet lawns.

Surface presentation practices

The emphasis placed on the three primary cultural practices in sportsturf maintenance, mowing, fertilisation and irrigation differs between lawn bowls, croquet and tennis. The two primary requirements of high shoot density and fine leaf texture result from a low mowing height and high mowing frequency. Specific requirements of degree of surface resilience, amount of verdue and impact resistance are crucial presentation criteria. Rolling is a critical practice for bowls and tennis. Player wear patterns are a critical managerial problem for lawn bowls and tennis.

Maintenance regimes

Mechanical mowing is a practice unique to grassed sportsturf surfaces. The emphasis on mowing height and mowing frequency of croquet lawns is less compared to that on lawn bowling greens and tennis courts. Detrimental effects of low mowing as listed by Beard (1973) are more pronounced on lawn bowling greens and lawn tennis courts compared to croquet lawns.

Low and frequent mowing is fundamental for the presentation of high quality lawn bowling greens but varies significantly between countries. Two significant factors which have contributed to the high degree of uniformity and presentation of bowling greens in Australia are bowling green mower design and the frequency of mowing. Since the advent of the lawn bowling green mower in the late 1920s by Scott Bonnar, the design of bowling green mowers has been highly modified and copied by several Australian companies. The Australian bowling green mower is electric or hydraulically driven for vibration-free mowing, possesses a modified bottom blade and an eighteen-bladed cylinder for extremely precise mowing.

In Australia, a mowing height of 1.5 millimetres on Tifdwarf bowling greens in south east Queensland is not uncommon. Bentgrass bowling greens throughout Australia are typically mown between 2 to 3 millimetres. A mowing height of between 3 to 5 millimetres is advised in Britain depending on the season (Evans, 1992). In New Zealand, a slightly lower mowing height is common on bowling greens of non-gramineous species. For *Leptinella* spp. and *Plantago* spp. bowling greens a height of less than 2 millimetres is used (Ormsby, 1993). Bentgrass bowling greens are mown at a height of between 2 to 3 millimetres for championship play. Lawn bowling greens in South Africa are normally mown at a height not less than 3.5 millimetres (Louw, 1996). In the United States, a mowing height of between 4 to 6 millimetres on bentgrass and 2 millimetres on bermudagrass is common (Henry, 1977). The frequency of mowing on lawn bowling greens is dictated by seasonal growth patterns and the required standard of presentation. In Australia, the higher standard greens are mown daily during the growing season. In northern Australia the variation to the frequency of mowing does not vary significantly between seasons. Winter dormancy of Tifdwarf bowling greens may require only weekly mowing. In Britain, as the playing season commences in April to May, it is advised to increase the mowing interval to three times per week (Evans, 1992).

Due to the surface performance demanded of high quality lawn tennis courts, particularly for tournaments, the mowing height must be relatively low and with a minimum of verdue. The mowing height of Australian bermudagrass tennis courts varies between 2 to 4 millimetres. The traditional tennis court mower used in Australia and New Zealand has been of British manufacture. While these mowers are still used in Australia for tournament play, some use is made of putting green triplex mowers to mow multiple courts for club play. The mowing frequency on bermudagrass tennis courts during the growing season is on alternate days. The requirement of mowing is considerably less for croquet lawns compared to lawn bowling greens because of a lesser demanding playing surface. In Australia, croquet lawns are mown at a height of approximately 4 to 5 millimetres two to three times per week during the growing season depending on grass species. Tennis courts in New Zealand are recommended to be mown at a height between 3 to 6 millimetres (NZTCI, 1971).

The fertiliser requirements of lawn bowls, croquet and tennis is a function of frequent clipping removal during mowing, nutrient leaching losses and a limited nutrient absorptive ability due to a restricted root growth. The emphasis of presentation of lawn bowling greens in Australia is not aesthetics but of playing surface quality. The overall fertiliser strategy shown among bowling greenkeepers in Australia during the growth season of bentgrass or bermudagrass is to produce a hard wearing playing surface with minimal verdue, by the correct timing of soluble nitrogen fertilisers, and to attempt to encourage maximum root growth, by the use of phosphorus and potassium fertilisers. The emphasis of fertiliser formulations during the growing season on lawn bowling greens is based on soluble types so as not to interfere with the roll of the bowl. The results of long-term nutritional analysis in Australia suggest that phosphorus applications are excessive on bowling greens. Bowling greenkeepers in Australia typically use a combination of straight soluble fertilisers, largely based on nitrogen and potassium, and the numerous range of complete fertilisers which contain nitrogen, phosphorus and potassium in varying ratios and forms. Organic fertiliser forms, largely based on processed fowl manures, have also been widely used for many years Australia wide. A dramatic increase in the use of controlled-release fertilisers during the renovation period of lawn bowling greens and tennis courts has been largely market driven. In Britain, a wider range of organic fertilisers are used on lawn bowling greens (Evans, 1992). Adams and Gibbs (1994) have given a guide to the annual fertiliser requirements of lawn bowling greens in Britain but with some provision for test results on sand greens. In New Zealand, the annual requirements of cotula bowling greens have been stated as 50 to 60 kilograms of ammonium sulfate, 10 to 15 kilograms of superphosphate and not in excess of 40 kilograms of potassium chloride (Grant, 1990).

Throughout the world, the seasonal irrigation requirements, irrigation water source and chemical quality, and irrigation technology vary widely. The application and control of soil moisture probably represents the single, greatest challenge among bowling greenkeepers. In Australia, South Africa, southern California and Florida, the irrigation requirements for lawn bowling greens during summer are relatively high in comparison to countries such as Britain and Canada. The irrigation schedule during summer is particularly acute in Australia and in southern California for lawn bowling greens constructed on near pure-sand profiles.

Throughout Australia on bermudagrass bowling greens during the growing period when greens are in play the irrigation schedule is typically twice weekly. In the absence of an automatic sprinkler system, lawn bowling greens are irrigated almost exclusively during the early and late daylight hours. It is not uncommon in Australia for a bowling club to nominate one day per week as a day of no play devoted to maintenance. For Australian bowling greenkeepers following the nominated days of irrigation complaints of moist surfaces and inconsistent or reduced green speed may be common. Additional irrigation along the edges of lawn bowling greens in Australia by manual methods is commonplace during days of high evapo-transpirational loss and to counteract problems of hydrophobicity. Soil wetting agents are commonly used during late spring and mid-summer on high sand content bowling green soils.

The objective of rolling of lawn bowling greens is to enhance the roll of the bowl by reducing the friction between the bowl and surface, and it is practiced at the highest level where bowls is played professionally. In Australia, the design and use of dedicated bowling green rollers is the highest among all countries where lawn bowls is played. Ron Kaye, a retired Australian aeronautical engineer, has single-handedly influenced the design of Australian bowling green rollers more than any other person. From the large diameter, walk-behind single-roller designs of the 1950s, the modern Australian bowling green roller from the 1970s utilised multiple-rollers to maximise the rolling efficiency, and are ride-on designs for speed of operation. The rolling of lawn bowling greens in Australia may be on a daily basis or at least on the day of any bowls competition. During the non-growing season in Australia, lawn bowling greens may be rolled in preference to mowing. In Britain, the use of rollers on lawn bowling greens is not as common as in Australia, and the use of motorised rollers for lawn bowling greens is relatively recent.

The rolling of lawn tennis courts has been traditionally practised to increase the level of soil resistance against physical deformation to influence the bounce of the ball. The design of the rolling machines has been relatively large, petrol-driven units often weighing in excess of several tonnes. The frequency and emphasis of rolling lawn tennis courts is generally much less than adopted for lawn bowling greens largely due to differing soil types. Croquet lawns are not rolled to the extent of lawn bowling greens because of the lesser demanding playing surface required.

Marking out for play

Marking out of the playing surface is normally the final input into presentation for the day's play. The primary objective of marking out is to define the area of play. Marking out presents the unique opportunity of not only controlling where the day's play is concentrated, but more importantly protecting the playing surface by rotating player wear.

Marking out of lawn bowling greens is based on the width of a rink. Marks on the bowls rest are typically used to define the width of a rink. In Australia, it is commonplace to use a mirror device placed on the green and chalk to define the rink centreline. The location of the rink can be moved as required. The marking out of lawn tennis courts is critical so that the court is marked out with lines at right angles and of equal length. Tennis court marking equipment typically utilises white calcium carbonate lime. Tennis court outlines can be moved laterally along the baseline or longitudinally along the sideline. Once to twice weekly marking out of lawn tennis courts is commonplace. The marking out of croquet lawns is a relatively straightforward procedure and is based on the correct distance of each peg.

Management of turfgrass pests

Numerous pests throughout the world affect the aesthetic uniformity and playing surface quality of lawn bowling greens, croquet courts and lawn tennis courts. The pests include numerous species of terrestrial and semi-aquatic plants, such as freshwater algae and moss, arthropods, pathogenic fungi and nematodes.

Management of arthropod pests

Numerous species of phytophagous insects and mites cause damage to sportsturf. The greatest number of arthropod pest species occur in Australia, New Zealand, North America and South Africa. Related insect and mite pest species occupy similar ecological niches within different countries. The pest status of each species varies within and between countries and includes endemic and introduced species. There are no effective turfgrass insecticides or miticides against some specific pest species in some countries. Arthropod pests most common to lawn bowling greens, croquet courts and lawn tennis courts are various species of scarabs, weevils, moths, grass mites and some species of flies (ATRI, 1996; Baldwin, 1990; Beard, 1973; Brockett, 1982; Liffman, 1984; NZTCI, 1971).

In Australia, the principal insect pest of bentgrass is the larvae of the Argentine stem grass weevil (*Listronotus bonariensis*). The principal insect pests of bermudagrass are the larval stage of a dipterous pest, the couch tip maggot (*Delia urbana*) and various species of grass webbing mites (*Oligonychus* spp.). These pests are active during summer. Less important pests of bermudagrass are the ground pearl (*Eumargerodes* spp.) and couch grass scale (*Odonaspis ruthae*). Sporadic foliage pests of bentgrass and bermudagrass are various species of armyworms (*Persectania* spp., *Pseudaletia* spp. and *Spodoptera* spp.), cutworms (*Agrotis* spp.), sod webworm (*Herpetogramma licarsisalis*) and underground grass caterpillars (*Oncopera* spp.). Soil inhabiting pests include the African black beetle (*Heteronychus arator*), Argentinian scarab (*Cyclocephala signaticollis*), black-headed cockshafer (*Aphodius tasmaniae*), red-headed cockshafer (*Adoryphorous couloni*) and the lawn scarab (*Sericesthis geminata*) (ATRI, 1996).

Significant insect pests of lawn bowling greens in New Zealand are the native grass grub (*Costelytra zealandica*) and the introduced black beetle (*Heteronychus arator*). The introduced Argentine stem grass weevil (*Listronotus bonariensis*) damages the turf during spring and autumn. Less important pest species are the armyworm caterpillars (*Pseudaletia aversa* and *P. separata*) and the native porina moths (*Wiseana cervinata, W. signata* and *W. umbraculata*). The major pests of cotula (*Leptinella* spp.) bowling greens in New Zealand are the larval stages of armyworms, black beetle, porina moth and the Argentine stem grass weevil (NZTCI, 1971).

In South Africa, soil scarabs including the black maise beetle (*Heteronychus arator*), the lawn caterpillar (*Spodoptera cillium*) and the ghost moth (*Dalaca rufescens*) are the most common insect pest species of lawns and courts (Brockett, 1982). In North America, pests of bermudagrass lawn bowling greens include mites (*Aceria* spp., *Oligonychus* spp.), bermudagrass scale (*Odonaspis ruthae*) and larvae of various cockshafers (Watschke *et al.*, 1994). On bentgrass greens various species of armyworm and cutworm larvae are known to cause some damage. In Britain, the common insect species on lawn bowling greens are the crane fly (*Tipula paludosa*) which is referred to as a leatherjacket. Lesser common fly species are the fever fly (*Dilophus febritis*) and the St. Marks fly (*Bibio marci*). Other insect pests which cause sporadic damage throughout Britain are various species of cutworm larvae (*Agrotis* spp.) and the larvae of the cockshafer (*Phylloperta horticola*) (Baldwin, 1990; Evans, 1992).

Management of insect and mite pests commences with an understanding of the basic biology and ecology, recognition of the pest and feeding habits. The use of insecticides and miticides for turfgrass application must be justified based on the threshold population and degree of damage caused by the pest species. Once the accurate identification of the pest has been confirmed as the actual cause of damage, the selection and application of a turfgrass insecticide or miticide must be made in accordance with label recommendations.

Management of terrestrial and semi-aquatic weeds

Numerous monocotyledonous and dicotyledonous species of terrestrial plants possess the morphological and structural adaptations to persist under the low mowing regimes of lawn bowling greens, croquet courts and lawn tennis courts. The most challenging plant species to control are the annual and perennial grassweeds. Throughout the world in temperate climates pereniated ecotypes of *Poa annua* (winter grass) remain the predominant grassweed of fine turfgrass. Its reproductive biology has been widely reviewed and studied (Lush, 1990). Numerous herbicides are used to control *Poa annua* in cool-season and warm-season turfgrass with varying degrees of short-term success (Baldwin, 1990; Beehag, 1994). Bensulide, endothal, ethofumesate, fenarimol and pronamide are the most common herbicides used for *Poa annua* control. In the United States, the plant growth regulators paclobutrazol and flurprimidol are now available (McCarty and Murphy, 1994). Other annual grassweed species with a cosmopolitan distribution in tropical and semi-tropical climates include *Digitaria sanguinalis* (summer grass), *Eleusine indica* (crowsfoot), *Eragrostis* spp. and *Setaria* spp. (McCarty and Murphy, 1994). Pre-emergent and post-emergent herbicides are used for their control; these include atrazine, dithiypor, DSMA, MSMA, diclofop-methyl, fenoxaprop-ethyl, fluazifop, hexazinone, metrabuzin, oxadiazon, pendamethalin, oryzalin and sethoxydim (McCarty and Murphy, 1994; ATRI, 1996).

Perennial grassweeds of bowling greens in Britain include cultivars of bentgrass (*Agrostis* spp.), perennial ryegrass (*Lolium perenne*) and Yorkshire fog (*Holcus lanatus*) (Evans, 1992). In Australia, perennial grasses include the turfgrass species *Agrostis* spp., and *Cynodon* spp., *Digitaria didactyla* (Queensland blue couch) and *Paspalum distichum* (water couch). In New Zealand, Yorkshire fog (*Holcus lanatus*) exists in bentgrass lawn bowling greens (NZTCI, 1971). The maintenance of cotula bowling greens in New Zealand presents considerable limitations for the selective removal of non-gramineous species (Sullivan, 1990). Active ingredients used include 2,2-DPA, terbacil, bromacil, oxadiazon and pendamethalin (NZTCI, 1971; Harrington, 1997). In Australia, common broadleaf weed species include *Oxalis corniculata* (creeping oxalis) and *Ranunculus repens* (creeping buttercup).

Semi-aquatic plants also persist under the specific conditions of saturated soils and partial shade. Species include freshwater algae, moss, sedges and perennial, broadleaf weeds. Moss and algae are common in countries which have cooler climates and a relatively frequent rainfall. Numerous species of cyanobacteria and eukaryotic algae are known in sportsturf swards in several countries including Australia, Britain, Canada and New Zealand (Baldwin, 1992). Management practices for algal control have typically included scarification and, with limited success, using specific copper-based algacides and fungicides such as thiram (Baldwin, 1992). A relatively common broadleaf weed species encountered in moist regions of lawn bowling greens in Australia, Britain and New Zealand is *Sagina procumbens* (pearlwort). *Cyperus brevifolius* is a persistent and widespread sedge in Australia in *Cynodon* spp. bowling greens.

Management of fungal diseases

Numerous fungal diseases are known to cause damage to bentgrass and bermudagrass cultivars throughout the world because of their widespread use as fine turfgrass. The causal agent and taxonomy for certain disease symptoms is not yet well understood in some countries. Disease symptoms vary between bowling greens, croquet lawns and tennis courts depending on the relative mowing heights. Fungicide efficacy varies widely between active ingredients and, for certain ectotrophic soil fungi, cultural practices cannot be precluded from the control strategy.

Fungal diseases of most significance in Australia are dollar spot (*Sclerotinia homeocarpa*), brown patch (*Rhizoctonia* spp.), damping off

(*Pythium* spp.), fusarium patch (*Fusarium nivale*), summer patch (*Bipolaris* spp., *Drechslera* spp.) and spring dead spot (*Leptosphaeria* spp.) (ATRI, 1996). Spring dead spot is the most widely distributed and damaging fungal disease on closely mown bermudagrass in Australia. South African bermudagrass (*Cynodon transvaalensis*) and Tifdwarf bermudagrass appear to be the two most susceptible *Cynodon* spp. cultivars to *Leptosphaeria* spp. in Australia. Fungicides registered in Australia against spring dead spot disease include bitertanol, iprodione, procymidone, propiconizole and TMTD. On the north coast of New South Wales and the coast of south east Queensland, two unknown fungal diseases are known on Tifdwarf bowling greens (Beehag and Wong, 1997). An unknown fungal condition has also been observed on lawn tennis courts of *Digitaria didactyla* (Queensland blue couch) in Sydney. Fusarium patch is the principal fungal disease on bentgrass greens in southern Australia. The principal fungal diseases in New Zealand on bentgrass bowling greens are fusarium patch (*Fusarium nivale*), brown patch (*Rhizoctonia solani*) and red thread (*Laetisaria fuciformis*) (NZTCI, 1971). On cotula bowling greens in New Zealand the most significant fungal diseases are brown patch (*Rhizoctonia solani*), Rolf's disease (*Sclerotium rolfsii*) and phytophthora (*Phytophthora crypogea*) (Ormsby, 1990).

In Britain, the most common fungal diseases on lawn bowling greens and occasionally on tennis courts are fusarium patch (*Microdochium nivale*) and dollar spot (*Sclerotinia homeocarpa*). Less widespread pathogenic diseases are anthracnose (*Colletotrichum graminicola*), red thread (*Laetisaria fuciformis*) and take-all (*Gaeumannomyces graminis*) (Baldwin, 1990; Evans, 1992). In North America, the main fungal diseases of bentgrass bowling greens in Canada are dollar spot (*Sclerotinia homeocarpa*) and brown patch (*Rhizoctonia solani*) in summer and fusarium patch (*Fusarium nivale*) and snow mould (*Typhula* spp.) in winter (Anon, 1986). In the northern United States, during the cooler months, pink snow mould (*Microdochium nivale*), grey snow mould (*Typhula incarnata*) and cool-temperature brown patch (*Rhizoctonia cerealis*) are relatively common (Watschke *et al.*, 1994). During the warmer months in the southern United States on Tifgreen bermudagrass bowling greens, helminthosporium leaf spot (*Bipolaris* spp. and *Drechslera* spp.) has been observed. The extent of ectotrophic soil-borne fungi on bermudagrass bowling greens in North America is unknown. In South Africa, the range of fungal diseases are brown patch (*Rhizoctonia solani*), dollar spot (*Sclerotinia homeocarpa*) and helminthosporium (Brockett, 1982). Spring dead spot and fairy ring diseases are common on bermudagrass bowling greens (Louw, 1997).

Management of parasitic nematodes

In Australia and the southern United States, numerous species of parasitic nematodes have long been recognised as disease-causing organisms on fine turfgrass (Stynes, 1971; Smiley, 1983). The only nematicide registered for turfgrass application in Australia is fenamiphos (ATRI, 1996). Fenamiphos has been shown to undergo enhanced biodegradation at several bowling green sites in Australia as well as in Florida (Beehag, 1995). The significance of parasitic nematodes in Britain is unknown (Baldwin, 1990). In New Zealand, parasitic nematodes have been recorded on bentgrass and cotula greens (Knight, 1990). Circumstantial evidence suggests that biological control agents and certain organic products, such as molasses and seaweed extracts, may provide an integrated approach to nematode control.

Deterioration and restoration of playing surfaces

Natural turfgrass surfaces suffer deterioration over time in the absence of proper cultural management. The components of surface deterioration are abrasive wear of the surface (Canaway, 1980; Green, 1980; Adams and Gibbs, 1994) and densification of the underlying soil medium (Carrow and Petrovic, 1992). The pattern and type of surface deterioration is normally unique for each sport and is visibly expressed by surface symptoms, such as partial loss of verdue due to direct player wear and the normal accumulation of excessive thatch. Deterioration of the physical properties of the growing media may be not readily recognised and it is therefore insidious.

Abrasive wear and excessive soil compaction on lawn bowling greens primarily occurs on the edges due to the crushing effect on the verdue by players, as well as the turning of the mower. One of the principal reasons for the widespread adoption of improved *Cynodon* spp. selections in Australia on lawn bowling greens during the early 1970s was due to the relative intolerance of wear by *Agrostis* spp. cultivars. The location and type of physical wear caused by tennis players along the service and base lines can be relatively extreme particularly on bentgrass. Wear is due to the crushing and sliding motion by tennis players when serving and receiving the ball. The sliding motion displayed by tennis players to receive a ball is unique to lawn tennis courts. Soil compaction on lawn tennis courts results primarily from the fine texture of the underlying soil and influenced by the mowers and rollers during times of high soil moisture content.

Restoration of playing surfaces

Restoration of the sward and the underlying growing media to meet management objectives and playing surface requirements are in response to deterioration resulting from play. The degree and type of surface restoration practices are governed by the ways in which the playing surface and growing media fall short of being acceptable and not meeting the surface requirements. Minimisation of wear on lawn bowling greens and tennis courts can be achieved through the selection of wear-tolerant species such as *Cynodon* spp. cultivars in warm climates. Compaction resistance of soil media is possible to a large extent by the use of uniformly graded sands for bowls and croquet. Thatch accumulation may be minimised by use of less aggressive species. Excessive soil compaction effects and remedial action have been reviewed by Carrow and Petrovic (1992). Mechanical equipment used for the relief of excessive soil compaction without removal of the playing surface includes hollow and solid tynes, solid drills, vibratory deep slicers and more recently, high-pressure water injection. The practice of thatch control involves scarification (vertical mowing) and topdressing when the playing surface is to be retained. Rejuvenation of abnormally low levels of density and verdue may involve overseeding, stolonisation or re-sodding (returfing) (Beard, 1973).

Hollow tyning and scarification are routinely used on lawn bowling greens during the growing seasons. Inorganic and organic amendments such as lime, gypsum, processed organic manures and peat have long been used. The use of zeolite is relatively new in Australia. In Australia and New Zealand, the dramatic approach of 'shaving' for thatch management has long been used (Adams and Gibbs, 1994). The technique of 'shaving' *Cynodon* spp. lawn bowling greens in Australia commenced with the importation of a Godwin shaver unit in the 1970s from South Africa by the mower manufacturer, Scott Bonnar. Other equipment now used to shave greens in Australia normally to a depth of 5 to 10 millimetres includes a modified sod cutter, utilising an angled and wide thin blade, and a purposely-designed, self-propelled rotary shaver. The use in Australia of 'grooming reels' to replace the normal cutting cylinder on electric bowling green mowers is commonplace as an additional management procedure of thatch control. The use of 'mini-tyning', 'drilling' and 'vibratory deep slicing' of lawn bowling greens in Australia is additionally practised to relieve soil compaction. Recent use is made of the Verti-Drain and the Hydroject units in Australia. Lawn bowling greens may also require levelling by top-dressing to a depth of 5 to 10 millimetres to restore surface playing levels. This practice traditionally used a 'rail' system but has largely been superseded by mechanised laser-guided equipment. Oversowing of *Agrostis* spp. bowling greens in southern Australia is commonplace during the renovation period. In Australia, a lawn bowling green may be 'out of play' for periods of between 4 to 10 weeks to allow for renovations depending on the extent of the restoration work.

Hollow tyning and scarification are typically used on lawn tennis courts. Scarification during the summer growing season in Australia of *Cynodon* spp. lawn tennis courts is required to minimise surface resilience. The Verti-Drain machine is ideally designed for relatively deep penetration of lawn tennis courts given the required degree of soil compaction. Topdressing of lawn tennis courts is required due to the need to restore surface levels along the base lines. Topdressing by manual means may be required for isolated sections of a court. Restoration practices for croquet lawns normally include hollow tyning and scarification. Topdressing and re-levelling may be required.

Preparation for championship play

All sporting surfaces require peak performance for tournament and championship play. This is particularly true for international tournaments being sponsored and televised worldwide. Preparation for championship play involves forward planning and adequate surface preparation during the tournament. Items include the forward planning of seasonal renovations to ensure that the required surfaces will meet the designated championship performance during events. Amount of thatch accumulation and the degree of level are the two primary playing surface factors requiring strict attention during a restoration program well in advance of the actual tournament. Championship tournament surface preparation involves the key practices of mowing, irrigation and fertiliser regimes. Mowing heights may be reduced slightly for the event depending on the specific directions demanded by tournament organisers. The frequency of mowing may be increased to twice daily, morning and afternoon. Irrigation application needs to be reduced to ensure a highly playable and true surface. In Australia, during bowls championships, hand irrigation during the late afternoon after play may be practised along the edges should abnormal soil drying be a factor. Fertiliser application needs to have ceased immediately prior to the championship so as not to stimulate a 'lush' and therefore a wear-intolerant sward. The rolling of lawn bowling greens and tennis courts is an additional key cultural practice during championship play. In Australia, the rolling of lawn bowling greens is normally conducted twice daily, before and after play. During some international events rolling may be practised during the day between events to ensure a true rolling surface.

References

Adams, W.A. and Gibbs, R.J. 1994. 'Natural turf for sport and amenity—Science and Practice', *CAB International*, Wallingford, Oxon, 404 pp.

Anon. 1986. *Bowling green maintenance handbook*, Lawn Bowls, Canada.

Arnold, E.H. 1957. 'Soils for greens renovations', *N.Z. J. of Agriculture*, vol. 94, no. 2, p. 165.

ATRI. 1996. *Disease, insect and weed control in turf*, 4th edn, Australian Turfgrass Research Institute Ltd, Concord, Australia.

Baker, S.W. 1990. *Sands for sports turf construction and maintenance*, Sports Turf Research Institute, Bingley, UK.

Baldwin, N.A. 1990. *Turfgrass pests and diseases*, 3rd edn, Sports Turf Research Institute, Bingley, UK.

Baldwin, N.A. 1992. 'Cyanobacteria and eukaryotic algae in sports turf and amenity grasslands: a review', *J. Appl. Phycology*, vol. 4, p. 39–47.

Beale, R. 1924. *Lawns for sports—their construction and upkeep*, Simpkin, Marshall, Hamilton, Kent and Co., London.

Beard, J.B. 1973. *Turfgrass: science and culture*, Prentice-Hall, Englewood Cliffs, N.J., 658 pp.

Beehag, G.W. 1994. 'A review of *Poa annua* (L.) management', *Proc. 1st ATRI Turf Research Conf. (Sydney)*, Australian Turfgrass Research Institute Ltd, p. 120–6.

Beehag, G.W. 1995. 'A review of enhanced biodegradation of fenamiphos in Australia', *Proc. 1st ATRI Turf Research Conf. (Sydney)*, Australian Turfgrass Research Institute Ltd, p. 61–5.

Beehag, G.W. and Surrey, R. 1992. 'Couchgrass culture in Australia', *ATRI Turf Notes*, Australian Turfgrass Research Inst., Vol. 11, no. 3, p. 10–11.

Beehag, G.W. and Wong, P.T.W. 1997. 'Leopard spot and frog eye symptoms—the knowns and unknowns', *ATRI Turf Notes*, Australian Turfgrass Research Inst., vol. 16, no. 3, p. 3–6.

Bell, M.J. and Holmes, G. 1988. 'Playing quality standards for level bowling greens', *J. Sports Turf Res. Inst.*, vol. 64, p. 48–62.

Bowls Australia. 1998. *Laws of the Game of Bowls in Australia*, Bowls Australia, 30 pp.

Brockett, G.M. 1982. (ed.) *Grass for turf and revegetation*, N 21/1982, Grassland Research (Cedara), South Africa.

Canaway, P.M. 1980. 'Wear', in *Amenity grassland: An ecological perspective*, I.H. Rorison and R. Hunt (eds), John Wiley and Sons, p. 137–54.

Carrow, R.N. and Petrovic, A.M. 1992. 'Effects of traffic on turfgrass', in *Turfgrass*, D.V. Waddington, R.N. Carrow and R.C. Shearman, (eds), American Society of Agronomy, Monograph no. 23, p. 285–330.

Davis, W.B. 1977. 'Soils and sands for bowling greens', *Cal. Turf Culture*, vol. 27, no. 3, p. 19–20.

Dawson, R.B. 1939. *Practical lawncraft*, Crosby Lockwood & Son Ltd, London.

Evans, R.D.C. 1992. *Bowling greens–their history, construction and maintenance*, Sports Turf Research Institute, Bingley, UK.

Gibbs, R. 1997. 'Further comparisons of natural and synthetic bowling greens', *New Zealand Turf Management J.*, vol. 11, no. 4, p. 25–9.

Grant, A.S. 1990. 'Basics of a fertiliser program', *Proc. 4th New Zealand Sports Turf Convention*, Massey University, p. 36–8.

Green, B.H. 1980. 'Management of extensive amenity grasslands by mowing', in *Amenity grassland: An ecological perspective*, I.H. Rorison and R. Hunt (eds.), John Wiley and Sons, p. 155–61.

Haley, E.R. 1979. *Better greens*, Escondido, California, USA.

Harrington, K. 1997. 'Controlling weeds in new cotula greens: Some new options', *New Zealand Turf Management J.*, vol. 11, no. 4 p. 24–5.

Henry, M. 1977. 'Introduction to lawn bowling', *Calif. Turfgrass Culture*, vol. 27, no. 3, p. 17–19.

Holmes, G. and Bell, M.J. 1987. 'Other sports', in *Standards of playing quality for natural turf*, The Sports Turf Research Institute, Bingley, UK, pp. 50–2.

Knight, K. 1990. 'Nematodes in bowling greens', *Proc. 4th New Zealand Sports Turf Convention*, Massey University, pp. 72–3.

Liffman, K. 1984. *Bowling greens—a practical guide*, RMIT Tafe Publications Unit, Morphett, 130 pp.

Louw, C. 1996. 'Bowls in South Africa', *ATRI Turf Notes*, Aust. Turfgrass Res. Inst. (Sydney), vol. 15, no. 2, pp. 4, 5, 10.

Lush, M. 1990. 'Biology of *Poa annua*–the secret of success', *N.Z. Turf Management J.*, vol. 4, no. 3, p. 5–8.

McAuliffe, K.W. and Gibbs, R.J. 1993. 'A national approach to the performance testing of cricket grounds and lawn tennis courts', *International Turfgrass Research Journal 7*, R.N. Carrow, N.E. Christians, R.C. Shearman (eds), Intertec Publishing, Overland Park, Kansas, p. 222–30.

McCarty, L.B. and Murphy, T.R. 1994. 'Control of turfgrass weeds', in *Turf Weeds and Their Control*, A.J. Turgeon (ed.), Am. Soc. Agronomy, Wis., p. 209–48.

McMaugh, P. 1968. *Bowling green construction*, 1st Impression, Grass Research Bureau (NSW), Sydney.

Murphy, J.W. and Nelson, S.H. 1979. 'Preliminary investigations of sand growth media for cotula', *N.Z. Journal of Experimental Agriculture*, vol. 7, p. 257–62.

Neylan, J. and Robinson, M. 1993. *Comparison of natural turf and synthetic bowling greens*, Horticultural Research and Development Corpn Project no. TU112.

NZTCI. 1971. *Turf culture*, 2nd edn, New Zealand Institute for Turf Culture, Palmerston North, N.Z.

Ormsby, D. 1990. 'Diseases of cotula in the North Island', *Proc. 4th New Zealand Sports Turf Convention*, Massey University, p. 55–6.

Ormsby, D. 1993. 'Developments in New Zealand bowling green surfaces—alternative surfaces', *New Zealand Turf Management J.*, vol. 7, no. 3, p. 13–15.

Pierce, S.J. and Rigney, C.B. 1946. *Bowling greens—their construction and maintenance*, NSW Bowling Assn, Sydney, 5 pp.

Stynes, B. 1971. '*Heterodera graminis* N. sp., a cyst nematode from grass in Australia', *Nematologica*, vol. 17, p. 213–18.

Sullivan, J. 1990. 'Common pests of cotula', *Proc. 4th New Zealand Sports Turf Convention*, Massey University, p. 70–1.

Thorpe, J.D. and Canaway, P.M. 1986. 'The performance of tennis court surfaces. I. General principles and test methods', *J. of the Sports Turf Research Institute*, vol. 62, p. 92–100.

Watschke, T.L., Depnoeden, P.H. and Shetlar, D.J. 1994. *Managing Turfgrass Pests*, Lewis, USA.

CHAPTER 17

Prescription surface development: Sportsfield and arena management

J.J. NEYLAN, D.J. McGEARY and **M.R. ROBINSON,** Turfgrass Technology Pty. Ltd., Sandringham, Melbourne, Victoria, Australia

Introduction

Turf quality and turf survival determine the playing and use characteristics of all grassed areas and reflect how well the grass is surviving in its particular environment. The turf environment includes soil type, fertility, grass species/cultivar, management and wear, and each one separately or collectively can have a significant effect on turf quality. The foundation of all turf areas is the soil used in construction; and soil selection is a critical factor in determining the long-term quality of the turf. The use of inferior soils with poor drainage and aeration or soil profiles that are too shallow does not provide an environment for satisfactory growth, particularly under conditions of high use. The use of poor quality soils is common and it often occurs in sportsfields subjected to, or intended for, heavy use. The long-term maintenance needs of such areas are high, however sufficient resources are often not available to provide an adequate maintenance level and a poor quality surface is the result.

Turfgrass areas are subjected to two main traffic stresses, soil compaction and wear. Soil compaction is the pressing together of soil particles, and as the particles are pushed together, the spaces between them are reduced in size. The spaces or pores in a soil play an important role in determining the drainage and water holding characteristics of the soil. Soil pores also affect the oxygen supply to the roots and the ease with which roots can grow. As the pore spaces are reduced in size excess water does not move freely through the soil, and the ease with which water can be removed from the soil is reduced. As the pores become smaller, a greater proportion of the available space contains water to the exclusion of oxygen which is essential for healthy root growth. Wear is a direct injury to the plant tissues caused by pressure, scuffing, abrasion and tearing (Beard, 1973). Wear tends to crush the leaves, stems and crowns of the plant, reducing the vigour of the plant and increasing the susceptibility to disease infections. Wear tolerance depends on turfgrass species, intensity of culture, environment, and intensity and type of traffic. Wear and compaction often occur at the same time, but one is normally the dominant stress. With sandy or dry soils that can withstand compactive forces, then wear is dominant. Compaction is usually dominant on soils high in silt and clay, particularly under moist conditions.

Key construction issues for sportsfields and arenas

Typical regulation playing surface dimensions are shown in Table 17.1.

Improving soil conditions, that is drainage and porosity, improves the environment for turf

Table 17.1 Regulation playing surface dimensions.

Sport	*Dimensions (m)*	*Typical area of field*
Rugby Union	max: 122 × 69	0.83 ha
Rugby League	max: 144 × 68	0.98 ha
Soccer	max: 120 × 90	1.08 ha
	min: 90 × 45	0.41 ha
Australian	max: 185 × 155	2.87 ha
Rules Football	min: 135 × 110	1.49 ha
Hockey	max: 91 × 55	0.50 ha
Cricket pitch	20.1 × 3	60 m²
Polo	274 × 183	5.01 ha

growth. Growing conditions can be improved by selecting soils for construction with high drainage rates (such as sands) or by improving the drainage of sports turf through the use of agricultural drainage, sand slitting, subsoil aeration or a combination of techniques.

Trials conducted by the Sports Turf Research Institute (STRI), England, compared the efficiency of various construction and drainage methods (Canaway and Baker, 1993). The best surfaces were those with a sand playing surface, that is slit drainage with a 25 millimetre sand top layer, sand carpet (100 millimetres of sand over topsoil with slit drains) and sand profile. Where slit drainage alone was used the playing surface rapidly degenerated to a level similar to pipe drain plots. The reduction in performance was due to capping of the slit drains with soil from the surrounding area when the plots were subjected to wear. To improve the efficiency of sand slitting it was recommended that an essential part of maintenance is the regular topdressing with sand at a rate of 16 kilograms per square metre per year (160 tonnes per hectare per year). A study by Baker and Gibbs (1989) and Gibbs and Baker (1989) showed that soccer fields on soils with no drainage or pipe drainage only were prone to waterlogging and gave only 2.0 to 3.7 hours of play a week, while on sand profiles the amount of usage was 5.3 to 10.9 hours per week. In work carried out by Gibbs *et al.* (1989), it was calculated that 7.8 hours of usage per week over a 35 week soccer season (during winter) was the maximum usage on a sand profile. At this stage there were 30 to 40 unstable (i.e. bare ground) wear areas on the playing surface.

Gibbs *et al.* (1993) analysed the cost effectiveness of various types of sportsfield construction which included the cost of construction, hours of usage, operating costs and maintenance hours. Based on operating costs only, the pipe drained field was the least cost effective. The undrained pitch was very cost-effective due to the good natural drainage of the soil and mild winters, however the surface was poor during wet weather. Therefore good natural drainage can provide a field that is cheap to run, particularly when winters are mild, although there is still the risk that the quality of play will decrease dramatically in severe weather. When including the construction plus running costs, the slit drainage fields were most cost effective, providing that the field had a life of at least seven years. Slit drainage fields can deteriorate rapidly under high usage which then makes the sand carpet fields more cost effective. Sand-based, perched water table constructions are the most expensive and have a low cost effectiveness. Sand-based fields can also incur greater maintenance demands under high usage because as the turf cover is lost, the surface becomes eroded and results in reduced stability. Consequently, greater efforts are required to repair the surface. The main advantage of sand-based fields is that they provide a high quality surface under most rainfall conditions.

Traction is an important characteristic of turf surfaces, as many sports involve running and sharp turning. If the sport is to be played at the highest level, then the surface must be stable. This requirement not only applies to ball sports such as soccer, Australian Rules football and rugby but also to horse racing. Traction is the amount of horizontal force the player can apply to the surface without slipping or falling (Canaway, 1985). It is often referred to as shear strength but there are forces other than shearing taking place and traction is a more accurate term. On sports turf it is a combination of the sward and the soil which govern traction. The soil affects traction principally through its effects on drainage, moisture content and binding strength, while the sward has an effect through root reinforcement. Soils with a high clay content maintain surface stability when devoid of grass compared to sand, which becomes very loose and provides poor traction unless it can be kept at a high moisture content and rolled. Adams *et al.* (1985) demonstrated that on a well grassed sand rootzone, plant roots increase the shear resistance by a factor of two to three times. The effect of the roots is much greater than the increase in traction achieved by increasing the silt and clay content. In a soil with a fine fraction (< 50 millimetre) of 12 per cent there was an increase in shear resistance of about 50 per cent over that of pure sand. Trials carried out by Lemaire and Bourgoin (1981) showed that wear and compaction modified the weight and distribution of roots with depth. This depended on soil type, plant species and the amount of wear. As compaction increased, such as on heavier soil types, there was a reduction in root weight with the majority of roots being confined to the surface five centimetres. On sandy soils, the compaction had less effect and there was less root injury. Gibbs *et al.* (1989) also emphasised the importance of root material in stabilising the surface of sand-based rootzones, particularly those subjected to high use. Even when the above ground parts of the plant are worn away a well established root system, including the thatch–rootmat component, will maintain surface stability.

Construction of sportsfields and arenas

Sportsfields have been constructed using a wide variety of soils and construction techniques, often with little thought given to the amount of use, the required quality, and ongoing maintenance demands. Sportsfields have often been the dumping ground for high clay content soils and building rubble, as well as being the final product of land fill areas. The demands for better quality playing surfaces, player safety and the aesthetic requirements for televised sport have necessitated a more scientific approach to sportsfield construction. If satisfactory turfgrass growth is to occur and a good quality surface is to be produced, then the growing media must balance the two divergent needs of drainage (including aeration) and moisture retention (Adams and Gibbs, 1994). From an agronomic point of view, the ideal sportsfield soil is one that has sufficient large, stable pores to allow rapid drainage and good aeration under heavy or persistent rainfall, with the remaining space taken up by small pores for moisture retention to meet the requirements of turfgrasses.

Using local soils for sportsfield profiles

Most local soils will only meet the requirements of good drainage and aeration if the sand, silt and clay particles are bound together in a fragile system of water stable aggregates (Adams and Gibbs, 1994). The successful use of the local soil in its natural state depends on the level of usage, climate and soil type. Where natural sandy and loamy sand (consists of at least 75 to 80 per cent sand particles) soils exist there is sufficient macroporosity to ensure rapid drainage and good aeration. However, the majority of local soils have a relatively high silt and clay content (> 30 per cent) and it is very difficult to maintain a high level of drainage and aeration. These finer textured soils are susceptible to compaction, particularly when they are subjected to traffic under wet conditions. In this situation the soil aggregates break down into smaller particles which results in a loss of macroporosity. When local soils are used, they are often modified by the inclusion of slit drainage, sand carpets and frequent subsoil aeration so that macroporosity is maintained.

Sand-based sportsfields

Given the requirements of sportsfields to have good drainage, adequate aeration and to resist compaction, sands have become the preferred rootzone medium. There are several standard sand profile specifications that have been used including the United States Golf Association (USGA) method for greens construction which has been adopted for sportsfields, the Prescription Athletic Turf (PAT) system, and the California method of construction. In Australia, McIntyre and Jakobsen (1993) have developed a sand-based sportsfield specification based on the moisture release characteristics of the selected sand.

The first rootzone construction system developed on soundly based scientific principles and supported by extensive field and laboratory research was the Texas–United States Golf Association (USGA) method of rootzone construction (Ferguson *et al.,* 1960). The research involved more than developing a sand profile, requiring a comprehensive assessment of soil physics, moisture retention, nutrition, infiltration rate and plant growth requirements. The result was a specific method of construction that has been refined over the years, culminating in a major review and refinement of the system by Hummel (1993). The features of this method include a shaped subgrade, a network of agricultural drains at a spacing of about five metres, a 100 millimetre thick gravel drainage layer over the drainage pipes, and an intermediate layer of coarse sand/fine gravel. Where the gravel meets particular criteria for permeability, bridging and uniformity as related to the rootzone sand (Table 17.2), this layer is eliminated. The intermediate layer is only 50 millimetres in thickness and needs to be installed by hand: it is therefore impractical for sportsfield construction. The 300 millimetre thick rootzone mixture consists of a sand with a specific particle size distribution (Table 17.3) which is amended with organic matter, usually peat moss. The rootzone mix has to meet particular criteria for porosity, moisture retention and drainage (Table 17.4). The presence of the gravel drainage layer has two important effects; it speeds the flow of water to the drains and forms a perched water table which increases the water holding capacity of the profile.

Table 17.2 Size recommendations for gravel when intermediate layer is not used in a USGA profile construction (Hummel, 1993).

Performance factors	*Recommendation*
Bridging factor	• D_{15} (gravel) ≤ 5 × D_{85} (root zone)
Permeability factor	• D_{15} (gravel) ≤ 5 × D_{15} (root zone)
Uniformity factors	• D_{90} (gravel) ≤ 5 × D_{15} (root zone) • No particles greater than 12 mm • Not more than 10% less than 2 mm • Not more than 5% less than 1 mm

Table 17.3 Particle size distribution of USGA rootzone mix (Hummel, 1993).

<table>
<tr><th>Name</th><th>Particle diameter (mm)</th><th colspan="2">Recommendation (by weight)</th></tr>
<tr><td>Fine gravel</td><td>2.0–3.4</td><td colspan="2" rowspan="2">Not more than 10% of the total particles in this range, including a maximum of 3% fine gravel (preferably none)</td></tr>
<tr><td>Very coarse sand</td><td>1.0–2.0</td></tr>
<tr><td>Coarse sand</td><td>0.5–1.0</td><td colspan="2" rowspan="2">Minimum of 60% of the particles must fall in this range</td></tr>
<tr><td>Medium sand</td><td>0.25–0.50</td></tr>
<tr><td>Fine sand</td><td>0.15–0.25</td><td colspan="2">Not more than 20% of the particles may fall within this range</td></tr>
<tr><td>Very fine sand</td><td>0.05–0.15</td><td rowspan="2">Not more than 5%</td><td rowspan="3">Total particles in this range shall not exceed 10%</td></tr>
<tr><td>Silt</td><td>0.002–0.05</td></tr>
<tr><td>Clay</td><td>less than 0.002</td><td>Not more than 3%</td></tr>
</table>

Table 17.4 Physical properties of the USGA rootzone mix (Hummel, 1993).

Physical property	*Recommended range*
Total porosity	35–55%
Air-filled porosity (at 30 cm tension)	15–30%
Capillary porosity (at 30 cm tension)	15–25%
Saturated conductivity	
Normal range	(15–30 cm/h)
Accelerated range	(30–60 cm/h)
Organic matter content (by weight)	1–5% (ideally 2–4%)

The Prescription Athletic Turf (PAT) system, described by Daniel *et al.* (1974), consists of a sand profile enveloped in an impermeable membrane where the moisture levels are controlled by drainage pipes, suction pumps and sub-irrigation. The main features of the PAT system are a flat subgrade so that the water levels in the profile are consistent over the entire sportsfield, an impermeable plastic membrane that encloses the entire profile, a network of pipes at four to six metre spacings that are used to control drainage and sub-irrigation, and a 250 to 350 millimetre sand rootzone. The critical aspect of the PAT system is the control of the water table by utilising pumps to either remove or add water. Under high rainfall conditions, particularly before a match or during play, the pumps can be activated to increase the hydraulic gradient within the sand rootzone and prevent ponding.

The California method of construction was developed at the University of California in the early 1960s as a means of constructing golf greens (Davis, 1973). The California method, or more particularly the principles behind this method, have been used in sportsfield construction. The key to the success of this method of construction is the narrow distribution range of the sand particle sizes. The dominant particle size is the medium sand fraction (0.25 to 0.5 millimetre) with a minimum of 60 per cent of particles in this range. Up to 10 per cent is permitted in the very coarse sand to fine gravel (1 to 2 millimetre) range, and 2 to 8 per cent in the very fine sand (0.05 to 0.1 millimetre) silt and clay (< 0.05 millimetre) fractions. The depth of sand specified is 300 millimetres. The California method may or may not incorporate an agricultural pipe drainage system depending on the permeability of the subgrade. Where the subgrade permeability is less than 12 millimetres per hour, a drainage system is recommended with pipe spacings up to 3 metres apart. The drainage system involves the use of pea gravel around the pipe with no drainage blanket. There have been many sportsfields constructed based on principles similar to those outlined in the California method.

The Technical Services Unit (TSU) in Canberra, Australia, has successfully based its method for sand-based sportsfield construction on soil physics and engineering principles (McIntyre and Jakobsen, 1993). The key to this method of construction is to produce a high drainage rate field with a perched water table. The principles involved are not dissimilar to the USGA method, however a more rigid specification for the sand and gravel types and the depth of these layers has been developed. The profile consists of a shaped and compacted base, agricultural drainage system, gravel drainage blanket and a sandy rootzone soil. The selection of the topsoil is considered to be the most important aspect of constructing a perched water table field. The sand for the topsoil must meet the specification in Table 17.5.

Table 17.5 Technical Services Unit (TSU) topsoil and gravel specification for a sand profile sportsfield (McIntyre and Jakobsen, 1993).

Type 'P' topsoil		*Drainage gravel*	
> 2 mm	0	> 5 mm	< 10%
0.25–0.5mm	> 60%	1–5 mm	> 80%
0.25 mm total fines	< 25%	1–2 mm	< 10%
Total silt/clay	< 5%	< 1 mm	0

The sand specification is similar to that of the USGA specification, however it is much tighter and does not allow as many coarse particles. The sand specification also requires a moisture release curve on a compacted and uncompacted sample to determine sand suitability. The bulk density of the compacted sample must be less than 1.7 grams per cubic centimetre and there should be only minimal differences in the bulk density between the compacted and uncompacted samples. The moisture release curve of the sand is also used to determine the critical depth of sand. TSU recommends the amendment of the top 100 to 150 millimetres of the profile with organic matter if the sand contains no silt or clay. The recommended source of organic matter is either composted pine bark or peat moss added at 10 to 15 per cent by volume, resulting in 1 to 2 per cent organic matter. The main difference in regards to organic amendments between the TSU and USGA specifications is that the TSU recommends amending only the top 100 to 150 millimetres compared to the full profile (300 millimetres) for the USGA specification.

Improving rootzone stability

Turfgrass injury and reduced playing surface quality are increasing problems on heavily used turf areas. Even employing the 'best' maintenance program on sand-based grounds with the hardest wearing grasses, will not necessarily guarantee a 'non-wear' turf under all conditions. The problem that often occurs is that as the quality of the turf improves, there are increased pressures to use it more frequently. The other issue is the multi-use of turf facilities, particularly those that are non-turf related such as concerts, motocross and other events that require large areas for the performers and audience. In recent years, sand reinforcing materials have been introduced to improve turf wearability and to cope with this non-sport use. Several reinforcement materials have been used and there are a number of mechanisms by which these materials may improve the wear tolerance and quality of turf (Baker *et al.*, 1988), such as by load-spreading, therefore reducing the rate of soil compaction; by reducing the effects of shearing forces, which helps to preserve the continuity of large pores at the soil surface; by protection of the crown tissue of the grass plant; and by increasing traction through the interaction between the fibres in the reinforcement material and the studs on the players' footwear.

Adams and Gibbs (1989) used a needle punch geofabric in an attempt to stabilise sand rootzones. The material was placed near the surface and covered with 5 millimetres of sand to provide a seed bed. As the turf sward deteriorated, they found that there was improved surface stability and traction. There was some restriction in root growth through the material but the improvement in surface stability more than compensated. It was concluded that synthetic fibres could be used to stabilise high wear areas such as goal squares. One major disadvantage is the restriction on the range of cultivation techniques that can be used.

Baker *et al.* (1988) investigated the use of several reinforcement materials, including needle punched geotextiles, a semi-rigid polyethylene mat, a rigid polyethylene mat and a material which was a mixture of polypropylene fibres and sand. The trials indicated that some reinforcement materials, such as the needle punched geotextile, could improve the quality of playing surfaces subjected to heavy use. It was suggested that these materials would be best used in high wear areas such as goal mouths. Another method of reinforcing sand profiles is the use of randomly orientated interlocking mesh elements. Trial work carried out in the USA showed that the mesh elements reduced divoting and lateral cleat tear which resulted in quicker recovery of the affected area (Beard and Sifers, 1989). The mesh elements have now been used in the Sha Tin and Happy Valley race tracks in Hong Kong, the Moonee Valley race track and the Melbourne Cricket Ground, in Melbourne, and the Parramatta Stadium in Sydney, Australia. Another reinforcement agent being used is Turfgrids®, small grids of polypropylene which are mixed with the rootzone soil to increase soil strength and stability. More recently, a new reinforcement system has been developed by Desso, which involves the injection of synthetic fibres into a natural turf system. The fibres are injected at 2 centimetre intervals to a depth of 20 centimetres with 2 centimetres of fibre visible at the surface, and make up 5 per cent of the surface. There is evidence that it dramatically increases the wearability of the playing surface. Research is also being conducted into natural turf transportable systems in which modular units of instant mature natural turf can be configured for indoor use.

Utilisation of sports turf species and cultivars

The rootzone media and level of management provide the growing conditions for turfgrasses and play a large part in determining their survival. However, there is another factor to consider: that of wear tolerance of individual species and cultivars within species. The factors that favour wear tolerance are above ground biomass, and high lignin and high cellulose contents in the shoot. These are characteristics typically found in ryegrass (*Lolium* sp.) and bermuda or couchgrass (*Cynodon* sp.), the grass species most often used for turf sportsfield. In a series of trials, Canaway (1983) assessed the wear tolerance of several turfgrass species on both sand (87 per cent particles in the 0.125 to 0.5 millimetre range) and a high clay content soil (35 per cent particles less than 0.05 millimetre). On both rootzone materials the wear tolerance was rated as *Poa annua* > *Lolium perenne* > *Poa pratensis* > *Festuca arundinacea* > *Agrostis castellana* > *Festuca rubra*. However on the soil, wear caused the playing conditions to deteriorate greatly, that is turn to mud, while the sand provided good conditions throughout the winter.

Another important factor related to grass selection is the effect on the playing conditions, that is traction and ball bounce. Even though winter grass (*Poa annua*) has a high wear tolerance, it has a very low shear strength and ball bounce, resulting in poor playing conditions. The result is often observed on turf swards dominated by *Poa annua* both in sportsfields and arenas. The low ranking of tall fescue (*F. arundinacea*) is surprising and was due in part to the use of a non-turf variety, which produced a low density sward and possibly a sward that was not sufficiently mature. However Lemaire and Bourgoin (1981) found that both tall fescue and perennial ryegrass rate highly in their ability to withstand wear and compaction. The superior performance of *L. perenne* in handling wear and providing a good playing surface is not surprising. Similiar results have been reported by Evans (1988). It has been suggested by Carrow and Wiecko (1981) that future plant breeding research should look at selecting grasses that have a greater tolerance to low oxygen levels and can overcome the high mechanical impedance of compacted soils. This applied at both the species and intra-species level, and requires a greater understanding of the plant physiological mechanisms involved.

Turf grass selection has often been seen as the solution to the problems caused by poor drainage and excessive wear. However, grasses can only tolerate so much wear, even under the best growing conditions, before they deteriorate and lose density. When turfgrasses are growing under ideal conditions, particularly climatic, they have excellent recuperative potential and will generally tolerate high levels of wear. However, when the conditions are less favourable, such as in cool weather, low light intensities and high moisture conditions, the turfgrasses are less able to recover before being subjected to more traffic. As a consequence, the effects of the wear are cumulative and rapid turf deterioration can occur. Of the warm-season grasses, couchgrass is the hardest wearing species and has greater recuperative potential than zoysiagrass (*Zoysia* sp.), and hybrid couchgrass (*C. dactylon x C. transvaalensis*) (Cockerham *et al.*, 1993). Turf subjected to heavy wear is often invaded by inferior grass species such as *Poa annua* and summergass (*Digitaria sanguinalis*) which are very invasive and well adapted to compacted soils.

In cool, temperate climates, where cool-season grasses are grown, perennial ryegrass has proven to be the best grass for year-round growth and recovery. Kentucky bluegrass and tall fescue are particularly good during the summer, however under cold conditions there is very little growth and very poor recovery potential. Given that many of the high wear sports of rugby, Australian Rules football and soccer are winter sports, then a turf with a high proportion (> 50 per cent) of ryegrass is desirable. Bermuda or couchgrass is the dominant species in warm humid, warm subhumid and warm semi-arid climates and are utilised to varying degrees in the transitional zones (Beard, 1973). Their preferred optimum temperature range is 27 to 32°C and under these conditions they provide a hard wearing, vigorous high quality turf. Couchgrass is tolerant of low mowing and provides a fast, firm surface that gives excellent traction. Couchgrass and other warm-season grasses are generally unsuitable in transitional climates for heavy winter traffic unless they are overseeded. At temperatures less than 15°C, most warm-season grasses become dormant or dramatically slow down in their growth and consequently have no recovery under moderate to heavy traffic.

In transitional climates where warm-season grasses, such as couch and kikuyugrass (*Pennisetum clandestinum*) are dominant during the summer but dormant in the winter, a base of these grasses overseeded with ryegrass (*Lolium* spp.) will provide the hardest wearing surface for winter sports. In mixed swards that have a high proportion of ryegrass it is difficult to transition out or weaken the ryegrass sufficiently to allow the warm-season grass to dominate during summer. If there is sufficient time between the different sporting seasons, e.g. football and cricket, the ryegrass can be chemically removed in the spring, the couchgrass allowed to grow in over summer and then overseeded with ryegrass in the autumn. The reality is that the ryegrass is often allowed to persist and this is eventually to the detriment of the couchgrass. In time high wear areas, such as the goal to goal line and goal squares, are best returfed with couchgrass and then overseeded.

Establishment of the playing surface

Preparing a playing surface suitable for an intended use or sport is both an art and a science. The ability of the curator to produce a high quality surface not only depends on the structure (soils, grass type) they have to work with, but also how they manipulate the growing conditions. Many management practices are employed including fertilising, cultivation, irrigation, pest control and renovation/repair techniques to produce the optimum playing surface conditions. Conversely the failure to implement the correct management strategies will quickly result in a deteriorating surface, irrespective of how well a sportsfield is constructed.

Pre-establishment surface conditions

The preparation of a playing surface before establishment is critical because it is the best opportunity to set the standards and conditions for the sportsfield, as well as to improve on various problems and deficiencies that may exist with its initial construction. It will also determine the amount of work needed to maintain a good playing surface, and governs the inputs required to achieve this. If the effort is not made to prepare the surface properly before establishment then a less than desirable playing surface may eventuate, which cannot cope with high levels of usage or poor weather conditions. As a result it may be extremely expensive and labour intensive to repair and maintain the sportsfield in a good condition. Before work commences on preparing a surface, it is important to ascertain the problems and issues that exist, and may include some or all of the following: grass cover and turf composition, weed species and population, depth of thatch–rootmat, type and quality of topsoil, depth of topsoil, presence of any compacted layers, depth of water table, surface contours, evenness of surface, training practices and facilities, level of usage, surface and subsurface drainage, and irrigation.

Some of these factors are inherent to the site, however all of these factors will affect the amount of work needed to improve or prepare a good playing surface. There are several critical aspects in the preparation of a sporting surface for a particular use, and these include removal of weeds and unwanted plant material, removal of thatch and rootmat, break up of any soil layers, amendment of the soil to improve soil structure, soil cultivation to prepare a uniform soil profile, supply of sufficient and balanced nutrients to sustain healthy turf growth, grading, contouring and levelling to provide a smooth and even surface, preparation of a consolidated surface to avoid unwanted settling and an uneven surface, and provision of adequate soil moisture.

The first step in preparing an existing surface prior to reestablishment is to remove unwanted plant material. It is often possible to kill all plant material using a non-selective herbicide, such as glyphosate, but this will obviously depend on the plant species present. Plants such as *Poa annua* are prolific seeders and a large seedbank of a future potential *Poa annua* problem is likely to exist in the soil. The use of a non-selective herbicide may not resolve this problem and further expensive post- and pre-emergent control of *Poa annua* may be needed if insufficient control is achieved. Soil fumigation will ensure that any unwanted weed or grass seeds that exist on-site are destroyed, however this practice is generally uneconomical especially on large turf areas. The removal of the top 25 millimetres of soil from the surface can greatly reduce the seed bank population in the soil. Mouldboard ploughing is also a useful practice, providing that a sufficient depth of topsoil exists. This helps to bury the seed bank and unwanted plant material deep in the soil profile where it cannot germinate and re-establish. However soil types and depth of topsoil often limit this practice.

Many turf surfaces in need of renovation have a layer of thatch–rootmat which will affect both the quality of the playing surface and the behaviour of the turf environment if it remains. A small layer of thatch–rootmat can be mixed into the soil profile without any harmful effects. However where a thick layer of thatch exists the only practical solution is to remove this thatch and rootmat layer, either with a turf cutter or using a grader blade, before cultivating the soil. This will also help to reduce the weed and plant seed population. If the reconstruction or rehabilitation of an existing sportsfield is to take place this offers an excellent opportunity to modify the existing soil conditions. At this time both soil structure and soil fertility can be improved to provide good grass growing conditions which will be reflected by improved turf quality and vigour. It is important to test the soil to determine the nutrient status, pH and salt levels, and the soil physical characteristics (drainage, aggregate stability etc) to make adjustments according to these results. At construction, soil amendments such as gypsum, lime and organic matter can be incorporated through the profile.

The main reasons for cultivating are removing existing plant species, breaking up compacted layers, preparing a suitable seedbed, and levelling of the soil surface. Cultivation of the soil increases soil porosity, aeration and infiltration of water into the soil, and decreases runoff and erosion. The overall effect is to produce better conditions for plant growth. Though there are many benefits of cultivation, the over-cultivation of fine textured soils can destroy the soil structure and must be avoided. A seed bed of fine tilth is required for good soil–seed contact and should be as uniform as possible without any depressions. The soil must be levelled and graded to sufficiently consolidate to avoid unwanted settling and the creation of an uneven surface. Final levelling and consolidating needs to be conducted using low ground pressure equipment for a long period to ensure that the soils are evenly consolidated without being excessively compacted. A light roll will help to provide sufficient consolidation. Bulk density measurements can be undertaken to determine optimum consolidation and evenness of consolidation of the topsoil, while maintaining satisfactory soil conditions for turf growth.

Seed and vegetative establishment

Establishment of the surface includes several phases: preparing the seedbed for sowing or planting, seeding, sprigging, sodding and post-care management. Turf establishment can be achieved by seeding, sprigging or sodding, and there are a number of factors that will determine the method of establishment including time restraints and costs.

Seeding is the ideal way of establishing most turf areas because the turf manager has control of grass species, density and thatch levels. However, the greatest disadvantage is that seeding requires the longest establishment period and is more susceptible to washouts, wind erosion and climatic stress (e.g. heat, cold, drought). Some important points to consider to ensure a high quality turf are uniformity of seeding, and using the correct seed rate and seed depth. A high seeding rate may lead to the development of spindly, immature plants less able to cope with wear or environmental stress, whereas low seeding rates will provide greater opportunity for weeds to invade any thin or bare areas. Similiarly, while seed must be placed deep enough in the soil to protect it from disease, pests and weather, they must be shallow enough to be able to reach the surface on the seed reserves. Small seeds such as bentgrass need to be sown closer to the surface for successful establishment than ryegrass which is a large seed.

Sprigging is where plant material is used to vegetatively propagate and establish a new turf surface. This is commonly used for establishing stoloniferous warm-season grass species such as bermuda or couchgrass (*Cynodon dactylon*), St. Augustine grass (*Stenotaphum secundatum*), and other grass species that cannot be established from seed.

Sodding or turfing is the laying of mature turf that is taken from a sod nursery. Sod is available with a layer of soil or the soil can be removed which provides a clean, bare rooted turf. If the soil is left on the sod it is very important that it is compatible with the rootzone soil on which it is to be laid. On sand profiles, washed turf or turf grown on sand is strongly recommended. Cool-season turf often consists of a mixture of grasses with Kentucky bluegrass (*Poa pratensis*) as the base grass providing the sod strength due to its stoloniferous and rhizomatous growth. The bluegrass is most often then sown with perennial ryegrass or tall fescue. Warm-season species include couchgrass, kikuyugrass and St. Augustine grass (buffalograss). Sportsfields sodded with couchgrass have been ready for use within four weeks of laying and within six to eight weeks for cool-season grasses. Where the surface is used immediately after sodding, a deeper sod with a layer of soil must be used. This will minimise any transplant shock and ensure a stable surface that can be played on immediately.

Post establishment care

Newly sown turf requires special care for about four to six weeks after establishment. In this early stage the new turf requires frequent watering to promote root penetration and active growth. Ideally the soil should be kept moist until the seedlings are at least 2.5 centimetres high. Watering can then be tapered off, but avoid stressing the turf as this will delay the maturing process of the turf. Where sprigs or instant turf are planted, the soil must be kept wet for the first five to seven days while new roots are initiated. As the sprigs or sods become established the watering regime can be reduced to allow the soil to go through more regular wetting and drying patterns that will in turn stimulate deep root growth. Once the sprigs of warm-season grasses have established new roots and are relatively safe from desiccation, the drying out and heating of the surface stimulates stolon growth and increases the rate of establishment. The young grass must receive adequate fertilisation during this period for active growth and to promote a healthy, vigorous and mature turf. Once the grass has reached the second true leaf stage, regular applications of nitrogen and potassium are required. Prior to mowing, the newly established turf will benefit from a light roll. This will promote tillering and lateral growth by slightly crushing the growing points of the grass. This process will also improve surface stability.

The first mowing should be carried out when the new turfgrass plants are about 7.5 centimetres in height for cool-season grasses established from seed, and 2.5 centimetres for warm-season grasses established from sprigs. Sodded turf must be mown as soon as it can tolerate a mower without dislodging the turf. The cutting height on the sodded turf should be at the height of cut at which the sod was maintained prior to delivery. If possible, do not apply herbicides for weed control during the first three months when the grass is still young and tender and susceptible to chemical damage. If the seed bed has been poorly prepared and there is a high population of weed seeds, then a decision may have to be made to spray while the grasses are very young. It is a matter of balancing the detrimental effects of the herbicide versus the competition from the weeds which may smother the new grass. Most annual weeds will be controlled in the first mowing. Herbicides for broadleaf weed control should be delayed as long as possible because they can affect new root initiation in turf grasses. Minimise the usage on a newly established area for the first six months.

Playing surface standards and performance testing

The playing quality of a sportsfield is of paramount importance in order to enable a high standard of play to be achieved and to provide a high degree of player safety. A sports turf surface should be judged by how well it plays rather than by its overall appearance, turf cover and colour. These aesthetic characteristics are important but are not directly related to the playing quality of a surface. The playing quality is dependent upon the physical properties of both the immediate surface layer and the underlying material, that is soil and sward characteristics. The two main factors affecting playing quality are player–surface interactions and ball–surface interactions which are made up of several components including player–surface interactions of friction and traction, and hardness and resilience, and ball–surface interactions of ball bounce, rolling resistance, and friction and spin. There are certain objective measurements (i.e. ball rebound, ball roll, traction and hardness) that can be used which correlate with the players' perception. Bell *et al.* (1985) and Baker and Canaway (1993) provide an excellent review on these interactions and associated testing procedures.

Player–surface interactions

Friction and traction allow the necessary player movements without excessive slipping. Friction applies to smooth soled shoes and traction to shoes having studs, spikes or cleats which provide extra grip. Excessive friction/traction is undesirable as there is a risk of knee and ankle injuries and too little friction/traction is also undesirable as there will be increased slipping and falling. There are numerous test procedures used to measure surface friction and traction ranging from pendulum tests to measure translational friction, towed sledges to measure friction on artificial surfaces, the distance a trolley with a test foot slides after being released down a ramp, and the measurement of the torque required to cause slippage of a studded disc. The studded disc apparatus is widely used in the measurement of turf surfaces and consists of a 150 millimetre diameter disc with six studs similar to those found on a football or soccer boot and weighing 42 kilograms. It is dropped from a height of 300 millimetres and then rotated using a torque wrench until the turf fails. Traction measurements have ranged from 36 to 83 newton metres for various turfgrass species, and 32 to 41 newton metres for bare ground (Canaway, 1981).

The hardness of a surface is defined as the ratio of an applied force to the amount of surface deformation (i.e. stiffness), while resilience is a measure of the amount of energy returned to the player from the surface after impact as a proportion of the energy put in before impact. Both these factors are impc
falling and injury pote
cause jarring of limbs anu
increase the risk of injury fro
face that has low resilience c
fatigue. The Clegg Impact Soil Tes
measures the deceleration of a we
from a fixed height is the most com
used for testing surface hardness. In 1996 the American Society for Testing Materials (ASTM) adopted a standard procedure for determining the shock-attenuation characteristics of natural turfgrass surfaces using the CIT (ASTM F1702-96). Other test methods employed use the penetrometer, and the Stuttgart and Berlin Artificial Athlete, both of which measure surface deflection under a falling weight. The Clegg Impact Soil Tester has been used extensively on cricket pitches. Clegg (unpublished, 1982) used a 0.5 kilogram hammer released from a height of 300 millimetres at the Western Australia Cricket Association (WACA) Ground in Perth, Australia, during a three-day match and produced the following values as a guide for grounds preparation (Table 17.6).

Table 17.6 Clegg Impact Test (CIT) values on a cricket pitch at the West Australian Cricket Association (WACA) grounds.

Period of measurement	*Range of values (gravities)*
2 days prior to game	320–390
1 day prior to game	400–500
1st day of game	520–590
2nd day of game	640–730
3rd day of game	660–740
Day after game (pitch watered)	310–370

These results were related to soil moisture content and ball bounce and are similar to those observed by Lush (1985). In New Zealand, the penetrometer has been used as an aid to preparing cricket pitches by measuring soil hardness and relating this to soil moisture content and ball bounce (McAuliffe and Tuohy, 1987).

Ball-surface interactions

Ball bounce is important in many sports including soccer, Australian Rules and cricket. The ideal height of bounce varies for different sports with the consistency of bounce from one part of the field or pitch to the other being of most interest. Ball bounce resilience is used as a measure of bounce and is the ratio of the height the ball bounces to the height from which it is dropped. The actual drop height is not important provided that the rebound height is expressed as a percentage of the drop height, however a drop height of 3 metres has been found to be convenient and is widely used for soccer. Ball bounce resilience has often been shown to be highly correlated to surface hardness measurements. The type, condition and age of the ball can have an effect on ball bounce characteristics, especially with cricket balls. Stewart and Adams (1968) produced a scale which related ball bounce to cricket pitch pace. They dropped a cricket ball from a height of 4.88 metres and measured the rebound height (Table 17.7).

Table 17.7 Ball bounce resilience (BBR) and its relationship to cricket pitch pace.

Ball bounce resilience (%)	*Pitch pace*
> 15.6	very fast
13.0–15.6	fast
10.4–13.0	moderately fast
7.8–10.4	easy paced
< 7.8	slow

Canaway et al. (1990) developed performance standards for the playing quality of natural turf soccer pitches (Table 17.8). These were produced by analysing player responses shortly after they had played on the surface being measured. It was noted by the authors that these standards are somewhat subjective and liable to revision.

Table 17.8 Playing quality standard for soccer (adapted from Canaway *et al.*, 1990).

Quality	*Minima*	*Maxima*
Rebound resilience		
Preferred	20%	50%
Acceptable	15%	55%
Distance rolled		
Preferred	3 m	12 m
Acceptable	2 m	14 m
Traction		
Preferred	25 N m	—
Acceptable	20 N m	—
Surface hardness		
Preferred	20 gravities	80 gravities
Acceptable	10 gravities	100 gravities
Surface evenness		
Preferred	—	8 mm standard deviation
Acceptable	—	10 mm standard deviation

The rolling resistance is a significant factor in sports where the speed of the surface is important such as golf, bowls, hockey, cricket (outfields) and to a lesser extent soccer. It can be measured in terms of ball deceleration or the distance rolled by the ball. The two main measurement techniques involve either propelling a ball with a standard force or releasing a ball down a ramp and then measuring the distance the ball rolls. A more sophisticated apparatus involves the use of several infra-red timing gates set at standard intervals, which are used to measure the deceleration or the change in velocity. Any rolling measurements are usually conducted in opposing directions to compensate for any effects of slope, wind, etc. Golf and bowls are the only sports that have incorporated the measurement of ball roll into standards for rating greens for tournament or membership play. For example, the stimpmeter is used in golf for rating green speed, while in bowls various testing ramps have been used for timing the speed of bowling greens. The friction between a ball and surface is responsible for the variations in speed, direction and rate of rotation of a ball after contacting a surface. These are important properties in tennis, golf and cricket.

There are many complex interactions involved between horizontal velocity, spin, bounce and friction, and these are used by players to produce subtle variations in the pace and direction of the ball to deceive their opponent. Because of these complex interactions, the measurement of friction and spin has generally relied on video analysis or electronic detecting equipment.

Players' perception

If these tests are to be useful, the results must relate to the user's opinion of the surface so that the results can be interpreted in a meaningful way (Baker and Canaway, 1993). Canaway *et al.* (1990) used player questionnaires to interpret the results from the test measurements. They describe the players as being notoriously variable in their responses, and that a large sample is very important. The results of the questionnaires then provided acceptable limits for measurements related to ball rebound, ball roll, traction, hardness and surface evenness. In relation to interpreting the physical measurements in terms of injury potential, Baker and Canaway (1993) suggest that player questionnaires are not particularly useful. Even though there are numerous testing procedures available, there has been only minimal adoption of playing quality standards for particular sports. Test methods and performance requirements have been established for soccer pitches in the United Kingdom and standards based on performance have been derived for hockey, rugby and lawn bowls.

Surface preparation practices

Maintenance regimes

The standard of maintenance depends on the usage of the sportsfield and includes the following practices: fertilising, soil amendments, mowing, weed and insect control, topdressing, oversowing, soil renovation, thatch removal, irrigation scheduling, rolling, training practices, and line marking. The importance of fertilising is often overlooked in sportsfields. It is important to maintain good nutrient levels. The aim of any fertilising program is to provide sufficient, regular and even supplies of nutrients to the plant to avoid a weakening of the turf, and hence damage from wear and pests. A heavily used sportsfield will require more frequent fertilising than a low use sportsfield because of the need to be able to repair it from the wear it receives before irreparable damage occurs. Typical fertiliser requirements for various grades of sportsfields are listed in Table 17.9.

Table 17.9 Quantity of nutrients required per year

	N (*kg*)	*P* (*g*)	*K* (*g*)
High use sand-based sportsfield	6.0	800	3300
Low use sportsfield	1.6	520	590

Soil nutrient tests and leaf tissue tests must be conducted at least once a year on a sportsfield to determine any imbalances in the level of fertility. This will help to accurately formulate a suitable fertiliser program, and avoid unwanted problems or poor turf wearability and recovery which often occurs on sportsfields. Soil amendments are mainly used before the establishment of the turf area, however there may be an ongoing need to amend the soil with various materials, such as gypsum or organic matter, depending on the type and requirements of the topsoil and subsoil layers. Mowing determines the appearance, durability and health of the turf. It is important to mow little and often to help maintain a dense, healthy and vigorous turf, and avoid weakening of the sward. The cutting height depends on the intended use of the turf area and the grass species. In particular, it is important to avoid low mowing. Mowing heights less than the optimum for the grass species being maintained require greater inputs to maintain a good playing surface. Increased watering and fertilising are just some of the maintenance practices that must be used to compensate for low mowing or high levels of usage. Mowing height can affect the composition of the grass sward and depends on grass species, time of year, and type of sportsfield sport. Typical mowing heights for Australian rules football and soccer are 20 to 40 millimetres, cricket 15 to 30 millimetres and hockey 15 to 30 millimetres. The frequency of cut will depend on the growth rate of the plant and the time of year, but it is generally at least once a week during the cool seasons and at least twice a week during warm seasons.

Topdressing of sportsfields needs to be conducted for a number of reasons including maintaining a level surface, dilution of the thatch layer, maintaining a firm surface, and improving the soil conditions. Whether topdressing is necessary will depend on the situation, but it should be considered where surface levels are critical or the integrity of sand slit drainage is to be maintained. Topdressing can be conducted at any time during the year providing the turf is actively growing and it is not applied heavily. Another thought to consider with topdressing is the material used. Fine sand is likely to blow around and may become a hazard to the eyes of the players. Therefore an appropriate sand, or otherwise an appropriate time of topdressing, must be considered.

The main wear areas on a sportsfield caused by match play are the goal to goal line, especially the goal squares, centre half forward and centre square. The extent of the damage is influenced mainly by the number of games played and the prevailing weather conditions during each game. The main complication to the quality of the surface is when grounds are also used for training. Training is the single most destructive practice on a sportsfield because it is generally restricted to a couple of areas on the ground. Most of the damage caused near the pavilion and goals closest to the pavilion is due to training. While it is sometimes difficult to avoid damage from matches, it is important to minimise the damaged caused by training. The ideal situation is to remove training altogether from the sportsfield and utilise other facilities or areas of less importance or priority. If there is no alternative than to use the ground, then training must be spread as much as possible to spread the wear, to allow areas a chance to recover and to minimise the damage to the surface. To achieve this, training lights are needed around the entire ground, and discipline is also required. Commonsense must also prevail during periods of wet or inclement weather since greater damage will occur during wet weather.

Marking out for play

In sportsturf management, the final presentation of the playing surface is of great importance and can be improved, or alternatively marred, by the standard marking out procedures. Uniform and positive line marking is not only pleasing in appearance, but it is essential to all participants and spectators in order that the rules of the game can be correctly interpreted. The persistency of chemical materials varies and some are better suited to harder surfaces than turf. In general, an ideal marking material should be waterproof, quick drying, not easily rubbed off or likely to flake or powder. The materials available include whiting or chalk (the long term persistence of the material can encourage weeds, worms and disease due to the modification of the pH in the treated area), proprietary marking materials (often semi-permanent which compensates for higher cost), proprietary dry mine materials (dry aggregate combined with a border compound), emulsion paint, tape, sawdust, thread, and road line paints.

Management of turfgrass pest species

Healthy turf is less likely to be affected by pests, however the incidence of pests may be unavoidable due to climatic stresses. It is therefore important to monitor the turf for pest types and numbers, and to implement appropriate control programs once the pest reaches an unacceptable threshold. Weeds are likely to invade a weakly grassed or poor growing turf surface and less likely to invade a dense vigorous sward. Once a weed problem exists, it is important to identify the weed species so that the correct herbicide can be used to eradicate the problem without affecting the turf species involved. *Poa annua* is a major problem in sportsfields, and in some cases becomes the major grass species that must be managed accordingly. Sportsfields are susceptible to all disease and insect problems that occur in turf. The incidence of these problems obviously depends on the grass species present, the stress on the playing surface and climatic conditions. Grass mown higher on a sportsfield is less likely to be affected by diseases that occur on lower growing turf areas, such as golf and bowling greens. Nevertheless they can occur, given the right situation. The major disease problems occurring in sportsfields are red thread (*Laetisaria fuciformis*), rust (*Puccina* spp.) and some of the leaf blight (*Bipolaris* spp. and *Drechslera* spp.) diseases. It is difficult to justify treating disease affected sportsfields with fungicides unless it is a high priority sportsfield because of the large area and the cost involved. It can generally be controlled through the application of nitrogenous or balanced fertilisers as well as other management practices without excessive expense. The main insect problems in sportsfields are the scarab beetle larvae (*Heteronychus arator*) and cockchafer grubs (*Aphodius tasmaniae* and *Adoryphorous couloni*). These can cause serious loss of turf cover if left unchecked and will warrant chemical control once they reach sufficient levels to cause turf damage.

Restoration of the playing surface

Renovation includes any cultural practice that is used to improve the conditions of the playing surface and to restore the environment in which the plants are growing. Renovation practices will depend primarily on the inherent problems of the turf area and the types of problems which develop. The common problems that occur include change in surface levels, loss of grass cover (whether uniform or patchy), density, texture, grass species, turf composition, weeds, disease, pests, excessive build up of thatch, inadequate soil depth, soil depth, degree of soil compaction, root growth, poor growth, poor drainage, and root development. The role of renovation is to assist in resolving the problems that have developed. Some of the renovation practices employed may be similar to the maintenance practices for the sportsfield in question. However they are normally regarded as operations over and above general maintenance.

Renovation practices conducted on a sportsfield to remedy these problems include scarifying (to remove thatch, improve water penetration, improve surface levels, provide a good seed bed for oversowing and establishment); surface aeration which includes spiking, hollow tyning, coring, slicing (to improve surface aeration and water penetration and remove thatch); subsurface aeration including verti-drain, mini-mole plough, vibra-mole (to alleviate compaction and improve soil drainage and aeration); topdressing (to restore surface levels); and overseeding (to maintain or improve turf density and/or improve or change turf composition).

References

Adams, W.A. and R.J. Gibbs. 1989. 'The use of polypropylene fibres (VHAF) for the stabilisation of natural turf on sportsfields', in Takatoh, H. (ed.), *Proceedings of the 6th Inter. Turfgrass Res. Conf., Tokyo, Japan*, pp. 237–9.

Adams, W.A. and R.J. Gibbs. 1994. 'Natural turf for sports and amenity: science and practice', *CAB International*, Cambridge, 404 pp.

Adams, W.A., C. Tanavud and C.T. Springsguth. 1985. 'Factors affecting the stability of sportsturf rootzones', *Proc. 5th Inter. Turfgrass Res. Conf., France*.

ASTM-designation: F1702-96. 1996. *Standard test for shock attenuation characteristics of natural playing surface systems using lightweight apparatus*, American Society for Testing Materials, 65 pp.

Baker S.W. and P.M. Canaway. 1993. 'Concepts of playing quality: criteria and measurement', *Inter. Turfgrass Soc. Res. J.*, vol 7:172–81.

Baker, S.W. and R.J. Gibbs. 1989. Levels of use and playing quality of winter games pitches of difference construction types: case studies at Nottingham and Warrington, *J. Sports Turf Res. Inst.*, 65, 9–33.

Baker, S.W., A.R. Cole and S.L. Thornton. 1988. 'The effect of reinforcement materials on the performance of turf grown on soil and sand rootzones under simulated football-type wear', *J. Sports Turf Res. Inst.*, 64, 107–20.

Beard, J.B. 1973. *Turfgrass: science and culture*, Prentice Hall, Englewood Cliffs, N.J., 658 pp.

Beard, J.B. and S.I. Sifers. 1989. 'A randomly orientated, interlocking mesh element matrices system for sport turf root zone construction', *Proc. 6th. International Turfgrass Research Conf., France*, p. 253–7.

Bell M.J, S.W. Baker and P.M. Canaway. 1985. 'Playing quality of sports surfaces: a review', *J. Sports Turf Res. Inst.*, vol. 61 p. 26–45.

Canaway P.M. 1981. 'Wear tolerance of turfgrass species', *J. Sports Turf Res. Inst.*, 57:65–83.

Canaway, P.M. 1983. 'The effect of rootzone construction on the wear tolerance and playability of eight turfgrass species subjected to football-type wear', *J. Sports Turf. Res. Inst.*, 59, 107–23.

Canaway, P.M. 1985. 'Playing quality, construction and nutrition of sports turf', *Proc. 5th Inter. Turfgrass Res. Conf.*, 5:45–6.

Canaway, P.M., M.J. Bell, G. Holmes and S.W. Baker. 1990. 'Standard for the playing quality of natural turf for association football', in Schmidt, R.C., Hoerner, E.F., Milner, E.M. and Morehouse, C.A. (eds), *Natural and Artificial Playing Fields: Characteristics and Safety Features*, American Society for Testing and Materials, Philadelphia, pp. 24–47.

Canaway, P.M. and S.W. Baker. 1993. 'Soil and turf properties governing playing quality', *Inter. Turfgrass Soc. Res. J.*, 7:192–200.

Carrow, R.N. and G. Wiecko. 1981. 'Soil compaction and wear stresses on turfgrasses: Future research directions', *Proc. of 6th Inter. Turfgrass Res. Conf., Japan*.

Clegg, B. 1982. personal communication.

Cockerham, S.T., Gibeault, V.A. and Khan, R.A. 1993. 'Alteration of Sports Field Characteristics using Management Practices', *Proc. of 7th Int. Turfgrass Soc.*, 7, p. 182–91.

Daniel, W.H., R.P. Freeborg and M.J. Robey. 1974. 'Prescription athletic turf system', *Proc. 2nd Inter. Turfgrass Res. Conf.*, 2:277–80.

Davis, W.B. 1973. 'Sands and their place on the golf course' *California Turfgrass Culture*, 23(3) 20–4.

Evans, P. 1988. 'Species composition and management of winter sportsfields', *Proc. 1st Aust. Turf Res. Sem., Bermagui, NSW*.

Ferguson, M., L. Howard and M.E. Bloodworth. 1960. 'Specifications for method of putting green construction', *USGA Green Section Record*, 3 (5).

Gibbs, R.J. and S.W. Baker. 1989. 'Soil physical properties of winter games pitches of different construction types: case studies at Nottingham and Warrington', *J. Sports Turf Res. Inst.*, 65, 34–54.

Gibbs, R.J., W.A. Adams and S.W. Baker. 1993. 'Playing quality, performance and cost effectiveness of soccer pitches in the U.K.', *Inter. Turfgrass Soc. Res. J.*, 7: 212–21.

Gibbs, R.J., W.A. Adams and S.W. Baker. 1989. 'Factors affecting the surface stability of a sand rootzone', in Takatoh, H (ed.), *Proceedings 6th Inter. Turfgrass Res. Conf., Toyko, Japan*, pp. 189–91.

Hummel, N. 1993. 'Rationale for the revision of the U.S.G.A. greens construction specifications', *USGA Green Section Record*, Mar./Apr., 7–21.

Lemaire, F. and Bourgoin 1981. 'Consequences of artificial wear on the root systems of five turfgrass species grown in three substrates for purr wick system', *Proc. 4th Inter. Turfgrass Res. Conf., Canada*.

Lush, W.M. 1985. 'Objective assessment of turf cricket pitches using an impact hammer', *J. Sportsturf Res. Inst.*, 61:71–9.

McAuliffe, K.W. and M.P. Tuohy. 1987. 'Cricket wicket research carried out by Massey University during the 1986/87 season', *New Zealand Sportsturf Journal*, July, 3–9.

McIntyre, K. and B. Jakobsen. 1993. 'Design, construction and maintenance of sand-based facilities using the principle of the perched water table', *Turf Craft*, 31: 48–64.

Stewart, V.I and W.A. Adams. 1968. 'Soil factors affecting the control of pace on cricket pitches', *Proc. 1st Inter. Turfgrass Res. Conf., Harrogate*, 1:533–44.

CHAPTER 18

Prescription surface development: Racetrack management

I.H. CHIVERS, Racing Solutions Pty. Ltd., Sandringham, Victoria, Australia

Introduction

Racing of horses has been a part of the culture of many countries for centuries and is enjoyed by many millions of people across the world. It occurs in the tropics, the sub tropics and the temperate zones. It is followed ardently by people of all races, colours and creeds. It is undertaken in various forms from the highly formalised to the unstructured, with participants from the wealthiest to the poorest of the society. It is colourful, fast and exciting with the image of horses racing against one another at breakneck speed featuring in many people's minds as strongly as any athletic performance by highly trained men and/or women. Vast amounts of money are wagered on the outcome of horse races and prize monies for the winning owners can be very substantial. In Australia, the entire nation stops work for the Melbourne Cup on the first Tuesday in November each year.

The platform for the contest of racing is the racecourse. If the platform, or stage, is not adequate then the contest cannot be adequate. While the platform for racing may not be grass but could be soil, sand or other artificial surfaces, this chapter is addressing only grass or turf racing. This chapter considers the platform for racing and how the management of the turf on this surface is different from that of other grass playing surfaces. It will do so from a background of looking at the impact of the regulations under which racing operates, the construction of racecourses, the requirements for uniformity, the measurement of the quality of the racing surface, the soils and grasses used, the maintenance practices, and finally, the renovation practices.

At the start, let the obvious be stated: Horses are unlike humans in their weight, speed and number of feet, and as a consequence they impose a distinctly different pattern of wear on a turfgrass surface. By inference therefore, racecourse management is a distinctly different form of turf management to any other.

Regulations regarding the racing surface

Throughout the world thoroughbred racing occurs under the code of regulations known as 'The Rules of Racing' (Victoria Racing Club 1997). These rules are descriptive of the manner of riding, the equipment to be used on the horses, and other factors which have an impact on both the legitimacy of racing and its perception by members of the public. Strangely, however, there is no mention in 'The Rules' of the surface upon which the racing occurs. There is no restriction on the length, width, shape, camber or slope of a racecourse. According to the rules, anything goes. Imagine a soccer match being played on a pitch with widely varying dimensions, say, one goal located 30 metres off the centre of the pitch. Under 'The Rules' this would be quite acceptable. Similarly, there is no mention in 'The Rules' of the uniformity of the surface. Thus one side of the track can 'legally' be saturated and sloppy, while the other could be dry and hard. Furthermore, there is no prescription of the grass cover, with no indications of preference of type, uniformity, density, fertility or any other agronomic factor of turf.

This lack of rules within 'The Rules' occurs despite the fact that the presentation of the surface for racing will have a substantial impact on many factors. Not only will the condition and type of surface in use affect the performance of the horses, it also affects the fairness of the racing and the way it is perceived. All are vital to the continuing interest of devotees to the sport. The solution to this dilemma has been for the various racing authorities throughout the world to adopt 'Local Rules'. These rules set out the range of satisfactory conditions that are suitable and that provide uniform and safe racing conditions.

It is within this last phrase that the quality parameters for racing surfaces are set. The emphasis is on the safety of the surface for the conduct of race meetings, that is the ability to race a number of horses, at substantial speed, in close proximity to each other with minimal danger to either the riders or the horses. If the track surface is of an unusual shape or not uniform, then it will have an impact on the safety of the racing and may be determined as unacceptable by the local racing authorities. In terms of the agronomy, the track could have a surface made of thistles or even conifers without any impact on the suitability of the surface for racing. The vegetation type is not important, as long as it is safe.

Soil profiles for racing surfaces

For centuries horse racing has been undertaken on informal racecourses with little special preparation and few considerations made to the layout and construction of the racing surface. Racecourses were conventionally located on the outskirts of major towns on parcels of land that were of relatively low agricultural value. The racecourse was often seen as a way of making use of some poor land that otherwise would have no value. Thus many racecourses have an unfortunate legacy to carry, that is they are located on poor soil, sometimes old swamps, other times on shallow soils. It is indeed the exceptional racecourse that is located on deep, self mulching, sandy loam soils.

This circumstance held for many years with racecourses making the most of their 'local' soils. As the number of race meetings increased and the failings of the local soils showed up through becoming too soft when wet, or too hard when dry, or both, then the need for improvements increased. Commonly the first step towards improvement was the addition of drainage into the 'local' soil to alleviate wet areas and to allow racing to continue on days of inclement weather. Since water on the surface of a racecourse usually means the abandonment of the races on that day, the addition of drainage helped to guarantee the continuance of racing, and hence the fortunes of the club (Field, 1994).

The second step was usually the irrigation of the local soil, not only to promote grass growth during the dry times of the year, but also to reduce the hardness of unirrigated racing surfaces (Catrice, 1993). In dry climates this allowed for the conduct of race meetings on grass surfaces rather than on soil or sand surfaces, or in many cases not at all. Further refinements to the 'local' soil construction were achieved on many racecourses by the addition of sandy materials in an effort to improve both the porosity of the soil and the total soil depth, as well as reduce the surface hardness. In many cases this has proved successful, although a number of other cases exist where this approach proved completely unsuccessful with soil layering and reduced topsoil depth resulting. Greater improvements of 'local' soil profiles have been difficult once the processes of drainage, irrigation and topdressing have been undertaken. These processes have normally improved an unsuitable soil type to the maximum, practical degree possible.

The next major surface construction improvements came by the replacement of the entire root zone with a loamy sand soil. This process was undertaken, and is still undertaken, on racecourses which receive intensive usage and require the benefits of higher porosity, greater traffic carrying ability and a reduced tendency to compaction. Constructions such as these usually combined the practices of full irrigation and drainage systems. Thus a totally new soil profile was constructed in place of the 'local' soil with a loamy-sand topsoil growing medium placed above a drained base layer. The inclusion of a blinding layer in this construction pattern is not common and has, in some cases, caused problems with infiltration through to the drainage system. The effect of these constructions has been a large increase in the possible number of race meetings per year and a greater capacity to race throughout the full year. These tracks are characterised by being quite uniform, having drainage capacities of around 25 millimetres per hour, and providing fast and true racing.

Their limitations are seen firstly when heavy rainfall occurs on or soon before race meetings, and secondly when excessive usage causes considerable damage as the result of soil compaction. In the event that these limitations were restrictive on the racing club and its desired program of races with the problems of slow drainage under heavy rainfall conditions and soil compaction under constant use, some clubs have opted for full sand profiles. The sands used have usually had infiltration rates of around 500 to 1000 millimetres per hour. An inevitable consequence of using sands with infiltration rates such as these, is that they do not have sufficient stability to tolerate galloping and turf is easily removed by galloping horses.

To overcome this problem, various physical structures have been incorporated into the profile to add vertical and horizontal stability. These vary from mesh elements to monofilament fibres to expanding filaments to implanted fibres, with registered names of 'Netlon', 'Fibresand', 'Turfgrids' and 'Grassmaster' respectively (Adams, 1997; Baker, 1997). The intention of all of these products is to provide some interconnection within the soil profile that allows for substantial soil strength while retaining the rapid infiltration rates of the pure sands. The performances of these profiles under racing conditions have been highly variable, reflecting the somewhat experimental nature of the earlier installations. Much more is known at this time about the preferred construction materials and rates of inclusion of the various amendments, however little comparative work has been undertaken on their performance under galloping, with no published data available at the time of writing. Installations of this kind are very expensive and have been undertaken only at main metropolitan venues where the returns to the club or the industry can justify the additional cost of the inclusion of these materials.

Racecourse uniformity

The uniformity of a racing surface is vital to all involved. The horse needs uniform conditions in order to be able to gallop freely. In this context 'uniform conditions' refers to the grass coverage, the soil type, the moisture content, and the slope or camber of the track. If any, or all, of these factors is not highly uniform then horses will not only race poorly, but may well suffer injuries. For example, rapid changes in surface conditions, such as in going from hard to soft soil within one stride length, can cause serious injuries to horses if they are not prepared for, or expecting, the different surface softnesses.

Uniformity of grass cover

While uniformity of grass species for the full length and width of a racecourse is probably an ideal, it is practically impossible for all except the most highly managed racecourses. Agronomic factors may require the use of species blends in order to provide satisfactory conditions for year-round racing in many climates. Thus the sward composition of a racecourse usually becomes a blend of several grass types. Uniformity of grass cover more importantly, relates to the presence of gaps within the grass cover as they create problems of exposed soil, differing moisture conditions and potential slipping points. Zebarth and Sheard (1985) showed that a Bluegrass turfed surface had a greater ability (274 per cent) to withstand shearing than did bare soil. Under galloping then, bare soil is to be avoided as it will be more likely to cause slipping than does a fully grassed surface. Thus the intention of racecourse managers is to have a grass cover with no gaps and, preferably, a uniform texture and density.

Uniformity of soil type

Variability of soil type is often a major cause of poor uniformity of racing conditions. Many racecourses built with 'local' soil types have highly variable soil conditions. This is seen especially on those tracks where the topography is highly variable and incorporates hills and hollows. Under those conditions, it is common to find quite different soils on the hills in comparison to the hollows. Even on some relatively flat tracks the soil type can vary significantly. With all the variations in soil type come variations in the water holding and infiltration capacities. In turn, these reduce the uniformity of the racing surface and can create problems for galloping horses.

Uniformity of moisture content

The moisture content of the track should be similar, if not identical, for all parts of the track. Highly variable soil moisture conditions provide great variation in the racing conditions and lower performance of the racing surface. Drainage aims to reduce the excess water content of the worst performing areas, or indeed the entire racecourse, and can reduce the gross variations in soil moisture within the surface. Drainage cannot, however, overcome all of the differences in soil moisture content created by varying soil types; this can only be achieved by providing a uniform soil type for the entire racecourse.

On the other side of the relationship, irrigation of dry parts of the track can help to create more uniform conditions by matching the moisture content of the drier and wetter areas of the track. To achieve this requires a highly flexible irrigation system with great precision in application and careful monitoring of soil conditions. The unfortunate fact in regard to fixed irrigation systems is that they usually create more problems than they solve. The most common occurrence is that the uniformity of the water application is low and subject to substantial distortion under wind. Given that most racecourses are sited on large open areas, they are subject to considerable wind effects and consequently suffer from poor uniformity of irrigation application. The wide nature of many racecourses also takes many sprinklers beyond their range of uniform performance. On the other hand, these sprinkler systems have the advantage of providing water to the track quickly and with little effort.

Better uniformity is achieved by the use of movable sprinklers, and they occur in a number of forms from single sprinklers on stands, through arrays of sprinklers on aluminium lines, to computer operated mobile irrigators that span the entire racecourse. All of these systems have the advantage of being able to accommodate for wind and provide uniform applications across the entire track width. They have the disadvantages of being slow in application and requiring greater amounts of labour than the fixed systems.

The experience of many race clubs therefore, is that they need to install both a fixed system and a movable system. The fixed system is used to provide small amounts of water in a short period in order to overcome an immediate water requirement. The movable system is used to apply heavy waterings of greater uniformity when more time is available.

Uniformity of layout

A further reason for reconstruction of a racecourse is the need to rectify an unsuitable or dangerous layout. While not immediately relevant in a text on the agronomy of turfgrasses, it is necessary to note that sometimes, despite an exceptionally good grass cover, the layout of a racecourse requires reconstruction to occur. A poor shape of a curve or incorrect cambers on the turns or straights is often sufficient reason to reconstruct part or all of the racecourse.

Playing quality properties and measurement

The measurement of a racing surface is a matter of some conjecture around the world. For the majority of countries, the track conditions are measured by the local management by a number of means, such as a boot heel or an umbrella, and on the basis of the 'feel' of the track it is given a subjective rating. While these ratings vary from country to country, they usually have four or five stages from a 'soft' or 'heavy' rating to a 'hard' or 'fast' rating. In general, the ratings attempt to assess the amount of damage or disruption to the surface which a galloping horse will cause.

The aim of all of these ratings is to provide a guide to the state of the track at the time of racing. The aim then is to be able to compare the performance of two horses racing on different tracks over the same distance in order to judge which is faster. Thus horse A racing at venue M with a track rating of 'good', could be compared directly with horse B racing at venue N which also had a track rating of 'good'. If horse A recorded a time of 59 seconds for a 1000 metre race and horse B recorded a time of 60 seconds for the same distance, then it could be anticipated that horse A would beat horse B if they were to contest the same race. The problem, of course, is that the ratings are subjective and dependant upon the racecourse management at the various venues. As no two people are alike, no two racecourse managers are alike, and the ratings they give a track will occasionally vary.

Some associations encourage the use of basic objective measurement of the surface. The most common is the penetrometer, an apparatus that comprises a one kilogram weight dropped from a one metre height with a one square centimetre blunt end. It measures the amount by which the weight enters the surface. Its use follows initial assessments by Cheney *et al.* (1973), La Font (1985) and Zebarth and Sheard (1985) that correlated surface hardness with the speed of galloping. Penetrometers have been used in a number of countries but are used in only one, New Zealand, as the basis for track rating. In that country, track ratings are based entirely upon the penetrometer reading on the day of racing. The individual tracks are equalised against an industry average and a rating issued for that track on that day (Field *et al.*, 1993). While it works adequately in New Zealand, this system has not met with widespread acclaim and has not been adopted elsewhere. It is generally viewed that the penetrometer is not the appropriate piece of equipment as it does not measure any lateral strength of the soil–turf and is therefore solely one-directional, whereas the action of a horse's hoof on turf and soil applies in many directions.

The performance of the penetrometer has been assessed by Chivers (1996) by relating the race winning times on four Melbourne, Australia, tracks that had been measured with the penetrometer in accordance with the techniques of Field *et al.* (1993). Over a total of 1148 races conducted over an 11-year period, divided into two distances and three age groupings on four racecourses, the correlation coefficients in Table 18.1 were obtained between winning time and penetrometer reading.

Table 18.1 Correlation coefficients between winning time and penetromer reading (after Chivers, 1996).

Racecourse	*1000 m 2-y-old horses*	*1000 m 3-y-old horses*	*1000 m open age*	*1600 m 3-y-old horses*	*1600 m open age*
A	0.703	0.797	*	0.671	0.685
B	0.549	*	*	0.787	0.806
C	0.579	*	0.633	0.686	0.682
D	0.646	0.769	0.753	0.820	0.848
A,B,C & D combined	0.373	0.433	0.312	0.648	0.719

* indicates insufficient numbers for reasonable statistical evaluation with less than 50 races conducted over this distance on this racecourse for this age group of horses.

There was some relationship between the penetrometer reading and the winning time for individual tracks for all of the age groups and distances. However, it was not an extremely high correlation, and use of the penetrometer as a means of comparing racing performances was not warranted. Chivers (1996) calculated that the margins for error implied by these correlation coefficients would exceed the time differences caused by different penetrometer readings. The combined track data indicates a lower correlation between the value of the penetrometer and the winning time across all tracks. This serves to emphasise the work of Field *et al.* (1993) that an implement such as the penetrometer cannot be applied across all racecourses without calibration, but must be manipulated to provide a standardised reading relative to an industry average.

The Clegg Impact Hammer has been used on one American grass track to measure the surface hardness and to give assessments of the condition of the surface for racing (Sifers and Beard, 1992). This hammer measures the deceleration of a set weight dropped from a set height as it contacts the ground. The faster the deceleration, the harder is the surface. Sifers and Beard (1992) developed a range of performance criteria for the Santa Anita turf track in Los Angeles, USA, relating the measure of surface hardness obtained from the Clegg Impact Hammer to the likely galloping conditions. While not developing a rating system to indicate the likely racing conditions, it was found to be a useful guide to whether the surface was too soft or too hard and likely to cause injuries to thoroughbreds (Figure 18.1).

The action of the penetrometer and the Clegg hammer is to simulate a vertical impact of a hoof on a turf surface. While this may be adequate for a walking horse, it is not sufficient for a galloping horse where both horizontal and vertical forces are involved (Fredricson *et al.*, 1972). Furthermore, under normal rules of racing, the stipendiary stewards (who carry the ultimate responsibility for the rating of the track), view the vertical hardness of a surface as only an indirect method of measuring the surface. They consider that the most important measure is that of the amount of divoting that is caused by horses when galloping, that is the rating can only be determined accurately by conducting races on the

Figure 18.1 Operation of the Clegg Impact Hammer to assess surface hardness

surface (Gleeson, 1997). As a means of measuring the likely amount of divoting of a surface, the shear strength of a soil–turf may be as relevant as the hardness of the surface against vertical forces (Puckeridge, 1994). Measurements of this kind are possible using equipment such as a shear vane, although their relevance to thoroughbred racing is unknown. Any test such as this must closely simulate the movement of a hoof on the ground and apply similar forces (Baker and Cannaway, 1993).

A study on two Sydney, Australia, racecourses before and after a series of four race days was conducted by Puckeridge (1994). In this work Puckeridge determined the penetrometer values along and across the track, assessed damage to the surface, measured hoof imprints and undertook a series of tests on the soils of the two racecourses. He concluded from this work that 'research should be directed to designing an instrument that is similar in size to a horse's hoof and that measures both horizontal and vertical forces'. To date no further studies have been undertaken.

Surface materials specifications

Of relevance in this section are specifications on only the more recent constructions, as 'local' soils and earlier constructions used the material that was available on the site of the racecourse or nearby. More recent constructions have normally involved the use of sandy materials variously described as sand, packing sand or loamy sand. The pure sand constructions have required the use of structural materials of the like of 'Netlon' or 'Turfgrids' to provide sufficient structure to support galloping horses.

The loamy-sand soils used in other constructions have had sufficient structure on their own to be able to successfully conduct racing and do not require the use of these structural materials. That is not to say that 'Turfgrids', 'Netlon', 'Fibreturf', or 'Grassmaster' would not operate effectively to enhance the performance of those soils, it is simply that their use is not essential in this case, as it is with very open sands. Some typical soil analyses from recent track constructions are given in Table 18.2.

The loamy sand materials, while not meeting the USGA guidelines, produce highly satisfactory racing surfaces and, with the correct construction and subsoil drainage, are capable of tolerating heavy racing within a medium rainfall zone. The pure or open sands usually fall inside the USGA envelope and are capable, with the addition of structural items in the soil profile, of tolerating heavy racing demands within all rainfall zones.

Table 18.2 Typical analyses of some recent track constructions.

Particle description	*Particle size range (mm)*	*Pure 'open' sand (% retained basis)*	*Loamy sand (% retained basis)*
Gravel	more than 2.0	0	2
Very coarse sand	2.0–1.0	1	6
Coarse sand	1.0–0.5	11	16
Medium sand	0.5–0.25	38	34
Fine sand	0.25–0.10	47	31
Very fine sand	0.10–0.05	3	7
Clay plus silt	less than 0.05	0	4
Saturated infiltration rate	mm/h	550	60

In terms of the reconstruction of a racecourse surface, the choice of which type of sand to use is commonly determined by considerations of the rainfall intensity and volume, the amount of racing to be conducted per year and its intensity, the capacity to move the running rail and hence spread the wear, and the budget allowance of the club for both construction and maintenance.

Turfgrass selection criteria

Across the world almost all commonly used turfgrasses have been used, and are being used, on racing surfaces. There are no edicts from racing bodies on the particular choice of grasses for racing and each club is allowed to make its own choice of preferred grass or grasses. The ideal turfgrass for racing would have the following characteristics: remains green and vigorous throughout the racing season, produces a medium density coverage of leaves at a 75 to 100 millimetre mowing height, produces strong rhizomes that are capable of both rapidly filling into divots and holding a sandy soil profile together, and produces little thatch.

Within any environment, there are very few, if any, grasses that meet these criteria. Either the best choice is dormant or near dormant for part of the racing season, or it is too thatch forming, or other factors mitigate against its use. An example could be the use of bermudagrass (*Cynodon dactylon*) on many North American tracks. This grass often passes into a dormancy period for some tracks and almost always produces a turf cover of too high a density and too heavy a thatch layer. The former factor limits its vigour for repair and the latter factor tends to promote slipperiness of a surface. A further example would be the use of perennial ryegrass (*Lolium perenne*) on southern Australian and New Zealand racecourses. This grass has good winter vigour and will remain green throughout the year, it produces little thatch and a medium density coverage at the nominated mowing height. It does not, however, produce rhizomes. This poses two limitations. First, it means that it is killed when removed and a new plant needs to be sown in its place. The new plant needs time to establish itself, or it will be too easily removed if used again. Thus the running rail needs to be moved to allow time for repair, and necessarily the track needs to be capable of providing adequate rail shifts. Secondly, without rhizomes, it has reduced ability to provide structural support to tolerate galloping horses so that many sandy surfaces with a pure ryegrass surface are easily cut up by horse racing.

As stated above there are very few environments that allow the choice of one grass to provide the racing surface and that meet all of the criteria given above. The solution is usually one of compromise within both the racing program (avoiding the worst times of year) and the track (with a blend of grass types to permit racing over a longer time span). Combinations such as ryegrass and bluegrass (*Poa pratensis*) in cooler environments, or kikuyu (*Pennisetum clandestinum*) and ryegrass (the latter as an oversown grass) in the warmer transition zones, allow for reasonably good racing surfaces for most of the year. The prospects for improvement in this circumstance look positive on several fronts. The recent release of tall fescue (*Festuca arundinacea*) cultivars with strong rhizome growth offers potential for the temperate zone (Anon, 1995). At the same time, work on improved vigour and winter hardiness of zoysiagrass (*Zoysia* spp.) may help with many tropical zone tracks (Fukuoka, 1990; 1992). On a broader front, the science of plant breeding is rapidly changing with transgenic grasses likely to be available for broad-scale turf use within the next few years (Asano *et al.*, 1997; Lee *et al.*, 1997). While most of the breeding work in this area is directed to the larger golf industry, some beneficial side-effects are likely to be seen for racing.

Surface preparation

While all racing surfaces differ, and therefore the preparation for racing varies from track to track, there are a number of practices that are commonly undertaken in preparing for race meetings.

Mowing

Front-deck mowers are now accepted as the best machines for mowing tracks as they are lightweight, use turf type tyres, leave no wheel impressions of uncut grass and can operate when the grass cover is slightly wet. These mowers have replaced the trailed rotary mowers on most racecourses for all of the above reasons. Some, though not many, racecourses mow with sets of reel mowers. While these mowers tend to cut cleaner than rotary-type mowers, they are difficult to adjust to the higher mowing heights of a racecourse. The grass on most racecourses is maintained between meetings at heights of around 75 to 100 millimetres. The grass is then allowed to grow slightly beyond this height for racing. Commonly, racing is then conducted on surfaces with around 100 to 120 millimetres of grass cover. The collection of clippings off the racecourse is becoming a part of the normal management of major metropolitan venues as managers attempt to reduce the accumulation of thatch within the surface. It has the further benefit of reducing the tendency of the clippings to form a layer of wet thatch around the crowns of the turfgrass plants. Normally the collection is undertaken using a vacuum attachment on the deck of the front-mounted mower, but it can also be carried out as a separate operation using a powerful vacuum or sweeper-vacuum.

Irrigation

As mentioned earlier, the uniformity of application of irrigation water is vital in establishing a uniform racing surface. Planning for the correct irrigation practice to establish this uniformity in the time leading up to the race meeting is the most important practice within the manager's control. The usual practice, where suitable systems are available, is to water heavily (say 20 to 25 millimetres per application) in the weeks leading up to a meeting with a movable sprinkler system, as this allows for deep and uniform irrigation. Within the few days prior to the meeting, irrigation will reduce in order to maintain adequate moisture within the profile without having it saturated. Often this is carried out with a fixed, in-ground system applying around 8 to 10 millimetres of water per application. Naturally, a pattern such as this needs to be adapted for the soil type, the irrigation system, the evapotranspiration rate and other factors that apply on the individual racecourse.

Fertilisation

Adequate fertility, particularly nitrogen, needs to be maintained within the grasses to allow for strong leaf and root growth, and for rapid recovery after damage. Excessive nitrogen levels will promote soft leaves and high thatch production. Normal turf maintenance practices apply in the management of fertility levels on racecourses, as they do elsewhere.

Rolling

It is not uncommon to use a flat-bed roller of around 3 to 5 tonne weight to help restore surface levels after racing and sometimes to assist in making final surface preparations. The rolling after a race meeting usually follows the patching of the track with a divot repair mix of soil and (where appropriate) seed. The roller is used to smooth the repairs into the general shape of the track. These rollers are used in the knowledge that they will cause compaction of the surface, but are required in order to help smooth a surface and, once again, provide more uniform conditions. The use of these rollers is compensated for in the maintenance program by the regular use of surface aeration equipment.

Movement of railing

One of the saviours of racecourse managers is the ability to move the area of wear across the racing surface by the use of rail movements. Since most racing occurs in a narrow band to the outside of the rail, it is possible to have several movements of the rail across a racecourse on fresh turf. This presumes that the racecourse has adequate width and cambers to permit safe rail shifts. Normally, under the close racing patterns of Australia and New Zealand, a four metre rail movement will be sufficient to present fresh, unused turf for the next race meeting. The area that benefits the most from this ability to shift the rail is the home turn, where most of the damage to the turf occurs as the horses are accelerating and most closely concentrated. On non-uniform racecourses, rail shifts can cause problems and are to be avoided, however on most courses the clever use of the rail can allow for a greater number of meetings to be conducted on the same racing surface.

Weed, disease and insect control

Normal turf maintenance practices are required to keep pests under control. No specific 'racing-related' pests occur with the normal range of pests of all types being prevalent on racing surfaces. With the mowing height being considerably higher than those of golf greens, bowling greens and most athletics fields, there is less tendency towards the pests that strike close cut turf. There is also less ability to detect many pests because of the long grass height, however the pests do occur and the remedies are usually the same.

Renovation practices

The renovation of racing surfaces occurs for many reasons, some peculiar to racing, others common to all turfgrass surfaces. In common with all other turf surfaces, racecourses develop thatch, become compacted, lose grass cover and suffer pest problems. All of these problems need to be addressed and where this is not possible as part of the normal maintenance, it needs to be undertaken as part of a renovation program. A specific concern that arises on a racecourse is related to the requirement for uniformity of a surface. As the galloping horse is very disruptive to the surface, the ideal, smooth levels of the soil are quickly lost with many divots of soil taken from the surface and soil pushed outwards to mound the sides of the divots. Under a 75 to 100 millimetre grass cover these divots are often disguised and not corrected by the patching of the track.

In some cases this can lead to quite severe pitting of the soil surface. Frequently it is not seen from the surface as the grass cover hides the indentations. On stiffer soils, these indentations can be very sharp and hold a hard edge upon which horses can stumble. Renovation is required to firstly detect these indentations and secondly to overcome them. Usually this is achieved by removing much of the grass cover by mowing to around 25 millimetres, dethatching the surface, and then topdressing with sand and oversowing where necessary.

A second concern is deep compaction. Under the weight of a horse, compaction occurs at greater depth than normal for a golf course or athletics venue. The only effective means available at this time to remove this compaction is by use of deep-tyne aeration using machinery such as the Verti-Drain or Terra-Spike machines. These implements are large, PTO driven machines that are capable of decompacting a soil to a depth of 250 millimetres. They are slow to operate but are able to fracture a soil profile and increase its air content quite substantially. The machines cause little surface damage and are generally highly beneficial. However they can cause problems if used prior to rainfall in cooler environments, as the soil is able to absorb much more water than normal and often can reach saturation through the entire depth of the soil profile. These soils then are likely to remain fully saturated for some time as the transpiration rate is insufficient to remove significant water from the profile. A general guideline for the use of these machines in cool environments is to avoid their use prior to the wet season, to use them prior to the dry season, and not to use them in the four weeks leading up to a race meeting if at all possible.

A typical annual renovation program for a medium- to high-use track would be: immediately after racing mow close to the surface and remove the clippings, dethatch or scarify the surface to a 15 millimetre depth at 80 millimetre spacing between blades, topdress low areas with an appropriate loamy sand, use Verti-Drain or Terra Spike machines at 250 millimetre spacing, oversow if required, and apply fertiliser as necessary. Irrigation should follow if necessary in order to re-establish a strong grass cover on the smoother surface.

References

Adams, W.A. 1997. 'The effect of 'Fibermaster' fibres on the stability and other properties of sand rootzones', *International Turfgrass Society Research J.*, 8: 15–26.

Anon. 1995. *Ceres 'Torpedo' Tall Fescue in Turf in New Zealand—a management guide*, Pyne Gould Guiness Ltd., Christchurch, NZ.

Asano, Y., Ito, Y., Fukami, M., Morifuji, M. and Fujiie, A. 1997. 'Production of herbicide resistant transgenic creeping bent plants', *International Turfgrass Society Research J.*, 8: 261–7.

Baker, S.W. and Cannaway, P.M. 1993. 'Concepts of playing quality: criteria and measurement', *International Turfgrass Society Research J.*, 7: 172–81.

Baker, S.W. 1997. 'The reinforcement of turfgrass areas using plastics and other synthetic materials: a review', *International Turfgrass Society Research J.*, 8: 3–13.

Catrice H. 1993. 'Turf for horse racing: conflicting requirements of the turfgrass plant and the racing horse', *International Turfgrass Society Research J.*, 7: 517–21.

Cheney, J.A., Shen, C.K. and Wheat, J.D. 1973. 'Relationship of racetrack surface to lameness in the thoroughbred racehorse', *American J. of Veterinary Research*, 34 : 1285–90.

Chivers, I.H. 1996. *Assessment of the value of the penetrometer*, unpublished data.

Field, T.R.O., Murphy, J.W. and Lovejoy, P.J. 1993. 'Penetrometric assessment of the playability of coarse turf', *International Turfgrass Society Research J.*, 7 : 512–16.

Field, T.R.O. 1994. 'Horse Racing Tracks' in *Natural Turf for Sport and Amenity: Science and Practice*, by W.A. Adams and R.J. Gibbs, *CAB International*, pp. 329–53.

Fredricson, I., Dreveno, S., Moen, K., Dandanell, R. and Andersson, B. 1972. 'Equine joint kinematics and co-ordination: photogrammetric methods involving high-speed cinematography', *Acta. Vet. Scan. Supplementum*, no. 37 : 1–136.

Fukuoka, H. 1990. 'Breeding *Zoysia* spp.', *J. of Japanese Society of Turfgrass Science,* 17 : 185–90.

Fukuoka, H. 1992. 'Inter and intraspecific variations in ecotypes of *Zoysia* spp.', *Japan J. of Breeding,* vol. 42 :282–3.

Gleeson, D. 1997. Chief Stipendiary Steward for the Victoria Racing Club, Melbourne, Victoria, Australia, personal communication.

Lafont, J. 1985. 'How to master the irrigation of horse race turfs: the use of a penetrometer', in *Proc. 5th International Turfgrass Research Conference,* 5: 851–6.

Lee, L., Laramore, C., Hartman, C.L., Yang, L., Funk, C.R., Grande, J., Murphy, J.A., Johnston, S.A., Majek, B.A., Turner, N.E. and Day, P.R. 1997. 'Field evaluation of herbicide resistance in transgenic *Agrostis stolonifera* and inheritance in the progeny', *International Turfgrass Society Research J.,* 8: 337–44.

Puckeridge, D.M. 1994. Assessing the surface hardness of turf racecourses and defining impact forces and wear caused by the galloping thoroughbred, Thesis for Graduate Diploma in Agricultural Science (Turf Management), Faculty of Agriculture, University of Sydney, 130 pp.

Sifers, S.I. and Beard, J.B. 1992. 'Monitoring surface hardness', *Grounds Maintenance,* 27 : 60, 62, 90.

Victoria Racing Club. 1997. *Australian Rules of Racing,* Victoria Racing Club, Flemington, Victoria, 300 pp.

Zebarth, B.J. and Sheard, R.W. 1985. 'Impact and shear resistance of turfgrass racing surfaces for thoroughbreds', *American J. of Veterinary Research,* 46 : 778–84.

CHAPTER 19

Prescription surface development: Cricket wicket management

KEITH W. McAULIFFE, New Zealand Sports Turf Institute, Palmerston North, New Zealand

Inroduction

Cricket is one sport where the surface has a major influence on the outcome of a match. The result of a contest can rest upon who wins the toss and how the quality of the pitch changes over the course of the match. It is little wonder that home sides have been accused of doctoring the pitch to suit their teams' strengths. For the curator, cricket pitch preparation offers as great a challenge as any job in the sports turf industry. The needs of the cricketer are exact and there are many variables to consider when preparing a pitch. The pitch can also be an 'easy target' for an out-of-form player (or team). It is not surprising to learn that pitches in virtually every country where cricket is played have come under criticism at some stage.

Requirements of the player, cricket administrator and curator

What does the player and cricket administrator want from a cricket ground in the way of a pitch, outfield and practice block? The required qualities of the pitch will depend on the type (e.g. limited overs versus five-day test) and level (e.g. club versus international) of cricket played. A good pitch should offer: a playable surface under a range of weather conditions such that cancellation or delays are minimised; uniform and consistent bounce; bounce of good height; acceptable pace with the 'ball coming on to the bat' as this is deemed to allow a wider array of shots to be played, resulting in more entertaining cricket; and an acceptable degree of sideways movement, the extent of which will depend on the type of cricket played (and the stage within a match). For test and three- or four-day matches sideways movement is accepted early in the match. For one-day cricket sideways movement is undesirable at all stages.

In a match of longer duration, e.g. a three- to five-day match, the pitch should slowly deteriorate to take spin later in the match. However wear should not be excessive such that bowlers' run-ups and foot mark bare out badly to create problems for the bowler. The surface should be safe to bat on, with no alarming variation in bounce. There must be a sufficient number of strips for the entire season and/or individual strips with good recovery potential so that they can be re-used throughout the season. The surface must offer good performance over the off-season (in a dual use ground situation). The outfield should have the size and dimensions which conform to international regulations, a surface gradient which is neither too steep nor variable, an even and level surface so that the ball rolls truly and that players do not incur injury (e.g. twisted ankles), adequate (generally fast) speed of ball roll, and good drainage such that interference to play is minimised. The practice block should offer playing performance that mirrors that of the main block, and sufficient size and number of strips to cater for the season's needs.

What does the turf curator want from a cricket pitch? In addition to meeting the players' and administrators' requirements of the game (as discussed above), the turf manager will also want to have a surface which is easy to grow turf in, easy to prepare, easy to dry, remains stable (doesn't disintegrate) over the course of a match, and easy to renovate.

Performance specifications for a first class cricket ground

A typical match day performance specification for the pitch and outfield in a first class New Zealand cricket ground may be as follows. The outfield shall be of a smoothness and levelness that results in no alarming deviation or bounce of a shot played along the surface. The outfield levelness at any point (minimum of nine readings taken at random) should be such that deviations below a 3 metre straight edge do not exceed 15 millimetres. Match day outfield speed should be such that the distance of ball roll from a standard inclined plane should be a minimum of 4.5 metres (reading is an average of locations as outlined in NZTCI Publication No. 93/IR). The outfield surface should exhibit a dense turf grass cover, with less than 5 per cent bare ground and less than 5 per cent weed.

The pitch should appear even along its entire length with no noticeable holes. Height deviation along the pitch shall not exceed 7 millimetres at any point as measured by a 3 metre straight edge. The turf cover should look even along the entire strip. There will be no noticeable weed content in the strip. Turf density on match day will ensure an even balance of exposed soil and plant cover. The percentage of bare soil should not exceed 30 per cent. Grass height on match day shall be no greater than 5 millimetres. The strip shall be mown to its playing height no later than 24 hours prior to play. All clippings from mowing will be collected. The pitch shall be marked out with a white acrylic paint in accordance with the rules of the game. Lines should be no less than 12.5 millimetres and no greater than 18 millimetres in width. The markings should be clearly visible from a distance of 22 metres.

The surface hardness on match day should be such that mean Clegg hammer readings (using a 0.5 kilogram weight) exceed 40 units (400 gravities) at both ends of the strip (10 readings taken at each end). The standard deviation of Clegg hammer readings when expressed as a percentage of the mean must not exceed 15 per cent. The surface of the pitch shall not visibly disintegrate or break up upon ball contact. Throughout the match the pitch shall provide even and acceptable bounce at both ends, as judged by umpires and other officials. A medium-paced delivery pitched approximately 3 metres in front of the popping crease should bounce to at least stump height. Pitch pace should be consistent with the ball 'coming on to the bat'. For one-day cricket there shall be no disconcerting sideways movement off the pitch. Cores taken from the strip on or around match day must not show layering or cleavage planes within the surface 75 millimetres. In addition, there shall be no noticeable (greater than 2 millimetres) thatch layer. The pitch and adjacent strips must be able to be completely and effectively covered in the advent that rainfall is forecast over the 72 hours leading up to the match.

There are many variables that come together to determine the performance of a pitch (McAuliffe and Tuohy, 1987). Some of these variables pertain to the pitch direct and others, such as bowler action or type and age of the ball, are not. Cricket balls, albeit hard objects, have elastic properties. A ball will deform on impact with the pitch surface, with the amount of energy retained by the ball determining its rebound qualities. A hard pitch with good base density will allow the energy in the ball to be preserved, whereas a soft pitch will absorb ball energy. Other features that affect the amount of ball energy absorbed include presence of layering (any layering will absorb ball energy and reduce bounce), poor base density, moisture content, and clay type (some clays get harder than others). Preparation has an over-riding influence on how a pitch plays (Cameron-Lee and McAuliffe, 1989).

Common problems experienced with cricket pitches around the world

Press reports illustrate that virtually every cricketing country has had problems with their pitches from time to time (often when the home side is not performing well). These problems include uneven and variable bounce, low bounce, slow pace, excessive rate of wear, excessive cracking, excessive turn, poor or uneven turf cover, and slow to dry.

Uneven and variable bounce

Batsmen require consistency of bounce. Variable bounce or pace will create uncertainty in their mind. Inconsistent bounce and pace can be due to a number of causes including: an unlevel surface (which in turn can be due to a number of causes, such as poor renovation); variation in grass density and species composition (for example a mature bunch-type grass can give rise to soft spots in a pitch at the point of the plant crown); variation in soil density along the pitch; variation in soil moisture content (a soft surface contributes to variable bounce and excessive sideways movement); and irregularities such as the development of cracks or sub-surface layers (Figure 19.1).

Figure 19.1 An example of layer in a cricket pitch, Eden Park No. 2, New Patamahoe clay, New Zealand

Low bounce

If bounce is consistently low, players will be restricted in the range of shots able to be played. Players will become 'front foot' players. Low bounce can be due to a number of things, including insufficient base compaction, excessive soil moisture (pitch too wet), layering or cleavage planes, and excessive thatch content.

Slow pace

Pitches which are on the slow side limit the range of shots able to be played by batsmen, particularly shots square of the wicket. Most countries currently appear to be seeking increased pitch pace in order to encourage more exciting cricket, particularly for one-day cricket. Lack of pace in a pitch can be due to a number of things including: inadequate preparation (inadequate base density and/or surface hardness); type of soil used (for example, the soil may have limited potential to be 'shined' or it could have a high sand content); or method of surface preparation (such as amount of grass cover or amendments used to improve pace through shine). Recent research carried out by the New Zealand Sports Turf Institute illustrates the importance of surface friction and condition of the ball on pitch pace (Gibbs *et al.*, 1995).

Excessively rapid surface deterioration

A pitch that does not last well because it dusts up or breaks up can be a nightmare to play on and to manage. There are a number of reasons why a pitch can wear excessively including: poor binding clay at the surface (resulting in rapid scuffing up and excessive bowler run up wear); inadequate turf cover and root binding (exposing the soil to more rapid wear); some clay soils tend to self mulch and disintegrate upon drying, causing premature breaking up; poor repair techniques; buried layers (e.g. thatch overlaid by recent topdressing); and inadequate surface preparation (rolling).

Excessive cracking

Cracking in a pitch is not necessarily a bad thing, provided cracks are not excessively wide, the crack edges remain stable, cracks are continuous to depth and the prisms or blocks do not become misaligned. Cracking can result from clay type used (as clays with montmorillonite as the dominant clay mineral have swelling and shrinking tendencies); the balance of nutrients within the clay (for example the application of lime to a pitch can alter the behaviour of the soil); a soil that is under-compacted in the base is more likely to crack excessively at the surface; layered soils will generally crack prematurely (with the cracks extending only as deep as the cleavage plane); and soils that have received treatments such as slicing or vibra-moling can crack along the planes of weakness created.

Excessive spin and/or seam

Spin (turn) is an important facet of the game of cricket. A pitch should offer turn to the bowler, especially over the latter stages of a match. However, excessive turn too early in the match is undesirable. Seam, or sideways movement of the ball off the pitch (in contrast to swing), is desirable in the early stages of a three- to five-day match. For one-day or limited overs cricket the preference is for minimal seam throughout the match. Excessive turn on a new pitch could be due to high levels of thatch which allows the ball to sink in and grip; clays with a high sand content producing a rough surface; an overly soft (perhaps too wet) surface, giving the flighted delivery a chance to sink in and grip; an overly-dry surface, which has become rough and abrasive; and poor preparation of the surface, giving rise to variable density of both soil and grass cover.

Thin or patchy turf cover

An absence of a turf cover on a pitch is again not always a bad thing, and many top quality pitches have been prepared by virtually rolling out mud. But for a balanced game of cricket a uniform cover of turf should be present. Poor or patchy turf density could be due to poor renovation techniques; wrong turf species present (e.g. some species have lower tolerance to mowing and heat stress); excessively harsh preparation (e.g. over-rolling); soil being overly-compacted, limiting turf root development and/or water entry; and insufficient grooming (verti-cutting, brushing) of the turf resulting in clumpy growth.

Difficulty in drying (and getting the surface sufficiently hard)

A pitch surface that is slow to dry out to match-day hardness will be difficult to manage. Extras, such as raised covers, will become more critical, especially in a high rainfall environment or where evapotranspiration rates are lower. Excessive softness or stickiness of a pitch could be due to: climate, for example preparing pitches early in the season when drying is slow; inadequate resources available (e.g. no covers); poor shaping of the block, such that surrounding water runs onto the block; and type of clay used, with some clays taking longer to dry and harden up than others,

It must be appreciated from the outset that countries vary greatly in their natural resources, particularly climate and available soil types. Climate is over-riding with respect to what character of pitch is produced. Climate will dictate how quickly a soil can be dried, whether couchgrass (bermudagrass) can be used, the requirement for covers, and so on. One could argue that with modern technology, and use of raised covers, climate is now controllable by management (Figure 19.2).

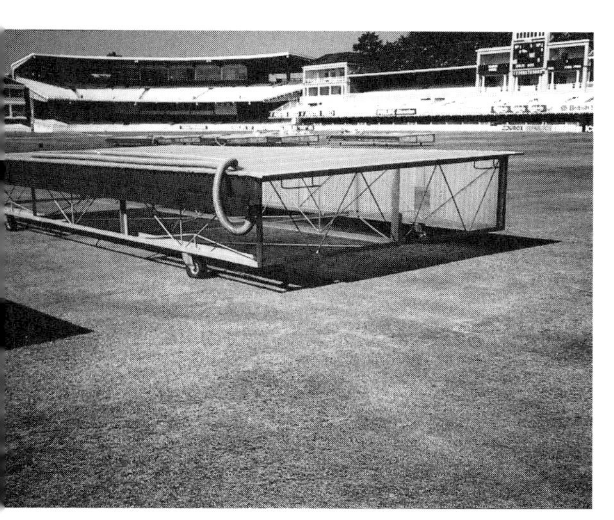

Figure 19.2 An example of a raised cover system for a wicket

The warmer temperatures and more predictable drying conditions (in places like Australia and India) allow heavier clays to be used. In wetter or cooler places, using a high clay content soil may be less practical without the use of quality raised covers. Furthermore, couch can only be grown in certain climatic zones of the world. For swelling/shrinking clays, couch is a great asset to help hold the pitch together during a match. Another consideration when comparing countries is the use made of the venue in the off-season. In places like New Zealand and Australia, cricket stadiums are used for sports such as rugby and soccer in the winter (McAuliffe, 1985). This places added requirements on grounds maintenance. Perhaps there is merit in individual countries and even individual provinces expressing their own unique characteristics through the way their pitch plays? Having all pitches around the world perform similarly may not necessarily be in the best interests of the game.

The role of soil type and turf in a cricket pitch

The selection of the clay soil is the single most important decision in wicket block construction. The ideal soil should offer: plasticity which allows moulding and compaction; hardness upon drying to provide good bounce, pace and wear capability; good binding and stability such that the soil does not break up, crumble or dust prematurely; the ability to recover from compaction and allow regeneration of structure; ability to be 'shined' and finished; good turf growth; ease of drying; and affordability and continuity of supply.

Clay particles are very small (< 0.002 millimetre in size). The chemical makeup of the clay mineral (i.e. the dominant ion, such as Al^{3+} or Si^{4+}) will determine its properties. Three common clay types used in cricket pitches are:

- Kaolinite. Because of strong hydrogen bonding, kaolinitic clays remain strongly bound together when wet, with the result that the soil doesn't swell or shrink.

- Illite. This clay mineral contains potassium (K^+) ions which help strong binding to limit the extent of swelling and shrinking.
- Montmorillonite. These clay minerals are also called smectites. They have poor interlayer binding, so tend to absorb water and swell when wetted and shrink when dried. Montmorillonite clays can potentially become the hardest of the three groups. They tend to develop cracks readily and are less prone to dusting up.

In addition to clay percentage and mineralogy, the soil selection process should also consider the sand and organic matter contents in a sample. The soil type used should not have excessive amounts of coarse sand, gravel or limestone, as this can enhance ball wear and spin. Soils used for cricket in Australia and South Africa are invariably heavy clays, commonly selected from lower horizons of a 'black earth' soil group. These soils are likely to contain more than 50 per cent clay; have montmorillonite as the dominant clay mineral; are plastic when wet; produce a brick-like surface when dry; swell on wetting and shrink on drying; be free from excessive sand and large aggregates; and have total dissolved salt levels less than 500 ppm (Figure 19.3) (Table 19.1).

Figure 19.3 Self mulching cricket wicket clay

Table 19.1 Properties of clays used at some major Australian cricket grounds (adapted from Lush *et al.*, 1985).

Ground	*Melbourne Cricket Ground*	*Western Australian Cricket Ground*	*Sydney Cricket Ground*
Clay %	53	82	52
Silt %	21	6	22
Fine sand %	20	6	15
Coarse sand %	1	2	7
Organic matter %	5	2	6
Clay mineralogy	smectite kaolinite	smectite illite	

There are practical ways of testing soil suitability for use in a cricket pitch. One of the standard tests is the 'motty' or ASSB test for soil binding strength (Stewart and Adams, 1969). In addition to binding strength, an examination should also be carried out on the swelling/shrinking properties, ease of drying, 'shine' capability and ability to grow grass. Testing should be mandatory before any soil is used. Experience has demonstrated that it is practical to mix soils for improved pitch performance. In some instances a 'strong' clay soil may be mixed with a lighter clay in order to make the former more manageable. If this practice is carried out it is important to get complete mixing of the two materials. A common practice in New Zealand is to introduce a veneer of an improved clay soil over the old soil layer. In this instance it is critical to get a good binding between the two materials, by thoroughly grooving and cleaning up the old surface before applying the topdressing.

To some it may seem strange that every effort is made to grow turf on a cricket pitch only for the turf cover to be decimated during the final preparation. The grass plant does, however, fulfil a number of important functions including its roots aiding in the deep and even drying of the pitch. Without turf cover, water removal will be by surface evaporation, which can create steep moisture gradients and bad surface cracking (Kirkman *et al.*, 1989). Roots, rhizomes and even rolled-in surface vegetation assist in holding a pitch together. Turf helps to rejuvenate the soil (e.g. provide drainage pores) over the course of time. The turf cover affects the performance of the pitch, including influencing pace, minimising damage (wear) on the ball and providing seam.

The 'ideal' turf species to use should recover well after close mowing or be adapted to the climate to grow when needed, and have good resistance to wear. The standard turf species used on pitches in warm regions is couch or bermudagrass (*Cydonon dactylon*). The desirable strains of couch have relative fineness in both leaf and stem, and low growth habit. A major advantage of couch is the ability to produce rhizomes. Rhizomes add to the tensile strength of a soil, helping to hold soil aggregates together. In cool regions of the world perennial ryegrass (*Lolium perenne*) is the main turf species used in wicket blocks. Ryegrass appears to best handle the stresses associated with preparing and playing a match (e.g. rolling). Furthermore, ryegrass germinates and establishes relatively rapidly. In some soil types and circumstances it may be possible to produce 'natural turf' pitches of adequate quality which are devoid of turf cover. Techniques such as rolling in clippings could be used to part-compensate for the lack of turf.

Construction of a wicket block

There are many theories around the world on how to construct a cricket wicket block. Historically, it was considered necessary to use a considerable depth of clay with appropriate underlying layers of materials. Today, many of our top grounds are built with a relatively shallow depth of clay soil laid over a firm base of free-draining material. Although specifications for the 'ideal' wicket block construction will vary from region to region, depending on characteristics such as climate, soils available and level of cricket played, some general construction guidelines can be provided.

Location of the wicket block

In most parts of the world the wicket table should run in a north–south direction to avoid problems of sun in batsmens' eyes later in the day. The square is generally placed in the middle of the ground, with the boundary ideally no further than 70 metres and not less than 45 metres from the corner of the wicket table. The minimum recommended distance from the stumps to the nearest boundary is 50 metres.Where the ground is used for multi-purposes (e.g. other sports play over the surface), it would be preferable to construct the wicket block between fields (Figure 19.4).

Gradient on the wicket table

In most cases the wicket table should be slightly elevated over the surrounding outfield. Some wicket tables are crowned in two or more directions, others tilted to fall in the general direction of the field (Figure 19.5). The objective of having a wicket table slightly elevated is to minimise run-off from surrounds onto the wicket block, which could interfere with play. Falls of 1:100 or less are common and are barely noticeable. The maximum gradient accepted for first grade cricket would be in the order of 1:80.

Size of wicket block

The number of strips on a wicket table will mainly depend on the projected usage of the ground and the number of matches to be played. The norm would be five strips per wicket block. Each strip measures approximately 3 metres wide, the standard recommendation is to use 3.04 metres per strip. In some countries a drainage or buffer strip is included between each of the playing strips (this practice is seldom used in New Zealand). A cricket strip is 20.12 metres between stumps. Most blocks are at least 25 metres in length, which caters for the last couple of delivery strides of the bowler.

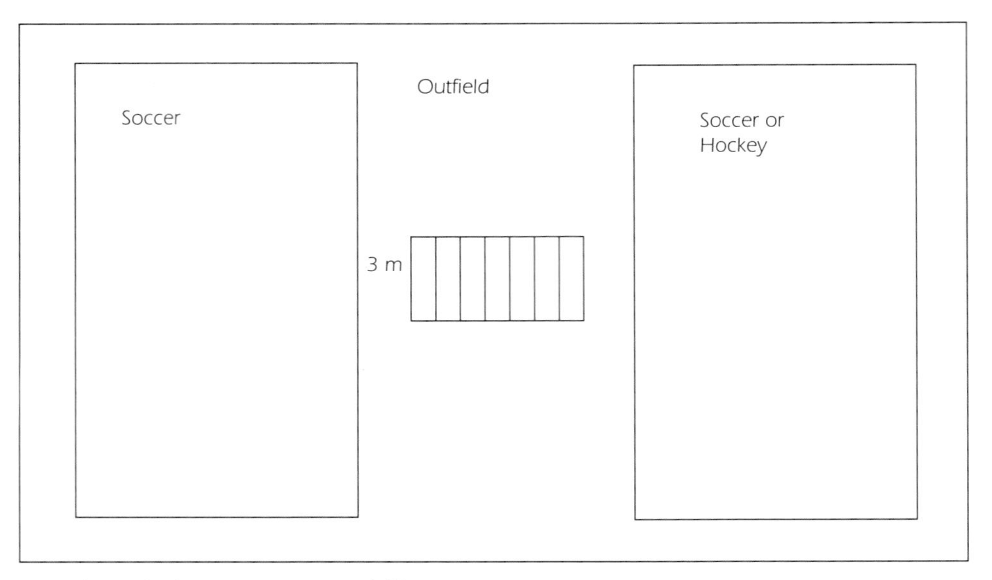

Figure 19.4 Cricket wicket between winter sports fields

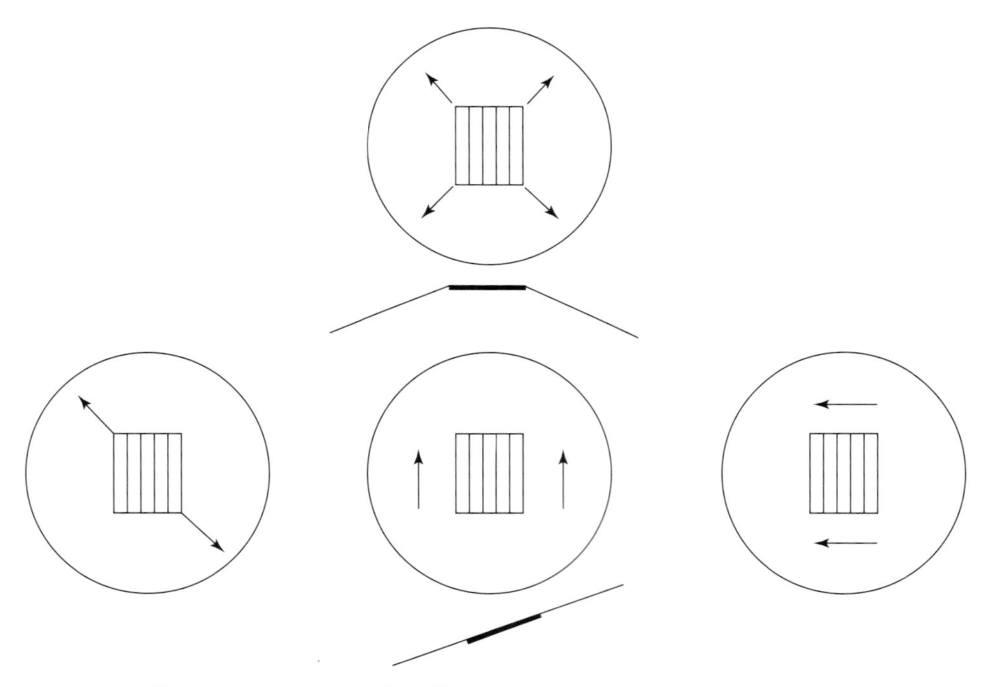

Figure 19.5 Different gradients on the wicket table

The wicket block profile

Each site will have different design requirements. At some sites it will be possible to establish a block simply and cheaply by building up a quality imported clay layer over the native soil. This approach could be used where the subsoil is free draining and stable. A typical profile construction using a reconstructed sub-base is shown in Figure 19.6.

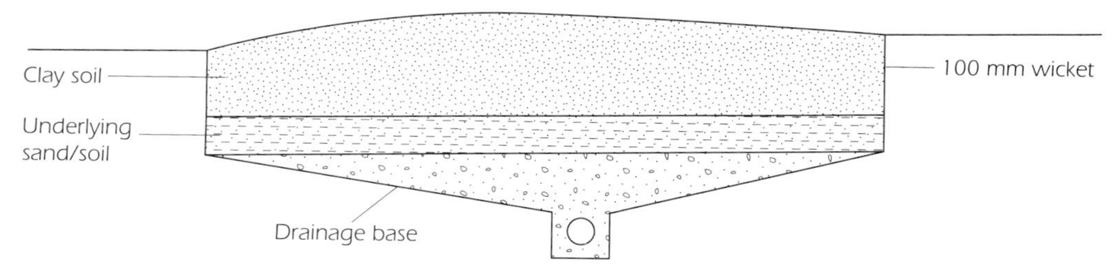

Figure 19.6 Typical profile using a reconstructed sub-base

Note that to create gradient on a block, it is best to shape the base to mirror the finished surface shape. Do not create a slope by varying the depth of clay (or any other profile layer) during construction.

Under drainage

Pipe drainage will generally be needed to the sides and below the block to provide an outlet for any surplus water that moves through the clay or comes up from underneath. Ideally any pipe drains installed should be positioned between strips and/or around the perimeter of the block. Note that since most clays used for wicket blocks have a very low permeability, especially once rolled, very little (if any) downward seepage of water through the system can be expected

Base drainage and sand layers

The drainage layer and intermediate sand layers are there to provide a well-drained and stable base. Choice of base materials should be such that the drainage and sand layers are both easily consolidated, yet remain permeable after compaction. Typically angular-shaped particles are preferred. A 100 millimetre depth of 5 to 10 millimetre diameter chip would be suitable for the drainage layer. A 100 millimetre depth of 0.25 to 0.5 millimetre diameter angular or sharp sand would typically be used for the sand layer. Sand should be completely consolidated, preferably by wetting then using a vibrating roller, during the laying process.

Clay soil layer

The clay soil used for the final stages should be crushed to pass through a 15 millimetre sieve. It should be free of stone, roots and other impurities. Clay depths in the range of 100 to 200 millimetres are typically used. When using the upper end of this range, it is important to get good consolidation of the clay base from the outset by applying the clay in layers and consolidating each layer well. Deeper clay (or clay–soil) depths are preferred in drier areas and/or where matches of longer duration are played. The clay soil should be placed in layers of no more than 80 millimetres (unconsolidated depth). Each layer must be well consolidated before adding the next layer, and a harrow device or similar used to loosen the underlying clay before adding the next layer. The final surface layer of clay should be accurately laid to level. Bear in mind that there will be further consolidation when the soil is rolled. One accepted method of levelling a wicket block is to use screeding boards and rails, although more frequent use is being made of laser-controlled, tractor-mounted levelling gear.

Turf establishment

The method of turf establishment will depend on the grass type grown. Where couch or bermudagrass is to be used, lay thoroughly washed turf, sprigs or stolons of an approved single strain. Washed sod is used by some curators to give rapid re-establishment of turf cover. When using washed sod it is imperative to get all foreign soil off the roots. It should also be expected that intensive verti-cutting will be required to thin out the turf mat if using washed sod. Thus verticutting will be particularly important if fresh soil is to be topdressed. When sprigging the wicket block, plant sprigs on a grid pattern or in horizontal slots at approximately 50 to 100 millimetres apart. After planting the surface should be light-rolled. If stolonising, disperse stolons over the surface and use a light roller to press plants into the soil. The soil is best kept moist in order to aid the bedding-in of stolons.

It may be necessary to lay hessian or scrim over the stolons before pressing them in with the roller. With cool-season grasses (e.g ryegrass), seeding is the most common method of turf establishment. Seeding can be done either by using a specialist seed drill or by thoroughly cultivating the surface and broadcast seeding. Regardless of the turf establishment method used, it will be critical to maintain good soil moisture control over the first few weeks until the new turf has rooted. Soil stabiliser products, such as hessian or shade cloth, are commonly used by curators to aid germination and to protect the surface.

Practice block

Ideally the practice block conditions should mirror those of the main block, since players need to practice in similar conditions to what will be experienced during the match. Practice strips can be constructed the full length with clay, but commonly the construction is limited to half length in size. In this instance it is often a good idea to apply good binding clay to the bowler's delivery crease area in order to improve surface stability. The practice block layout should also bear in mind safety and practicality issues, such as providing opportunity for cross-rolling and allowing opportunity for wicket keeping practice.

Pitch preparation

The starting point for any curator preparing a cricket pitch is to establish goals. The desired outcome will vary according to the standard and type of cricket played (e.g. test versus limited overs match). Pitch preparation techniques will be venue specific, hinging on things such as localised climate, soil type and philosophies of the curator. To derive a standard recipe for pitch preparation could be likened to the medical profession offering one cure for all ailments! Some generalisations and guideline pitch preparation techniques can be derived however.

Typical guideline pitch preparation for a representative match in New Zealand

Choose the strip approximately 3 to 4 weeks prior to the match. Mow the turf short, to about 5 millimetres, when the surface is dry. Use a comb on the mower or stiff brush to stand up and remove all dead or flat grass. Carefully inspect the surface to check for marks such as sprig marks. Carefully fill any depressions by hand with a light dressing of dry clay soil, taking care to avoid burying the turf. It is important to ensure good soil to soil contact is achieved with any topdressing of a cricket pitch. Apply a light dressing of nitrogen and potassium fertiliser to stimulate growth. Keep the strip well-watered over the next week or two. Ensure the moisture is kept to an even depth. Keep the turf mown at around 6 to 8 millimetres. Avoid mowing in the heat of the day to minimise stress on the grass. It is highly beneficial if good compaction has been achieved by the start of preparation (pre-season rolling is considered of great benefit in this respect).

The key goals over the initial week or two of preparation are to keep the turf alive, to gradually thin out the turf density, and to slowly bring up soil density and hardness. Avoid over-stressing the turf cover too early in the preparation through rolling. Throughout the full preparation period good control of moisture is necessary. Covers should be on hand from at least seven days out from the match in case of rainfall. It may be necessary to light water in order to slow down the drying of the surface if the weather is hot. Start rolling when the surface is relatively dry and firm, yet the soil underneath is still plasticine-like. Look to compact the base first then work through to the surface. Heavier rollers can be used as the soil begins to firm up. Aim to maintain a balance between keeping the grass alive and achieving a good level of compaction. Soil sampling and core assessment should be the basis of deciding when and how much to roll.

Approaching the last week, the base should be well compacted and there should be even moisture throughout the profile. Never let the top become too dry. The turf cover can be rolled into the soil two or three days out from the match. Do this by re-wetting the top few centimetres of soil prior to rolling. By the final few days there should be an even grass cover, with some soil visible through the turf. The day before play cut the surface to playing height, say 3 to 4 millimetres. Give the wicket a very light watering and immediately speed roll to shine the surface. If the pitch is excessively green it may be necessary to continue rolling in the heat of the day or to use some other method to bruise the turf and remove greenness. Before preparation starts, the strip should possess the basic ingredients of adequate levelness, good base compaction, high moisture content to depth, and uniform and healthy turf cover.

Watering

Most first class grounds use a combination of automatic pop-up sprinklers (for overall watering of the block) and hand-moved sprinklers or soak hoses (for watering individual strips). Hand watering is also used generally for light watering approaching match day. The aim of the pre-preparation watering should be to thoroughly soak the soil profile to near saturation.

There can be real difficulty in getting water penetration to depth in a pitch due to the low rate of infiltration, especially after the soil has swollen or the surface has become sealed through rolling or algal slime build-up. Duration of watering will be a function of the initial moisture content and the rate of soil water infiltration. For example, if the infiltration rate is one millimetre per hour and there is a deficit of 30 millimetres, it will be theoretically necessary to keep water applied to the surface for up to 30 hours. Pulsed applications and mist-type sprinkler systems are options to achieve this. Light watering during the preparation may be needed in order to slow down the rate of surface drying, keep the turf cover alive and healthy, or to re-wet the surface for a final surface shining/rolling. It is important to ensure that the moisture content is kept relatively uniform throughout the profile during the entire pitch preparation process. Steep moisture gradients can create problems.

Rolling

Rolling is probably the single most important operation in pitch preparation. Rolling of a cricket pitch can be done to increase the level of compaction and surface hardness, assist with improving surface levelness or to provide a surface finish (shine). In non-swelling clays that have become compacted through the course of heavy rolling, it may be necessary to use devices such as the hand fork, verti-drain or hollow tyne corer to aid deep root development over post-season. Any drill holes in the pitch must be thoroughly filled with topdressing soil.

Promoting recovery of turf

If turf grass is slow or there is insufficient turf density in the wicket block, chances are that the main reason is limited nutrition. Topdressing with a nitrogen-based fertiliser will be required as and when the turf needs a boost. Excessive turf growth in the off-season should be avoided in order to minimise accumulation of thatch.

General

In order to determine renovation needs, the curator should collect and analyse soil cores from the wicket block (Wells and McAuliffe, 1990). Core samples will show whether roots are struggling to penetrate, if there is excessive thatch accumulation or layering, and so on. If cores show that there is a major problem in the wicket block (e.g. a deep layer), then a major renovation operation is likely to be necessary. On the other hand, if there is no observable reason to renovate, then why bother? A thorough clean up of the surface (e.g. by brushing), coupled with a light soiling, may be all that is needed when the surface is well-grassed and unblemished.

When using a heavy roller on the pitch some ridging at the edge of the roller may show up. These ridges should get progressively smaller and less noticeable as rolling proceeds. Rolling in the early part of the season should be carried out in several directions, cross ways as well as lengthwise. This will reduce the wave action effect that can sometimes occur with pitches that are not perfectly flat. Be careful not to over-roll. Stop rolling if at any stage the surface starts to become sticky (with soil sticking to the roller) or if the plants start to show undue premature stress. Wait until the soil has dried out (or the turf vigour recovered) and then repeat the rolling process. A point worth noting about the swelling-type clay soils is that they contract and consolidate as they dry. This contraction can mean that very high soil densities are achievable in such soils. Rolling for shine is carried out over the last few days of pitch preparation. The objective is to lightly sprinkle the surface then speed roll in order to shine the soil or turf cover.

With respect to roller weights, most grounds are likely to only have one, perhaps two rollers. There is often the chance to alter the roller weight by increasing the amount of water ballast or by attaching removable cast iron weights. The compacting effect of a roller does not depend on its dead weight alone. The roller diameter and roller width as well as roller weight must be considered when determining the ground pressure and effectiveness of rolling. For example, a roller of smaller diameter can exert a greater pressure than a larger diameter roller of equal weight. The effectiveness of rolling is determined as roller weight, in relation to roller width × roller diameter. Categories of roller weights include the light roller, up to 500 kilograms, the medium roller, 500 kilograms to 1.5 tonnes, and the heavy roller, from 1.5 tonnes up. Most top grounds in New Zealand now have access to, and occasionally use, rollers of between 4 to 10 tonnes in weight.

Rolling for compaction is all about using the right weight and size of roller at the right soil moisture content in order to break down and mould 'loose' soil aggregates to form a dense stable mass. The optimal moisture content when rolling for compaction is around 80 per cent pore saturation. Figure 19.7 can be used to determine the soil moisture content this 80 per cent pore saturation figure represents once the soil bulk density is known. In most instances the lowest weight of roller (< 1 tonne) should be used at the outset, with increasing roller weight being used as the soil dries and density improves. Experience in New Zealand has shown that early season (pre-season) rolling when the soil is at optimal moisture content has achieved excellent base compaction. Rolling when the soil is too dry (notably when the soil is in crumbly and not plasticine-like condition) is unlikely to bring about improvement in soil density, and may only serve to place the turf under undue stress.

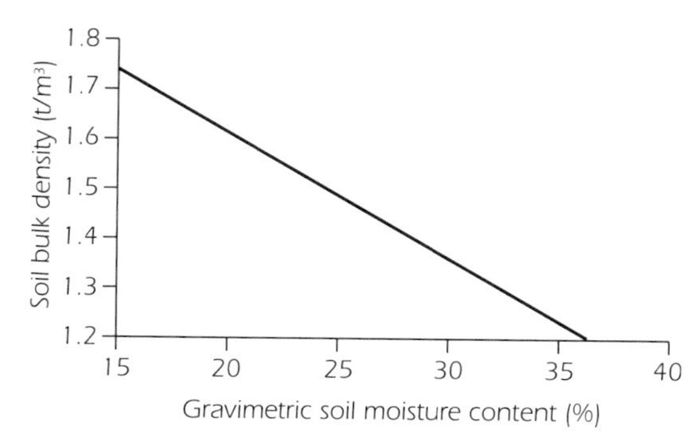

Figure 19.7 Standard curve showing the soil moisture content at which 80% pore saturation occurs for a range of soil bulk density levels (after Weaver, 1994)

Mowing and grooming

Various approaches are taken by curators around the world with respect to mowing, particularly bench and match day mowing heights. Typically a bench height cut would be around 10 millimetres, with a gradual reduction over the course of preparation down to as low as 3 millimetres playing height. An alternative approach is to retain the bench mowing height until a day or two out from the game before cutting down to match height. Progressive lowering of cutting height is likely to place the grass under less stress and give the grass opportunity to form new growth at the lower cutting height. Pitches cut down progressively generally stay greener for longer and recover more quickly after use. It may be necessary to keep match day mowing height higher than optimum if the surface is uneven. Close mowing under such conditions can lead to scalping of ridges and patchy cover. In such cases it is often beneficial to roll in the turf or to carefully brush and bruise hollows in order to provide a uniform-looking surface.

The scarifier (groover) is an important tool in pitch preparation. It is used to stand up flat (prostrate) grass prior to mowing, break up and thin out dense plant crowns, control excessive organic matter (plant stems, roots and mat), and provide good binding between newly-applied topdressing soil and the underlying soil.

Pitch preparation should generally aim to keep the mat or turf cover from being too dense on match day. Accordingly it is often necessary to thin out the turf cover and stand up growth using verti-cutting or brushing equipment. Verti-cutting and/or stiff brushing will also help break up any dense crowns of older grasses, which could otherwise produce soft spots, resulting in variable bounce. By match day, turf mowing and grooming should be such that the soil should be visible through the turf cover. Exceptions to this would be where a weak-binding soil (which easily dusts) has been used for topdressing.

Patching

Bowlers run-ups need to be repaired at the end of each day's play and after a match. A procedure for patching follows. Prior to returning fill ensure the edges of the area to be filled are stable; a hammer to firm the edges could be used for this. When the hole is ready, lightly water to ensure the spot is dampened. Mix up a batch of fresh soil and water to form a stiff plasticine consistency. Additives, such as cement or PVA glue, have been used to improve the rate of binding of soil. Press in, then ram the material to firm. Use a hessian cover over the soft fill if it is to be rolled. Apply clippings (preferably dry) if the area is to be played on.

Marking

Marking can be done with a cheap acrylic paint. Be careful not to use paint with an algicide content as this may kill off the turf. Line markings need to be easily distinguishable from a distance. Marking out may be achieved with the use of a paint brush and a specially constructed marking frame.

Use of covers

All first grade cricket blocks should have access to covers, ideally both raised covers and flat covers. The raised covers would be used during the preparation, while the flats would be used just prior to and during the match. Flat covers laid just prior to or during a match must be fully waterproof and properly laid out. Any tracking of water could result in seepage beneath the cover and wet spots. Hessian is usually placed under flat covers to avoid condensation and wet spots. A pitch dries out via the processes of evaporation (free water loss from the soil surface) and transpiration (loss via the plant). Moisture can move through the soil as drainage (in response to gravity), by capillary action (blotting paper effect) or by vapour transfer (in response to temperature gradients). It is the last mechanism that is responsible for 'sweating'. During the day any moisture vapour reaching the soil surface will evaporate to the atmosphere. At night, under covers, any moisture vapour escaping from the soil will condense on the underside of a flat cover. Although this volume of water is not great in comparison to what is lost during the day, it can cause problems if it is returned to the pitch surface in localised places. Scrim or hessian is often placed under flat covers to catch any condensate.

Selection of strips

Strip allocation should ideally be done at the start of the season. The allocation should aim to minimise interference to preparation. One possible order of use for a six strip block is: 1-5-3-6-2-4-.

Renovation of cricket pitches

Before setting out on pre- or post-season renovation it is important to clearly define the program's goals. The renovation program can be planned once it is known what is needed to bring the block up to standard. Renovation goals could include controlling organic matter build-up; over-coming a layering problem; establishing or re-establishing turf cover; establishing or re-establishing surface levelness; loosening the soil at the end of the season to assist water and root penetration; and fertilising to stimulate grass growth in conjunction with raising the mowing height. In general, major renovation work should be targeted for the post-season, to allow the off-season for recovery of turf cover and/or consolidation (Evans, 1991).

Controlling organic matter build-up

Excessive thatch is a big problem in cricket pitches. Experience shows that efforts to control thatch are often insufficient. It is important to have a clean soil surface leading into the season and certainly at the time when any topdressing is applied. Pre-season renovation often involves intensive grooving or dethatching to remove mat that has built up in the off-season. In this respect, management of a block receiving little or no winter use should aim to avoid excessive growth rate and accumulation of organic material (e.g. hold back on the fertiliser). In cases where the level of thatch accumulation is excessive, or too deep to be reached by standard grooving equipment, taking the entire top layer off and reconstructing may be the best solution.

Overcoming layering

In most cases layering in a block is best corrected by thoroughly re-working the soil down to and slightly below the offending layer. If the problem lies within the surface 30 millimetres, it should be possible to overcome the problem by intensive (possibly up to eight passes) grooving. For deeper-set problems, deep ripping using tyne harrows or similar, or even a complete re-build, may be necessary.

Re-establishing turf cover

At the end of the season (or at the end of each match) badly worn areas need to be worked and filled prior to re-turfing. Broadcast seed ryegrass in cool-season areas or sprig stolons of bermuda grass in warmer areas on bare spots that develop during the season. Where there is already a reasonable turf base, direct drilling is the main method of re-establishing a ryegrass turf cover at the end of the season. Turfing or stolonising are the most common ways of re-establishing bermuda grass. Germination and early growth of re-sown turf can be helped through using shade cloth or scrim to control moisture and temperature. Shade cloth also helps protect the surface against washouts or raindrop impact and sealing.

Re-establishing levelness

If the block has been used in the off-season for other sports, it is likely that damage will have been done to the surface. Renovation of this damage will involve thoroughly cleaning the surface of all dead plant material to expose the soil, lightly loosening the soil surface, then applying fresh topdressing soil and screeding to level. If the depth of fresh topdressing soil to apply is considerable, ensure the soil is well consolidated (by foot heeling or similar) as it is applied. Pitches that are not used in the off-season are unlikely to need much in the way of re-levelling. Many of the cricket clays used around the world are 'self mulching', which enables the natural regeneration of soil structure upon soil drying. Self mulching clays form their own structure upon wetting and drying cycles.

References

Burdett, L. 1986. *Turf cricket pitches,* South Aust. Cricket Assn.

Cameron-Lee, S.P. and McAuliffe, K.W. 1989. 'Principles of pitch preparation', *NZ Turf Management J.,* 3 (1):15–19.

Evans, R.D.C. 1991. *Cricket grounds. The evolution, maintenance and construction of natural turf cricket tables and outfields,* The Sports Turf Research Institute.

Gibbs, R.J., McAuliffe, K.W. and Wilkins, B. 1995. 'Development of a method for the measurement of cricket pitch pace', *Proc. of the Coaching Science Conference, Wellington,* NZ, p. 20.

Kirkman, J.H.D, McAuliffe, K.W. and Kirkman, A. 1989. 'Effect of grass verses bare soil on moisture loss from cricket pitch soils', *NZ Turf Management J.,* 3(2): 17–19.

Lush, W.M., Cummings, D.J. and McIntyre, D.S. 1985. 'Turf cricket wickets', *Search,* 16 (5,6):142–5.

McAuliffe, K.W. and Tuohy, M.P. 1987. 'Cricket wicket research carried out in New Zealand during the 1986/87 season', *NZ Turf Management J.,* July, pp. 3–7.

McAuliffe, K. W. 1995. 'The portable pitch—what are we waiting for?' *NZ Turf Management J.,* 9(4):1: 18,20.

Stewart, V.I. and Adams, W.A. 1969. 'Soil factors affecting the control of cricket pitches', *Proc. 1st Int. Turfgrass Res. Conference,* pp. 533–46.

Weaver, M. 1994. 'Using science as a tool in pitch preparation', *Proc. 5th N.Z. Sports Turf Conf.,* pp 133–5.

Wells, D.J. and McAuliffe, K.W. 1990. 'A pitch monitoring program for first class cricket venues in New Zealand', *Proc. of the Royal Aust. Inst. of Parks and Recreation Admin. Conf.*

CHAPTER 20

Environmental issues in turf management

J.T. SNOW, United States Golf Association, Green Section, Far Hills, New Jersey, and
M.P. KENNA, United States Golf Association, Green Section, Stillwater, Oklahoma, United States of America.

Introduction

Golf has grown dramatically in popularity during the past 20 years, with more than 25 million golfers in the United States of America alone. It now faces one of its greatest challenges from people who believe that golf courses and course maintenance practices are having a deleterious impact on the environment. One of the greatest fears is that the fertiliser and pesticides used to maintain golf courses will pollute drinking water supplies, which could include both surface and groundwater sources. Many people are concerned about the effects of high nutrient levels on human health and the ecology of surface waters, and about the potential effects of elevated pesticide levels in drinking water on cancer and other human and wildlife health problems. Claims have been made by some people that up to 100 per cent of the fertilisers and pesticides applied to golf course turf end up in local water supplies, a claim with no basis in fact, but one that has generated emotional reactions from people who are unaware of what happens to chemicals in the environment.

Similarly, no issue is more likely to have a significant impact on the game of golf in the future than that of how golf courses and golf course maintenance will affect the environment. At best, the cost of golf course construction and maintenance will increase in reaction to complying with a host of new environmental regulations, making the game even more costly to play. At worst, extreme restrictions or high costs will cause maintenance standards to decline significantly, driving people away from the game. Already compliance with environmental regulations has greatly increased golf course construction costs and is forcing new construction projects to be built on degraded sites, including abandoned agricultural land, industrial sites and landfills.

This is not to say that golf courses should not concern themselves about their impact on the environment and the need to be good stewards of the land. The 15 000 golf courses in the United States consume approximately 1.5 million acres of land, so golf should be concerned about its potential impact. The question has been, just what is the effect of golf courses on the environment? That is the question the United States Golf Association (USGA) struggled with during the 1980s as it considered the opinions of many groups and individuals, some saying that the impact is catastrophic, and others stating that the effects are nothing but good.

It wasn't until the 1970s that golf courses began to receive attention from people who were concerned about how the game might be affecting the environment. A series of widespread droughts during the late 1970s and early 1980s, highlighted by a severe drought in California and other western states, resulted in extreme restrictions on the use of water by homeowners and businesses in hundreds of communities. Golf courses were among the first and most severely restricted operations in many areas, due in part to their visibility in their communities and because they were considered non-essential users of water. Similarly, during the golf course construction boom of the 1980s and 1990s, golf courses again were under attack because of how golf course construction affected natural areas and because of the use of pesticides on existing courses. In many cases, anti-development groups tossed around unsubstantiated claims about the negative effects of golf courses in an effort to kill housing developments or commercial real estate development. Nevertheless, some of the environmental questions that people have raised about golf courses are based on sincere concern and deserve investigation and action by the golf and turfgrass industries. Unfortunately the game of golf and the turfgrass industry had little research information to refer to when responding to these concerns.

In 1989 the USGA decided to sponsor a significant amount of research on environmental issues. The area identified to receive the greatest amount of work concerned the effects of fertilisers and pesticides on surface and groundwater resources. In sponsoring environmental research, the USGA's Turfgrass and Environmental Research Committee adopted a mass balance approach, studying what happens to fertilisers and pesticides applied to golf courses. In other words, studies were conducted on the major pathways of chemical fate in the environment, including leaching, run-off, plant uptake and utilisation, microbial degradation, volatilisation, and other gaseous losses. The studies were conducted at twelve universities throughout the United States, representing the major climatic zones and turfgrass types. It was felt that by generating lots of data, a solid foundation would be built for determining how golf course activities and turfgrass management practices affect the environment (Snow, 1995).

USGA environmental research program

Between 1983 and 1995, the USGA funded 98 research projects at 33 land-grant universities, at a cost to the USGA of more than $US 12.5 million. Each of these projects addressed one or more of the environmental issues noted below. This was accomplished within the context of two distinct research programs. The Turfgrass Research Program (1983 to present) has as its goal the development of new grasses and cultural programs that will reduce the need for water, pesticides and other inputs for golf course maintenance. The Environmental Research Program (1991–93 and 1995–97) was designed to determine the effects of golf courses on people, wildlife and the environment. Four major areas of public concern were identified, consisting of the following: (a) the use of scarce water resources for golf course irrigation; (b) the potential pollution of our water resources by pesticides, fertilisers and other materials; (c) the loss of natural areas due to golf course construction and associated development; and (d) the possible effects of golf course activities on people and wildlife.

Use of scarce water resources for golf course irrigation

In addressing this issue, research strategies consisted of developing improved grasses that use less water, and identifying cultural management practices that result in less water use. Research was also conducted on the use of effluent water for golf course irrigation, and the use of new irrigation technologies. The cornerstone of this strategy was to invest in turfgrass breeding programs that eventually would produce new grasses that use significantly less water. At the time these plans were being developed in the early 1980s, relatively little was known about the water use rates and drought resistance characteristics of many of our warm-season and cool-season species and cultivars. It was decided to investigate these characteristics and develop screening techniques to help turfgrass breeders screen germplasm for resistance to heat, cold, drought, salt, and other environmental stresses. Results of some of this work, showing relative water use rates, drought resistance and salt tolerance, is shown in Tables 20.1, 20.2 and 20.3 (Kenna *et al.*, 1993). An important result of these plant stress mechanism studies was the verification that there is great variation for stress tolerances within many of the turf species, suggesting that there are many opportunities for significant improvement through traditional breeding efforts.

Table 20.1 Summary of mean rates of turfgrass evapotranspiration

*Turfgrass species**		
Cool-season	*Warm-season*	*Relative ranking*
	Buffalograss	Very low
	Bermudagrass hybrids Centipedegrass Bermudagrass Zoysiagrass	Low
Hard fescue Chewings fescue Red fescue	Bahiagrass Seashore paspalum St. Augustinegrass	Medium
Perennial ryegrass Tall fescue Creeping bentgrass Annual bluegrass Kentucky bluegrass Italian ryegrass	Carpetgrass Kikuyugrass	High

*Based on the most used cultivars of each species; there is significant variability among cultivars within most species.

Table 20.2 Relative drought resistance of turfgrasses in region of climatic adaption and preferred cultural regime

Turfgrass species,**		
Cool-season	*Warm-season*	*Relative ranking*
	Bermudagrass** Bermudagrass hybrids**	Superior
	Buffalograss Seashore paspalum** Zoysiagrass Bahiagrass	Excellent
Fairway wheatgrass	St. Augustinegrass** Centipedegrass Carpetgrass	Good
Tall fescue		Moderate
Perennial ryegrass** Kentucky bluegrass** Creeping bentgrass** Hard fescue Chewings fescue Red fescue		Fair
Colonial bentgrass Annual bluegrass		Poor
Rough bluegrass		Very poor

*Based on the most used cultivars of each species.
**Variable among cultivars within species.

Table 20.3 Relative salt resistance of several turfgrass species used in the United States

*Turfgrass species**		
Cool-season	*Warm-season*	*Relative ranking*
Alkaligrass	Seashore paspalum Zoysiagrass St. Augustinegrass	Excellent
Creeping bentgrass	Bermudagrass hybrids Bermudagrass	Good
Tall fescue Perennial ryegrass	Bahiagrass Centipedegrass	Fair
Fine fescues Kentucky bluegrass	Carpetgrass Buffalograss	Poor

*Based on the most used cultivars of each species; there is significant variability among cultivars within most species.

An important goal of the turfgrass improvement program was to significantly reduce water use and relative maintenance costs, using 1982 as a basis for comparison. One breeding strategy to accomplish this goal is to work to improve specific characteristics within each species, such as water use or heat tolerance. This is a long-term approach however, and can be expected to produce results in small increments over a period of many years. An example has been with the creeping bentgrasses, beginning with the vegetative bents in the 1920s, and progressing to Penncross and to the current selection and development of new cultivars.

Another strategy, and one that has the potential to produce a bigger return in a shorter period of time, is to replace stress-susceptible cool-season grasses with stress-tolerant grasses that have improved turf characteristics. A good example of this strategy is buffalograss (*Buchloe dactyloides*), which in a short period of time has been significantly improved for turf quality. Buffalograss now is being used to replace perennial ryegrass (*Lolium perenne*) and Kentucky bluegrass (*Poa pratensis*) roughs and other areas, with a saving in water use of 50 per cent or more and a large reduction in maintenance costs. Soon there will be buffalograsses for fairways, producing corresponding savings on those areas.

Other non-traditional grasses being improved for turf characteristics are alkaligrass (*Puccinellia* sp.), blue grama (*Bouteloua gracilis*), fairway crested wheatgrass (*Agropyron cristatum*), and curly mesquitegrass (*Hilaria belangeri*). Yet another breeding strategy employed has been to expand the range of adaptation or use of existing stress-tolerant turfgrasses, with the hope of replacing less tolerant species. For example, if warm-season grasses such as bermudagrass (*Cynodon dactylon*) can be improved for cold tolerance, it could be used to replace bentgrass (*Agrostis* spp.) and perennial ryegrass fairways in the transition zone areas of the US. This would result in a tremendous reduction in water and pesticide use in these areas.

Our last strategy is to develop improved seeded-type warm-season grasses, such as bermudagrass, zoysiagrass (*Zoysia* spp.) and buffalograss, which could replace existing vegetatively propagated cultivars and save significant establishment costs. Availability of seeded types also would encourage courses in the transition zone to establish stress-tolerant warm-season species instead of high maintenance cool-season grasses.

The USGA has financially supported turfgrass breeding programs at Tifton, GA (G. Burton), Rutgers University (R. Funk), University of Rhode Island (R. Skogley), and Pennsylvania State (B. Musser and J. Duich) for many decades, with significant results. The new programs initiated in the 1980s have been in existence for 9 to 12 years, and a number of improved cultivars or selections have come to market or have been released to seed companies for development (Table 20.4). All exhibit improved turf characteristics, stress tolerance or pest resistance. During the next decade, the number of new introductions will increase significantly, and the golf industry will be in a much better position to conserve and protect our natural resources.

Another water conservation issue concerns the use of recycled water for course irrigation. Hundreds of golf courses in the US use recycled water, and in fact its use is mandatory in parts of California and Arizona. It is expected that hundreds more courses will begin using recycled water during the next few years. The USGA has sponsored research on, and encouraged the use of, recycled or effluent water. In 1993 the USGA sponsored a symposium on this topic, and in 1994 published the proceedings in book form, titled Wastewater Reuse for Golf Course Irrigation.

Potential pollution of our water resources by pesticides, fertilisers and other materials

Initial environmental research was conducted from 1991 to 1993 by the USGA, and was

Table 20.4 Summary of USGA turfgrass breeding projects

Turfgrass	*University*	*Status of Varieties*
Creeping Bentgrass *Agrostis palustris*	Texas A&M University	CRENSHAW (Syn3-88), CATO (Syn4-88) and MARINER (Syn1-88), CENTURY (Syn92-1), IMPERIAL (Syn92-5), BACKSPIN (92-2) were released. All are entered in 1993 NTEP trials.*
	University of Rhode Island	PROVIDENCE was released.
	Pennsylvania State University	PENNLINKS was released
Colonial Bentgrass *Agrostis tenuis*	DSIR-New Zealand and University of Rhode Island	A preliminary line, BR-1518, was entered in the NTEP trials. A new line is being evaluated at the University of Rhode Island.
Bermudagrass *Cynodon dactylon*	New Mexico State University	NuMex SAHARA, SONESTA, PRIMAVERA and other seed propagated varieties were developed from this program.
	Oklahoma State University	Two seeded types, OKS 91-11, and OKS 91-1 were entered in the 1992 NTEP trials. OKS 91-11 was approved for release in 1996.
C. transvaalensis	Oklahoma State University	A release of germplasm for university and industry use is under consideration.
C. dactylon x *C. transvaalensis*	University of Georgia	TIFTON 10 and TIFTON 94 (MI-40) were released; a TIFWAY mutant (TW-72) is under evaluation for release.
Buffalograss *Buchloe dactyloides*	University of Nebraska	Vegetative varieties 609, 315, and 378 were released. Seeded varieties CODY and TATANKA were released. Three new vegetative selections, NE 86-61, NE 86-120 and NE 91-118, are currently being processed for release.
Alkaligrass *Puccinellia* spp.	Colourado State University	Ten improved families are under evaluation and have been released.
Blue grama *Bouteloua gracilis*	Colourado State University	ELITE, NICE, PLUS and NARROW are under evaluation in anticipation of release.
Fairway crested wheatgrass *Agropyron cristatum*	Colourado State University	Narrow leafed and rhizomatous populations were developed; nothing was released.
Curly mesquitegrass *Hilaria belangeri*	University of Arizona	Seed increases of 'fine' and 'roadside' populations are available for germplasm release and further improvement.
Annual bluegrass *Poa annua* var *reptans*	University of Minnesota	Selections #42, #117, #184, #208, and #234 were released and are under evaluation for seed production.
Zoysiagrass *Zoysia japonica* and *Z. matrella*	Texas A&M University	Ten vegetative selections were entered in the 1991 NTEP trial. DIAMOND (DALZ8502), CAVALIER (DALZ8507), CROWNE (DALZ8512) and PALISADES (DALZ8514) were released in 1996.
Seashore paspalum *Paspalum vaginatum*	University of Georgia	Germplasm has been assembled and is under evaluation. Two green types (AP 10, AP) and one fairway type (PI 509018-1) are being evaluated on golf courses.

*National Turfgrass Evaluation Program, Beltsville Agricultural Research Center, Beltsville, MD 20705.

followed-up by studies conducted from 1995 to 1997. For the sake of providing an overview of the results of these studies, representative data from both periods will be presented. The subject will be divided into the following topics: nitrogen leaching, nitrogen run-off, pesticide leaching, and pesticide run-off.

Nitrogen leaching

Golf courses use a significant amount of nitrogen (N) fertiliser, and there is concern that nitrogen leaching is affecting groundwater supplies. Seven different universities investigated nitrogen leaching, most using bucket lysimeters to measure leaching potential. In general, very little nitrogen leaching occurred when nitrogen was applied properly, that is according to the needs of the turf and in consideration of soil types, irrigation regimes and anticipated rainfall (Kenna, 1994). Representative of these results was the work at Michigan State University, which utilised 1100 millimetre (4 feet) deep lysimeters with undisturbed soil profiles. After 2.5 years, less than one per cent of the applied nitrogen had leached through 1100 millimetre (4 feet) profile (Figure 20.1). Most of the rest was recovered in the clippings, thatch and soil, and it is presumed that some volatilisation also occurred (Branham *et al.*, 1995, Miltner *et al.*, 1996). A three-year study on nitrogen volatilisation and denitrification will begin in 1998.

Researchers at Iowa State University observed similar results when nitrogen was applied at moderate rates and lightly irrigated (one 22.5 millimetre (1 inch) versus four 5.5 millimetre (0.25 inch) applications) after application (Starrett *et al.*, 1995). However, up to 40 times the amount of nitrogen was leached after the single 22.5 millimetre (1 inch) application, perhaps in part due to macropore flow caused by earthworm activity. At the University of California at Riverside, nitrogen leaching from a USGA profile sand-based green was generally less than one per cent when nitrogen was applied lightly and frequently (Yates, 1995).

On the other hand, nitrogen leaching was significant when nitrogen was applied at heavy rates and under less-than-ideal circumstances. For example, during the first year of the study at Washington State University on immature turf grown on a pure sand rootzone medium, nitrogen leaching amounted to 7.6 per cent at an annual application rate of 5.45 kilograms (12 pounds) nitrogen per 93 square metres (1000 square feet) (Table 20.5) (Braun *et al.*, 1995). Leaching was significantly less when peat was added to the sand (USGA mix), occurring at a level of about 3 per cent. On pure sand, nitrogen concentrations exceeded Federal drinking water standards (10 parts per million NO_3) several times at the 5.45 kilogram (12 pound) rate

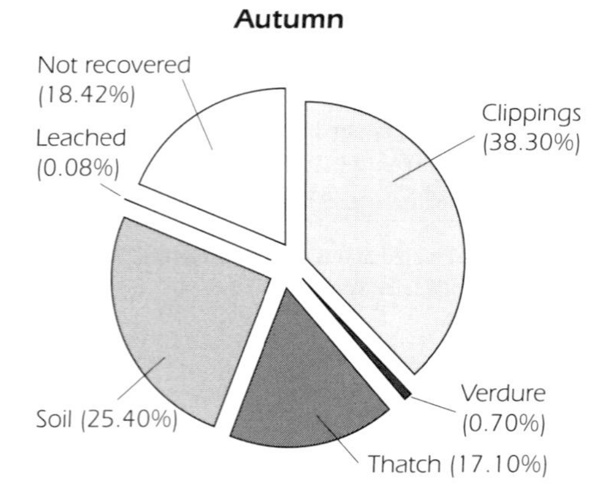

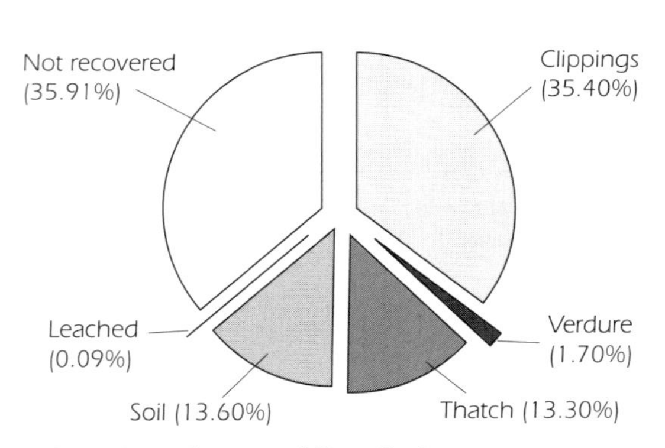

Figure 20.1 Percentage of total nitrogen applied recovered two years after spring and autumn (fall) applications

during the first year, whereas nitrogen concentrations never exceeded federal standards from the sand–peat mix. Significantly less leaching also occurred when less nitrogen was applied (1.8 kilograms and 3.6 kilograms (4 and 8 pounds)) and when application frequency was increased (22 versus 11 times annually). During years two and three, on mature turf, much less nitrogen leaching occurred for all treatments. Averaging results over all seven leaching projects during year one (establishment year), nitrogen leaching from pure sand rootzones was about 11 per cent; from sand–peat rootzones about 4 per cent; and from loamy sand, sandy loam, and silt loam rootzones about 1 per cent or less (Kenna, 1995).

Table 20.5 Percentage of total applied nitrogen leached as nitrate (Source: Braun *et al.*, 1995).

Rootzone Medium	*Annual N lb/1000 ft²*	*Annual N kg/93 m²*	*Year 1*	*Year 2*	*Year 3*
Sand	4	1.8	5.37	0.06	2.71
	8	3.6	6.31	0.04	3.17
	12	5.4	7.55	0.70	4.28
Amended sand	4	1.8	5.37	0.40	0.16
	8	3.6	0.91	0.02	0.17
	12	5.4	3.37	1.26	2.31

Generally, properly maintained turf allows less than 1 per cent of the nitrogen applied to leach to a depth of 1.21 metres (4 feet). When more nitrogen is applied than is needed, both the amount and the percentage of nitrogen lost increases. Sandy soils are more prone to leaching losses than clay soils. Nitrogen leaching losses can be greatly reduced by irrigating lightly and frequently, rather than heavily and less frequently. Applying nitrogen in smaller amounts on a more frequent basis also can reduce leaching losses.

Irrigating bermudagrass and tall fescue turf with adequate amounts (no drought stress) of moderately saline water did not increase the concentration or amount of nitrate leached (Bowman *et al.*, 1995). Higher amounts of salinity in the root zone, drought or the combination of these two stresses caused high concentrations and amounts of nitrate to leach from both a tall fescue and bermudagrass turf. This suggests that drought, high salinity or both impair the capacity of the root system of the turf, and that management modification may be needed to prevent nitrate leaching. In putting green construction, mixing peat moss with sand significantly reduced nitrogen leaching compared to pure sand rootzones during the year of establishment. Light applications of slow-release nitrogen sources on a frequent interval provided excellent protection from nitrate leaching.

Nitrogen run-off

Among the eleven universities investigating fertiliser and pesticide fate in phase one (1991 to 1993) of the USGA's environmental studies, only one looked at nitrogen run-off. The investigation was conducted at the Pennsylvania State University (Penn State) run-off plots, which were characterised by slopes of 9 to 13 per cent, good quality soil, and turf cover consisting of either creeping bentgrass or perennial ryegrass cut at a 12.5 millimetre (0.5 inch) fairway height. Typical of that part of the country, the fairway-type plots received 1.8 kilograms (4 pounds) nitrogen per 93 square metres (1000 square feet) per year. The irrigation water used to simulate rainfall itself contained a relatively high level of nitrate-nitrogen, ranging from about 2 to 10 parts per million. In no instance throughout the study did nitrate concentrations in the run-off or leaching samples differ significantly from the nitrate concentration in the irrigation water (Linde *et al.*, 1995).

The Penn State study was conducted on excellent quality turf and on soil with a relatively high infiltration rate. Although nitrogen run-off was not measured as part of the run-off studies at the University of Georgia, significant amounts of some soluble pesticides were found in run-off water, suggesting that the circumstances at the Penn State plots may not have represented the actual field conditions on many golf course fairways. Thus it was decided by the USGA Research Committee to expand the run-off studies during Phase Two (1995 to 1997) of the environmental program. One of these run-off studies was conducted at Oklahoma State University, where the effects of buffer strips and best management practices on pesticide and nitrogen run-off were investigated. Several interesting observations were made during 1995, the first year of the study (Baird, 1996; Cole *et al.*, 1997). During the first simulated rainfall event in July, soil moisture conditions were low to moderate. After a 50 millimetre (2 inch) rainfall event, less than 1 per cent of the applied nitrogen was collected in the run-off. In August, when the simulated rainfall occurred after 150 millimetres (6 inches) of actual rainfall the previous week (i.e. high soil moisture), the amount of nitrogen collected after the simulated rainfall averaged more than 8 per cent. When soil moisture was moderate to low in the Oklahoma study, the presence of an untreated buffer strip 2.42 to 4.85 metres (8 to 16 feet) significantly reduced nitrogen run-off, whereas when soil moisture was high, the buffer strips made no difference. In both cases less run-off occurred when sulfur-coated urea was applied compared to straight urea.

Results from USGA-sponsored run-off studies showed that dense turf cover reduces the potential for run-off losses of nitrogen, and significant run-off losses are more likely to occur on compacted soils (Kenna, 1995). Much greater nitrogen run-off occurred when soil moisture levels were high, as compared to moderate or low. Buffer strips reduced nitrogen run-off when soil moisture was low to moderate at the time of the run-off event, but not when soil moisture levels were high. Nitrogen run-off was significantly less when a slow release product (sulfur-coated urea) was used compared to a more soluble product (urea).

Pesticide leaching

Pesticide leaching studies were also conducted at eight universities throughout the US. Treatments were made to a variety of soils and turf species, and plots received varying irrigation regimes or rainfall events. During the first year of the studies, most turf areas were relatively immature. Results showed that very little pesticide leaching occurred with most products, generally less than 1 per cent of the total applied. However significant leaching occurred with certain products and under certain circumstances (Table 20.6).

Generally speaking, the physical and chemical properties of the pesticides proved to be good indicators of the potential for leaching, run-off and volatilisation(Kenna, 1995). Products that exhibit high water solubility, low soil adsorption potential, and greater persistence are more likely to leach and run-off. For example, fenamiphos (Nemacur), a commonly used nematicide, is highly water-soluble and has low adsorption potential, and its toxic breakdown metabolite tends to persist in the soil. As expected, losses of fenamiphos and its metabolite due to leaching were as high as 18 per cent from a sand-based green at the University of Florida (Cisar *et al.*, 1993; Snyder *et al.*, 1993; Snyder *et al.*, 1995; Cisar *et al.*, 1996), though when all studies are considered, the average loss was about 5 per cent. Soil type and rainfall/irrigation amount also were important factors in leaching losses. Table 20.7 shows the effects of soil type and rainfall on leaching of MCPP and triadimefon (Bayleton), two pesticides whose chemical and physical properties indicate a relatively high potential for leaching. Results show significant leaching from sand profiles, especially under high rainfall conditions, and much less leaching from sandy loam and silt loam soils (Petrovic, 1995).

Table 20.6 Summary of total pesticide mass (micrograms per square metre) and percent of total applied recovered in water effluent from putting green and fairway lysimeters.

Common name	*Trade name*	*GUS leaching classification*	*GUS score*	*n*	*Total Recovered* $\mu g/m^2$	*(Range, $\mu g/m^2$)*	*Percent recovered* %	*(Range, %)*
Fairway								
Chlorothalonil	Daconil	Non-leacher	1.27	1	0	(0)	0.00	(0.00)
Fenarimol	Rubigan	Intermediate	2.55	1	0	(0)	0.00	(0.00)
Metalaxyl	Subdue	Leacher	3.43	1	0	(0)	0.00	(0.00)
Propiconazole	Banner	Intermediate	2.00	1	0	(0)	0.00	(0.00)
Triadimefon	Bayleton	Intermediate	2.15	8	2,312	(27–11,160)	0.51	(0.01–2.44)
2,4-D		Intermediate	2.69	9	155	(0–329)	0.28	(0.00–0.60)
Dicamba	Banvel	Leacher	4.24	2	4,750	(3,700–5,800)	39.58	(30.83–48.33)
MCPP	Mecoprop	Leacher	3.51	6	44,236	(1,006–142,062)	19.34	(0.44–62.12)
Carbaryl	Sevin	Non-leacher	1.52	8	132	(24–75)	0.01	(0.00–0.02)
Isazofos	Triumph	Leacher	3.06	6	5,590	(5,590)	2.44	(0.00–10.40)
Trichlorfon	Proxol	Leacher	3.00	6	21,527	(5,763–40,341)	2.35	(0.63–4.41)
Putting green								
Chlorothalonil	Daconil	Non-leacher	1.27	4	2,961	(749–5,486)	0.08	(0.02–0.14)
2,4-D		Intermediate	2.69	7	871	(347–1,808)	2.25	(1.12–3.79)
2,4-D amine		Intermediate	2.00	6	46	(0–133)	0.12	(0.00–0.48)
Dicamba	Banvel	Leacher	4.24	7	201	(0–1,173)	3.07	(0.00–19.55)
MCPP	Mecoprop	Leacher	3.51	4	109	(0–329)	0.08	(0.00–0.25)
Carbaryl	Sevin	Non-leacher	1.52	4	372	(205–642)	0.04	(0.02–0.07)
Chlorpyrifos	Dursban	Non-leacher	0.32	4	92	(0–193)	0.04	(0.00–0.08)
Dithiopyr	Dimension	Non-leacher	0.97	4	139	(101–196)	0.24	(0.18–0.35)
Ethoprop	Mocap	Intermediate	2.68	1	41,138	(1,138)	0.05	(0.05)
Fenamiphos	Nemacur	Leacher	3.01	4	53,121	(419–199,038)	4.70	(0.04–17.61)
Fonofos	Dyfonate	Non-leacher	1.12	2	54	(4–103)	0.01	(0.00–0.02)
Isazofos	Triumph	Leacher	3.06	2	123	(41–204)	0.05	(0.02–0.09)
Isofenphos	Oftanol	Intermediate	2.65	2	43	(33–53)	0.02	(0.01–0.02)

Table 20.7 Effect of soil type and precipitation rate on the leaching loss of two pesticides from an immature turf, expressed as per cent of total applied (Source: Petrovic, 1995).

Pesticide	*Precipitation*	*Sand (%)*	*Sandy loam (%)*	*Silt loam (%)*
MCPP	Moderate	34.85	1.69	1.01
	High	73.76	0.10	1.26
Triadimefon	Moderate	1.00	0.06	0.24
	High	2.44	0.01	1.26

An interesting phenomenon occurred in the University of Florida study when fenamiphos was applied twice at a monthly interval. Although leaching from the first application amounted to about 18 per cent, leaching from the second application was just 4 per cent (Snyder *et al.*, 1993; Snyder *et al.*,1995; Cisar *et al.*, 1996). These results suggest that microbial degradation was enhanced due to microbial buildup after the first application, thereby reducing the amount of material available for leaching after the second application. Another interesting by-product of the studies occurred at the University of Georgia, where the actual leaching loss of 2,4-D was compared to that predicted by a computer model (GLEAMS) used by the U.S. Environmental Protection Agency (USEPA). It was found that the amount leached was a tiny portion of that predicted by the model (Smith *et al.*, 1993; Smith 1995; Smith *et al.*, 1996a). Generally the model over-predicted the actual amount leached by 10 to 100 times or more, for five of the seven pesticides screened by the computer.

Some conclusions observed from the pesticide leaching studies indicate that dense turf cover reduced the potential for leaching losses of pesticides; conversely, more leaching occurred from newly planted turf stands. The physical and chemical properties of the pesticides were good indicators of leaching potential. Current pesticide fate models used by USEPA over-predict the leaching loss of most pesticides applied to turf. Generally, sandy soils are more prone to leaching losses than clay soils. The average DT90 (days to 90 per cent degradation) in turf soils generally is significantly less than established values based upon agricultural systems (Horst *et al.*, 1995). Thus leaching potential for most pesticides is less in turfgrass systems. Turfgrass thatch plays an important role in adsorbing and degrading applied pesticides.

Pesticide run-off

The University of Georgia was the only site to carry out pesticide run-off investigations in the first round of studies (1991 to 1993). The studies were conducted on plots with a 5 per cent slope and a sandy clay soil (Table 20.8). Pesticides were applied, and 25 millimetre (1 inch) simulated rainfall events occurred 24 and 48 hours afterwards, at a rate of 50 millimetres (2 inches) per hour. Under these conditions, only very small amounts of chlorothalonil and chlorpyrifos could be detected in the run-off (Smith 1995; Smith *et al.*, 1996b). However, between 10 per cent and 13 per cent of the 2,4-D, MCPP and dicamba ran off the plots over an 11-day period, producing a relatively high level of contamination. About 80 per cent of this total moved off the plots with the first rainfall event. The significant loss of herbicides at the University of Georgia run-off project served to focus on the need for more run-off work in the next phase of the environmental research program. So, in addition to follow-up studies at the University of Georgia, new pesticide and fertiliser run-off investigations were initiated at Oklahoma State University. The purpose of the project was to develop best management practices by investigating how cutting heights and buffers of varying lengths can be used to minimise fertiliser and pesticide run-off. The effect of soil cultivation (core aerification) on run-off potential also was studied.

Table 20.8 The percentage of applied pesticide and concentration of pesticide transported from run-off plots during a storm event that occurred 24 hours after application (Source: Smith *et al.*, 1996).

Pesticide or fertiliser treatment	*Application rate (kg ha⁻¹)*	*Percentage transported (%)*	*Conc. at 24 h after application (µg L⁻¹)*
Nitrate-N	24.40	16.4	12 500
Nitrate-N (dormant bermuda)	24.40	64.2	24 812
Dicamba	0.56	14.6	360
Dicamba (dormant bermuda)	0.56	37.3	752
Mecoprop	1.68	14.4	810
Mecoprop (dormant bermuda)	1.68	23.5	1 369
2,4-D DMA	2.24	9.6	800
2,4-D DMA (dormant bermuda)	2.24	26.0	1 959
2,4-D DMA (pressure injected)	2.24	1.3	158
2,4-D DMA (2 m buffer strip)	2.24	7.6	495
2,4-D LVE	2.24	9.1	812
Trichlorfon*	9.15	32.5	13 960
Trichorfon* (pressure injected)	9.15	6.2	2 660
Chlorothalonil**	9.50	0.8	290
Chlorpyrifos***	1.12	0.1	19
Dithiopyr	0.56	2.3	39
Dithiopyr (granule)	0.56	1.0	26
Benefin	1.70	0.01	3
Benefin (granule)	1.70	0.01	6
Pendimethalin	1.70	0.01	9
Pendimethalin (granule)	1.70	0.01	2

*Trichlorfon + dichlorvos metabolite.
**Total for chlorothalonil and OH-chlorothalonil.
***Total for chlorpyrifos and OH-chlorpyrifos.

As reported in the section on nitrogen run-off, two experiments were conducted at Oklahoma State during 1995, the first when soil moisture was low to moderate prior to the simulated rainfall event, and the second when soil moisture content was very high due to previous heavy rainfall. From Table 20.9 it is clear that soil moisture content was a significant factor in determining how much of the pesticides ran off the plot areas (Baird 1996; Cole *et al.*, 1997). Where soil moisture was low to moderate, buffer zones were effective in reducing pesticide run-off; when soil moisture was high they were not effective except for the insecticide chlorpyrifos.

The follow-up run-off study at the University of Georgia also produced some interesting results in 1995. As much as 40 to 70% of the rainfall left the plots as run-off during simulated storm events. The collected surface water contained moderately high concentrations of treatment pesticides having high water solubility. For example, less than 1 per cent of the applied chlorothalonil, chlorpyrifos, benefin, and pendimethalin was transported from the plots in run-off water. On the other hand, as much as 9 to 16 per cent of the 2,4-D, dicamba, mecoprop and nitrate was transported in the surface water from the first two simulated storm events. Also the amount of the insecticide trichlorfon that ran off the plots was 5.2 times greater when broadcast as a granular compared to being pressure injected. Finally, the run-off loss of nitrate and several herbicides was much greater when applied to dormant turf as compared to an actively growing turf: 2,4-D 26.0 per cent versus 9.6 per cent; dicamba 37.3 per cent versus 14.6 per cent; MCPP 23.5 per cent versus 14.4 per cent; nitrate 64.2 per cent versus 16.4 per cent.

Among the conclusions or trends observed from the pesticide run-off studies were the following: (a) dense turf cover reduces the potential for run-off losses of pesticides; (b) the physical and chemical properties of pesticides are good indicators of potential run-off losses; (c) heavy textured, compacted soils are much more prone to run-off losses than sandy soils; (d) moist soils are more prone to run-off losses than drier soils; (e) buffer strips are very effective at reducing run-off of pesticides when soil moisture is low to moderate prior to rainfall events; and, (f) the application of soluble herbicides on dormant turf can produce very high levels of run-off losses.

Loss of natural areas due to golf course construction and associated development

During the past decade, more than 2000 new golf courses have been built in the US, bringing the total to about 15 000. Although many have been built on abandoned agricultural lands, commercial sites or other degraded areas, others have been built in natural areas, having an impact on wildlife and the local ecosystem. Local or national environmental groups concerned about potential negative impacts often contest construction in these areas. The USGA has no part in and no authority over selecting sites for new course construction, but we have been active in establishing programs to address some of these concerns. Our activities have included establishing the Audubon Cooperative Sanctuary Program for golf courses, publishing the *Landscape Restoration Handbook* (Harker, 1993), and establishing the Wildlife Links program.

In 1990 the USGA and the Audubon Society of New York State (now called Audubon International) teamed up to establish the Audubon Cooperative Sanctuary Program for Golf Courses. Among its objectives is to enhance

Table 20.9 Effect of low to moderate versus high soil moisture levels on pesticide and nutrient run-off losses from bermudagrass maintained as fairway turf, expressed as a percent of total applied (Source: Cole *et al.*, 1997).

Soil moisture	*Dicamba*	*2,4-D*	*MCPP*	*Chlorpyrifos*	*NH_4-N*	*NO_3*	*PO_4*
Low/moderate	0.35	0.79	0.81	0.04	0.2	0.09	0.2
High	5.4	8.7	9.3	0.025	5.1	3.1	7.7

wildlife habitat on golf courses and encourage active participation in conservation programs by golf course superintendents, course officials, golfers and the public. Participation in the program requires the completion of a resource inventory form, describing the property and its existing features. Audubon International then responds with ideas and technical information about what the participating course can do to enhance wildlife habitat and improve the environment. The course develops a plan of action and then acts on its plan. It can become certified in one or more of six categories by completing its plan and submitting documentation of its achievements to Audubon International personnel, who decide if the actions merit certification. Certification can be achieved in one or more of the following categories: environmental planning, member/public involvement, wildlife and habitat management, integrated pest management, water conservation, and water quality management. Since its inception nearly eight years ago, participation in the program has grown to more than 2700 courses. Most importantly, many thousands of people involved in golf are being educated about issues related to wildlife and the environment, and are participating in conservation programs that benefit both. Also in 1993 the USGA published its *Landscape Restoration Handbook* (Harker, 1993). This book discusses principles for establishing naturalised areas on golf courses, provides lists of native plants on an eco-region basis and gives extensive information about each plant, and lists nurseries where the plants can be obtained.

As part of its Environmental Research Program, the USGA sponsored a study of the effects of golf course activities on wildlife at the Ocean Course at Kiawah Island, SC, in cooperation with The Institute of Wildlife and Environmental Toxicology (TIWET) at Clemson University. This study led to a program of wildlife research that now is called the Wildlife Links Program, carried out in cooperation with the National Fish and Wildlife Foundation (NFWF), a Washington, D.C., organisation whose mission is to organise and help fund conservation projects that benefit wildlife and the environment. The NFWF provides technical expertise to help establish objectives and identify worthwhile research projects concerning wildlife and golf courses. As part of this program, the NFWF has established an advisory panel of experts representing several regulatory agencies and other organisations. The advisory panel establishes objectives, reviews proposals, and monitors the progress of the research projects. In addition, interested representatives of other environmental agencies and organisations are being kept abreast of the activities of the Wildlife Links Program, and are queried for suggestions.

Possible effects of golf course activities on people and wildlife

Before the Environmental Research Program was begun in 1991, the USGA's Turfgrass Research Committee commissioned a complete literature review on the topic of golf and the environment to help identify what was lacking in the scientific literature. The result was the 1992 publication of a 1000-page book titled *Golf Course Management and Construction: Environmental Issues* (Balogh, 1992). Concerns expressed about potable water used by golf courses come primarily from the arid parts of the US, but public concern about the potential effects of pesticides and fertilisers on water supplies can be found everywhere. Concern is not without good reason, since contamination of water supplies and destruction of wildlife has been associated with agricultural use of these materials in the past. Environmental groups and the media have broadly criticised golf courses for contributing to the problem, even though direct evidence of their contribution has been practically non-existent. Nevertheless, scattered incidents of bird and fish kills have occurred, and certainly the potential for water pollution is there if pesticides and fertilisers aren't applied with care.

In the late 1980s the USGA decided to begin investigating this issue, in part because there was little research with which to respond to criticisms. Among the strategies utilised were the following: (a) conducting and publishing an environmental literature review; (b) investigating what happens to pesticides and fertilisers applied to golf course turf; (c) supporting research to develop alternative (non-chemical) methods of pest control; (d) the use of breeding and biotechnology to develop new grasses that require less pesticide use; (e) the investigation of Best Management Practices and publish a book on the topic; and (f) modifying computer models to better predict pesticide and fertiliser fate.

In Phase Two of the USGA's Environmental Research Program (1995 to 1997), emphasis has been placed on following up on questions raised during the earlier studies, particularly concerning run-off and volatilisation losses, the effects of thatch and turfgrass soils on pesticide degradation, and pesticide and nutrient fate modelling. In the realm of alternative pest control methods, the rationale for funding these studies is that if non-chemical pest control methods can be developed, less pesticide need be applied and the potential for surface or groundwater contamination can be reduced. A majority of these studies have involved potential biological controls, generally concerning the search for natural antagonists in the soil that can control or suppress turfgrass diseases or insects. The virtues of biological controls are often extolled by environmental groups as a practical alternative to pesticide use on golf courses, but in our experience they are not (with few exceptions). Despite having spent more than $US1 000 000 in support of projects that sought biological antagonists for control of common turfgrass diseases and pests, not a single one made it to market. Major problems included product formulation, lack of field efficacy, or lack of commercial interest to develop the product because of high cost and small market potential.

Alternative pest control research projects that have shown some success and greater potential are those in the biotechnology category. Already creeping bentgrass plants have been transformed through genetic engineering to exhibit excellent resistance to glufosinate ammonium herbicide (Finale or Ignite), a non-selective herbicide. Through traditional breeding methods, it now may be possible to incorporate these herbicide resistance genes into putting green bentgrasses, allowing golf course superintendents to keep annual bluegrass (*Poa annua*) and other unwanted plant materials out of their greens. Potentially, this could reduce water and pesticide use on greens. Also two projects were initiated to introduced chitinase production genes into bentgrass, providing an internal mechanism to help control turfgrass diseases and thereby reduce pesticide use. In 1996, the USGA sponsored a symposium on turfgrass biotechnology, and in 1997 published the proceedings in book form titled, *Turfgrass Biotechnology: Molecular Approaches to Turfgrass Improvement* (Sticklen *et al.*, 1997). Another project underway is a book that will help educate golf course superintendents about how best to manage golf courses with the protection of the environment in mind. It should help educate regulators and critics of golf courses about appropriate, scientifically valid methods for minimising the potential negative impacts of course maintenance on the environment.

It isn't just environmentalists who are worried about golf course effects on the environment. Golfers, too, are concerned about the use of pesticides and their own exposure to these materials. As part of the USGA's pesticide fate studies, several research projects looked at pesticide volatilisation and dislodgeable residues. The volatilisation studies investigated how much of several pesticides volatilised into the air for several weeks after application, and how much exposure golfers or course workers might receive. From Phase One of the USGA's studies, the following are total volatile losses for several pesticides over a 4-week period, expressed as a percentage of the total applied: carbaryl < 1 per cent; 2,4-D < 1 per cent; MCPP < 1 per cent; triadimefon 7.6 per cent (Cooper *et al.*, 1995; Murphy *et al.*, 1996a; Murphy *et al.*, 1996b). At these levels, golfer exposure would be 1000 times below the EPA's no effect level (NOEL), and considered an insignificant risk. However, total volatile losses of trichlorfon and isazofos, two commonly used insecticides, constituted about 13 per cent of the amount applied. This is less than 100 times below the EPA's NOEL (No Effect Level), and although not considered a significant risk, it nevertheless suggests that a second look is prudent. Results of volatilisation studies showed that maximum loss occurred when surface temperature and solar radiation were highest, and that volatile losses were directly related to the vapour pressure characteristics of the pesticide. Thus checking the physical and chemical properties of the pesticide is a good way to determine if volatilisation losses are likely to occur under particular weather and application conditions.

In Phase Two of the Environmental Research Program (1995 to 1997), additional volatilisation studies were funded. It was shown that organophosphate insecticides that possess high toxicity and volatility might result in exposure situations that cannot be deemed completely safe as judged by the US EPA Hazard Quotient determination (Tables 20.10 and 20.11). Additional studies will be needed to determine the extent of the risk, if any.

People are also concerned about exposure they receive during a round of golf on turf that has been treated with pesticides. Exposure in this situation is caused by pesticide residues on the turf surface that rub off onto people or their equipment during a round of golf. At the University of Florida, a preliminary risk assessment was conducted on putting greens treated with insecticides. The assumptions for the study were as follows: the golfer kneels on every green to align putts, he handles golf grips that have been laid on the green, he contacts the soles of his

Table 20.10 Inhalation Hazard Quotients (IHQs) for turfgrass pesticides in the high, intermediate and low vapour pressure group (Source: Clark, 1996).

Group	*Pesticide*	*Day 1*	*Day 2*	*Day 3*
		Inhalation Hazard Quotients (IHQ)*		
Group 1: high vapour pressure (i.e. vapour pressures > 1.0 x 10^{-5} mm Hg)	DDVP	0.06	0.04	0.02
	Ethoprop	50.0	26	1.2
	Diazinon	3.3	2.4	1.2
	Isazofos	8.6	6.7	3.4
	Chlorpyrifos	0.09	0.1	0.04
Group 2: intermediate vapour pressure (i.e. 10^{-5} mm Hg > vapour pressures > 10^{-7} mm Hg)	Trichlorfon	0.02	0.004	0.004
	Bendiocarb	0.02	0.002	0.002
	Isofenphos	n/d**	0.02	n/d
	Chlorothalonil	0.001	0.001	0.0003
	Propiconazole	n/d	n/d	n/d
	Carbaryl	0.0005	0.0001	0.00004
Group 3: low vapour pressure (i.e. vapour pressure < 10^{-7} mm Hg)	Thiophanate-methyl	n/d	n/d	n/d
	Iprodione	n/d	n/d	n/d
	Cyfluthrin	n/d	n/d	n/d

*The IHQs reported are the maximum daily IHQs measured, all of which occurred during the 11:00 a.m. to 3:00 p.m. sampling period.
**n/d = non-detect.

leather shoes when cleaning them, he cleans his ball after every hole by licking it, the greens are sprayed every day with three insecticides and the golfer plays golf 365 days a year for 70 years. Under the outrageous assumptions of this study, after 70 years the golfer would have received about one-third of the lifetime reference dose considered safe by the USEPA (Borger *et al.*, 1994). This is not to say that golfers or workers could not receive unsafe exposure to pesticides on golf courses; under the conditions of this study, however, the golfer would not have been at significant risk. Other results of the dislodgeable residue studies indicated that less than 1 per cent of the pesticides could be rubbed off immediately after application, when the turf was still wet. Also, the 1 per cent could be reduced significantly by irrigating after the pesticide was applied. Finally, after the pesticides dried on the turf, only minimal amounts could be rubbed off.

Most media attention concerning golf courses tends to focus on the potential negative impacts of turfgrass and golf course management. Although people in our industry know intuitively that there are many benefits associated with turfgrasses and golf courses, the scientific basis for many of these benefits had not been documented or pulled together. Through a USGA grant, Beard and Green conducted an exhaustive literature search and documented many of these benefits. The results of their work were published (Beard and Green, 1994), and a summary is available from the USGA.

Future

Looking to the future, as golf addresses the environmental issues that threaten its growth and success, what is needed are more research information; greater participation in programs such as the Audubon Cooperative Sanctuary Program; greater industry support for the Wildlife Links Program; and ever higher environmental standards from golf course superintendents, architects and builders. There also needs to be greater understanding of golf's environmental issues among golfers; more constructive interactions with environmental organisations; and widespread publicity for what golf is doing to benefit wildlife and the environment (Snow, 1995). With respect to future research needs, the USGA has committed up to $US10 million from

Table 20.11 Dermal Hazard Quotients (DHQs) over a three-day post-application period for turfgrass pesticides listed by increasing the reference dose (Source: Clark, 1996).

Pesticide	*RfD*	*Day 1* 15 min.	*Day 1* 5 hours	*Day 1* 8 hours	*Day 2* 12:00 p.m.	*Day 3* 12:00 p.m.
Ethoprop	0.000015	16.0	1.64	1.35	0.23	0.34
Isazofos	0.00002	10.5	1.17	0.97	0.16	0.21
Diazinon	0.00009	3.0	0.28	0.22	0.04	0.05
Isofenphos	0.0005	0.32	0.05	0.05	n/d2	0.01
DDVP	0.0005	0.06	0.003	0.003	0.008	n/d
Trichlorfon	0.002	0.64	0.007	0.009	0.003	0.005
Chlorpyrifos	0.003	0.17	0.02	0.016	0.006	0.004
Bendiocarb	0.005	0.31	0.006	0.01	0.0005	0.0008
Propiconazole	0.0125	0.0002	0.003	0.0002	0.0006	0.0002
Carbaryl	0.14	0.03	0.0008	0.001	—	0.00002
Cyfluthrin	0.25	—	—	—	0.0004	—
Iprodione	0.61	0.0004	0.0003	0.0003	—	0.0003
Thiophanate-methyl	0.08	—	—	—	—	—

— = no data available at the time of this report.

1998 to 2002 to continue its turfgrass improvement research and to follow up on the results of its previous environmental studies. Topics for future environmental studies include integrated turfgrass management; site specific management; risk assessment; pesticide and nutrient fate/modelling; Best Management Practices (BMPs); sustainable land use; and wildlife and habitat management.

References

Baird, J.H. 1996. 'Evaluation of best management practices to protect surface water quality from pesticides and fertiliser applied to bermudagrass fairways', *1995 Environmental Research Summary*, United States Golf Association.

Beard, J.B. and R.L. Green. 1994. 'The Role of Turfgrasses in Environmental Protection and Their Benefits to Humans', *J. of Environmental Quality*, 23(3): 452–60.

Balogh, J.C. and W.J. Walker. 1992. *Golf Course Management and Construction: Environmental Issues*, Lewis/CRC Press, Boca Roton, Fla.

Branham, B., E. Miltner and P. Rieke. 1995. 'Potential groundwater contamination from pesticides and fertilisers used on golf courses', *USGA Green Section Record*, 33(1): 33–7.

Braun, S.E. and G. Stahnke. 1995. 'Leaching of nitrate from sand putting greens', *USGA Green Section Record*, 33(1): 29–32.

Borgert, C.J., S.M. Roberts, R.D. Harrison, J.L. Cisar and G.H. Snyder. 1994. 1994. 'Assessing chemical hazards on golf courses', *USGA Green Section Record*, 32(2): 11–14.

Bowman, D.C., D.A. Devitt and W.W. Miller. 1995. 'The effect of salinity on nitrate leaching from turfgrass', *USGA Green Section Record*, 33(1): 45–9.

Cisar, J.L. and G.H. Snyder. 1993. 'Mobility and persistence of pesticides applied to a USGA-type green. I. Putting green factors for monitoring pesticides', *International Turfgrass Society Research J.*, 8:971–7.

Cisar, J.L. and G.H. Snyder. 1996. 'Mobility and Persistence of Pesticides Applied to a USGA Green. III. Organophosphate Recovery in Clippings, Thatch, Soil, and Percolate', *Crop Science*, 36:1433–8.

Clark, J.M. 1996. 'Evaluation of Management Factors Affecting Volatile Loss and Dislodgeable Foliar Residues', *1996 Turfgrass and Environmental Research Summary*, USGA Green Section, p. 63.

Cole, J.T., J.H. Baird, N.T. Basta, R.L. Hunke, D.E. Storm, G.V. Johnson, M.E. Payton, M.D. Smolen, D.L. Matin and J.C. Cole. 1997. 'Influence of Buffers on Pesticide and Nutrient Runoff from Bermudagrass Turf', *J. of Environmental Quality*, 26:1589–98.

Cooper, R.J., J.M. Clark, and K.C. Murphy. 1995. 'Volatilisation and dislodgeable residues are important avenues of pesticide fate', *USGA Green Section Record*, 33(1):19–22.

Harker, D., S. Evans, M. Evans and K. Harker. 1993. *Landscape Restoration Handbook*, Lewis/CRC Press, Boca Roton, Fla.

Horst, G.L., P.J. Shea and N. Christians. 1995. 'Pesticide degradation under golf course fairway conditions', *USGA Green Section Record*, 33(1): 26–8.

Kenna, M.P. 1995. 'What happens to pesticides applied to golf courses?', *USGA Green Section Record*, 33(1): 1–9.

Kenna, M.P. 1994. 'Beyond Appearance and Playability: Golf and the Environment', *USGA Green Section Record*, 32(4):12–15.

Kenna, M.P. and G.L. Horst. 1993. 'Turfgrass water conservation and quality', *International Turfgrass Society Research J.*, 7: 99–113.

Linde, D.T., T.L. Watschke and J.A. Borger. 1995. 'Transport of N-NO_3 and nutrients from fairway turfs', *USGA Green Section Record*, 33(1): 42–4.

Miltner, E.D., B.E. Branham, E.A. Paul and P.E. Rieke. 1996. 'Leaching and Mass Balance of ^{15}N-Labeled Ureas Applied to a Kentucky Bluegrass Turf', *Crop Science*, 36:1427–32.

Murphy, K.C., R.J. Cooper and J.M. Clark. 1996a. 'Volatile and Dislodgeable Residues Following Trichlorfon and Isazofos Application to Turfgrass and Implications of Human Exposure', *Crop Science*, 36:1446–54.

Murphy, K.C., R.J. Cooper and J.M. Clark. 1996b. 'Volatile and Dislodgeable Residues Following Triadimefon and MCPP Application to Turfgrass and Implications of Human Exposure', *Crop Science*, 36:1455–61.

Petrovic, A.M. 1995. 'The impact of soil type and precipitation on pesticide and nutrient leaching from fairway turf', *USGA Green Section Record,* 33(1): 38–41.

Smith, A.E. and D.C. Bridges. 1996a. 'Movement of Certain Herbicides Following Application to Simulated Golf Greens and Fairways', *Crop Science,* 36:1439–45.

Smith, A.E. and D.C. Bridges. 1996b. 'Potential movement of certain pesticides following application to golf courses', *1995 Environmental Research Summary,* USGA Green Section, p. 84.

Smith, A. 1995. 'Potential movement of pesticides following application to golf courses', *USGA Green Section Record,* 33(1): 13–14.

Smith, A.E. and T.R. Tilloston. 1993. 'Potential leaching of herbicides applied to golf course greens', in K.D. Racke and A.R. Leslie (ed.), *Pesticide in Urban Environments,* American Chemical Society, Washington, D.C., p. 168–71.

Snow, J.T. 1995. 'The USGA's environmental strategies: what we've got and what we need', *USGA Green Section Record,* 33(3): 3–6.

Snyder, G.H. and J.L. Cisar. 1995. 'Pesticide mobility and persistence in a high-sand-content green', *USGA Green Section Record,* 33(1): 15–18.

Snyder, G.H. and J.L. Cisar. 1993. 'Mobility and persistence of pesticides applied to a USGA-type green. II. Fenamiphos and fonophos', *International Turfgrass Society Research J.,* 8:978–83.

Starrett, S.K, and N.E. Christians. 1995. 'Nitrogen and phosphorus fate when applied to turfgrass in golf course fairway condition', *USGA Green Section Record,* 33(1): 23–5.

Sticklen, M.B. and M.P. Kenna. 1997. *Turfgrass Biotechnology: Cell and Molecular Approaches to Turfgrass Improvement,* Ann Arbor Press, Chelsea, Michigan.

Yates, M.V. 1995. 'The fate of pesticides and fertilisers in a turfgrass environment', *USGA Green Section Record,* 33(1): 10–12.

INDEX